Digital Management of Container Terminal Operations

Ning Zhao · Yuan Liu · Weijian Mi · Yifan Shen ·
Mengjue Xia

Digital Management of Container Terminal Operations

SSTP

Springer

Ning Zhao
College of Logistics Engineering
Shanghai Maritime University
Shanghai, China

Yuan Liu
College of Logistics Engineering
Shanghai Maritime University
Shanghai, China

Weijian Mi
College of Logistics Engineering
Shanghai Maritime University
Shanghai, China

Yifan Shen
College of Logistics Engineering
Shanghai Maritime University
Shanghai, China

Mengjue Xia
College of Logistics Engineering
Shanghai Maritime University
Shanghai, China

ISBN 978-981-15-2936-8 ISBN 978-981-15-2937-5 (eBook)
https://doi.org/10.1007/978-981-15-2937-5

Jointly published with Shanghai Scientific and Technical Publishers
The print edition is not for sale in China (Mainland). Customers from China (Mainland) please order the print book from: Shanghai Scientific and Technical Publishers.

This Springer imprint is published by the registered company Springer Nature Singapore Pte Ltd.
The registered company address is: 152 Beach Road, #21-01/04 Gateway East, Singapore 189721, Singapore

Preface

At present, information technology has been involved in all aspects of the operation management of container terminals. With the further application of Big Data, Internet of Things, Mobile Internet, Artificial Intelligence and Cloud Computing technology, it is necessary to analyze and study the operation management of container terminals from a new perspective. In this book, the core technologies and concepts in operation management of container terminals are introduced as well as how to design and implement the key technical approaches around the terminal operation system (TOS) in the context of information age, aiming at providing valuable guidance through shaping the specific digital management logic and summarizing the previous experience in operation management.

The digital management in container terminals is selected as the main line of this book, and the related nodes in the information flow of various main operations as well as their relationship are systematically studied. The operation and management procedures in container terminals including information collecting and processing, resource planning and facility scheduling, and actual operations under the mode of Kanban management are elaborated in detail. A number of dynamic decision-making problems in the procedures of digital operation and management, including vessel unloading and loading, container delivery, container collection, etc, are figured out and analyzed as well.

The authors and their team have been engaged in the development of operation system and auxiliary decision support system as well as intelligent decision-making researches in container terminals for years, undertaken a great number of engineering projects of Tianjin Port, Shanghai Port, and Ningbo Port in China and overseas companies in the same industry, and accumulated solid research foundation and rich experience in port planning, terminal configuration, port logistics equipment design, process simulation, terminal operation and management, automatic loading and unloading operations, etc. Meanwhile, the research team have developed the terminal simulation system and terminal operation management system with independent intellectual property rights. A large amount of documents and videos accumulated in the engineering practice have provided a rich source of materials for the writing of this book.

Audience

The audience of this book are port management personnel; students and teachers of port management; and logistics software companies in the field of port and shipping logistics. With the constant improvement of the digital and intelligent level of container terminals and the increased automation degree of port equipment, the intelligent transformation of the container terminals has become the focus of attention. The digital and intelligent operation and management procedures and methodologies of container terminals are studied and analyzed elaborately and comprehensively in this book. For the audience concerning the digital management of enterprises and operations of container terminals, this book is a good reference.

Shanghai, China Ning Zhao

The original version of the book was revised: Figures have been corrected. The correction to the book is available at https://doi.org/10.1007/978-981-15-2937-5_7.

Acknowledgements

The publication of this book is funded by "MI Weijian workshop of intelligent port logistics, an innovation workshop for labor model of Shanghai education system".

Preparation of the manuscripts for this book has involved efforts and comments of my students, and appreciation goes to LI Ye, YANG Zhen, YANG Bin, GU Huajie, SHI Xuexin, LIANG Zhengguang, XUAN Beng, YE Zhilong, MIN He, FANG Shujie, ZHU Junmin, LI Youmei, and ZHANG Shuai.

Contents

Chapter 1
Introduction

1.1 Port and Terminal

A port, consisting of waters and lands, is an area constructed manually and equipped with infrastructures for vessel sailing and berthing as well as passenger and freight forwarding. The waters are the water areas within harbor line and have to meet two basic requirements, as one is to ensure the vessels safe entry and exit as well as berthing and unberthing, and the other is to ensure steady anchoring and handling operations. The waters include fore-dock water areas, entry and exit channels, turning basins, anchorages, navigation aids, etc. The lands are the land areas within harbor line including container handling areas and ancillary operation areas as well as some reserved areas. The container handling areas are equipped with infrastructures such as warehouses, freight stations, railways, roads, stations, passageways, etc. The ancillary operation areas are configured with garages, toolhouse, substations, repair houses, offices, firehouse, etc.

Port, as the gate of a country or a region, originates from the ancient Latin "port," which means the gateway to the coast with shields connecting the water and the land. In Chinese, port means the lanes near water, namely the passage of the land to the water (including rivers, seas, oceans, etc). Ports mentioned in this text refer to areas of waters and lands with clear boundaries equipped with infrastructures for vessels berthing, passengers embarking and disembarking, cargos handling, storing, transferring, and related services. Ports usually locate in or near to the cities and towns along rivers, lakes, and seas with frequent commercial trade activities. Ports are the hubs of water and land transportation, distribution centers of passengers and freights, transshipment joints of domestic and international trade as well as places for goods exchange.

The original version of this chapter was revised: Figures have been corrected. The correction to this chapter is available at https://doi.org/10.1007/978-981-15-2937-5_7

N. Zhao et al., *Digital Management of Container Terminal Operations*,
https://doi.org/10.1007/978-981-15-2937-5_1

A terminal is a part of the land area of a port, and it is used for vessels docking, cargos loading and unloading, and passengers embarking and disembarking. Generally, it also includes warehouses, yards, waiting rooms, handling facilities, railways, and roads. A terminal usually consists of the coastline and working area on the quayside. The coastline, namely the docking length along the bank, is the intersecting line of the vertical plane and horizontal plane of the terminal constructions on the quayside, which refers to hydraulic structures constituting the quay. The coastline is an important baseline determining the terminal horizontal position and its height, and usually falls into several categories, like deep coastline, shallow coastline, auxiliary coastline, etc, according to vessel draught and the functions. The overall length of all kinds of coastlines is an important index of the port scale, indicating the maximum docking number of vessels at the same time. The working area on the quayside refers to the area from the coastline to the front edge of the first row of warehouse (or yard). It is a location for cargos loading and unloading, transferring and temporary stacking, equipped with handling and moving facilities as well as lanes for mobile machines and vehicles, some even with railways for direct delivery. There is no specific standard defined for the width of the working area, which is usually determined according to the terminal function and the handling technology on the quayside. In China, the width of the working area ranges from 25 to 40 m. The surface course of the working area is generally paved with reinforced concrete and block stones, which can offer the strength needed for moving and handling operations.

The terminal is a hydraulic structure for vessels berthing, cargo handling, and passenger embarking and disembarking. Currently, most terminals are designed vertically, which improves the operating efficiency by facilitating the vessels berthing and direct distribution of the facilities to the coastline. The inclined terminals, preferred in areas with huge river water level difference, are equipped with pontoons in front of the inclined slope, which may lead to low operating efficiency due to complicated handling technologies and indirect distribution of the facilities to the coastline. Pontoon terminal can be adopted in areas with small water level difference, such as rivers, lakes, and harbor basin with natural or artificial shields, in which the pontoons are connected to the quays with mobile approach bridges. Pontoon terminal is usually designed for passenger transportation, fish handling, ferries, and other ancillary functions. The terminal structures include gravity quay, high pile quay, and sheet pile quay, which are designed in consideration of functional requirements, natural conditions, and construction conditions. ① The gravity quay, which can maintain stability with the gravity of the constructions and additives within, has excellent structure integrity and robustness and is easy to repair after damage. It includes integral type and prefabricated type and is suitable for terminals with good foundations. ② The high pile quay consists of foundation piles and upper structures. The down structures of the piles are buried in the soil and the upper structures above the water. The upper structures can be constructed with either beams and slabs, or large slabs without beam, or frames, or bearing platforms. The high pile quay, suitable for soft foundation, is an open structure, which allows the water to pass below the quay surface without reflection to the waves and interference to the flood discharge and decreases deposition. At present, the long piles with large span are popular, and the short piles with small sections are being replaced by large prestressed concrete piles or steel piles gradually. ③ The sheet pile quay consists of sheet pile walls and anchor fittings, which can support the side pressure caused by operation load and soil behind the walls. This simple structure results in rapid construction and can be adopted in most situations except in areas with extremely hard or soft foundations. However, the integrity and durability is barely adequate.

According to the purposes of the terminals, they can be classified as passenger terminals, freight terminals, auto terminals, oil terminals, yacht terminals, fishing terminals, navy terminals, and container terminals.

The main function of the passenger terminals is to embark and disembark passengers. Small passenger terminals can only be used for mini vessels like small ferries and yachts to berth, while those big ones can offer anchorage for big liners and cruises. Passenger terminals can be divided into public terminals, ferry terminals, and cruise terminals. The public terminals are available to all ships (depending on the depth of water, which means the vessels with larger draft cannot berth). The ferry terminals are usually occupied by special liners or shared by several ferries. In the ferry terminals adjacent to different countries or regions (like the Hong Kong—Macau Ferry Terminal), immigration facilities will be arranged. The cruise terminals are usually for cruises to berth, and configured with integrated ancillary facilities, like customs, immigration counters, health quarantine offices, luggage handling areas, tickets offices, parking zones for touring buses, and passenger pick-up and drop-off areas. Because of the huge volume and draft of the cruises, the cruise terminals are usually located in the broad ports with deep water. Most cruise terminals are not designated to a specific company. Some public passenger terminals are also used for freight handling in small volume, like the Wong Shek Terminal.

The freight terminals are designed for freight handling. According to their purposes and accessibility, the freight terminals can be divided into public freight terminals, container terminals, oil product terminals, mineral terminals, inland river freight terminals, and general freight terminals, etc.

The auto terminals are designed for some special vessels (usually large vessels) to berth, so the automobiles can roll on and off.

The oil terminals refer to those specialized terminals for crude and refined oil handling. They are usually located in an appropriate distance from the general freight terminals (or passenger terminals) in case of fire. Such terminals are configured generally with small freight loadings and simple handling facilities, and capable of handling most light tankers. Since the very large crude carriers (VLCC) are more and more popular in modern maritime transportation, the requirements on stable berthing for the terminals are not high due to the deep draft and strong resistance to wind and waves of VLCC. Now there are four kinds of deep terminals (or infrastructures) for crude oil handling, namely single point mooring terminals, multi-point mooring terminals, mooring island and open-type terminals. The first three kinds are not configured with constructions to resist wind and waves, and the last ones can be flexible according to their arrangements and local conditions.

The yacht terminals are designed for yacht to berth and usually owned by yacht clubs.

Fig. 1.1 An overview of a container terminal

The navy terminals, also known as military terminals, are designed for navy ships to berth and replenish and usually highly secured.

The container terminals are designed specifically for the container vessels to berth. They are the operating locations for containers handling, as well as the connections of water and land transportation of containers and the hubs of intermodal transportation. The container terminals are usually configured with berths, container yards, controlling offices, inspection gates, container machineries, and other special-purpose facilities (Fig. 1.1).

According to the geographic location, the container terminals can be divided into container terminals at seaports and container terminals at river ports.

① The container terminals at seaports are usually located at ports near to coasts and estuaries, where the water levels follow the tide variation. They are designed to handle ocean and coastal container vessels.

② The container terminals at river ports are located at ports near to inland rivers or lakes, where the water levels vary greatly due to the seasonal variation of rivers or lakes. Such terminals mainly handle inland river container vessels.

According to the degree of specialization, the container terminals can be divided into specialized container terminals and multipurpose container terminals.

① The specialized container terminals are constructed exclusively for containers handling with highly efficient handling and moving facilities, which can meet the requirements of rapid growth in container vessels amount and high-speed handling at the ports. At present, such terminals are being constructed at many ports worldwide and rapid development has been achieved especially in the last decade.

② The multipurpose container terminals can handle containers as well as other cargos, such as logs, steels, heavy general cargos, etc. They are generated due to the new potentials of containers transportation development and only temporarily transitional with relatively low loading and unloading capacity.

In this book, the specialized container terminals at seaports are mainly discussed and analyzed, and they will be mentioned in the text following as container terminals for simplification.

1.2 The Historical Development of Container Terminals

As the development of the society, the division of labor has changed constantly, the time and space interval of exchange has increased dramatically, and the organization of production has become more and more complicated, which gave the transportation (logistics) higher requirements that the conventional freight terminals could not satisfy, and containerization was just for these requirements. The container terminals appeared during the containerization of general cargos transportation in the world. When looking back to the history of container terminals, it is also the history of container transportation, which experienced the following periods.

(1) The origin period (1956–1966)
The most important event in this period is that the USA first adapted oil vessels and general cargo vessels into container vessels to conduct container transportation along its coasts and earned excellent economic benefits.
On April 26, 1956, the American Trans-Atlantic Steamship Company conducted a trial shipment from Newark, New York, New Jersey to Houston, Texas after adapted an oil vessel of T-2 modal and loaded 58 containers on deck. Three months later, the trial shipment earned a huge economic benefit by reducing the handling cost per ton from 5.83 dollars to 0.15 dollars, which is only 1/37 of the handling cost of the general freight vessels. After the big success of the trial shipment, the American Trans-Atlantic Steamship Company invested more interests on containerization and decided to fully and thoroughly promote it. In October 1957, this company adapted six freight vessels of C-2 modal into full container vessels with cells and named the first container vessel the "Gateway City." This vessel was configured with container crane and capable of loading 226 containers of 8 ft × 8.5 ft × 35 ft, which weighted 25 t, respectively. The "Gateway City" still sailed on the New York-Houston route, and its beginning of service symbolized the beginning of container transportation on the ocean.
The main characteristics of this period are that the container vessels were adapted from freight vessels; no special berths were available for container vessels; containers transported were not with standard sizes (namely 17 ft,

27 ft, or 35 ft); shipping routes for container transportation were only within the USA.

(2) The initiate period (1966–1071)

The most important event in this period is that the Land Sea Transport conducted sailings of adapted full container vessels with 226 35 ft containers loaded on the New York-Europe route in April 1966, which means the beginning of container transportation on international routes.

In September 1967, the Matson Liners sent vessels sailing on the Japan-Pacific North America routes, which initiated the container transportation in the Pacific liners. Enlightened by container transportation in the Atlantic and the Pacific, liner companies in Japan and Europe began to build container vessels of medium size, establish container vessels operation systems, and join the international activities of container transportation on routes connecting Japan, Europe, USA, Australia, etc. Until the beginning of 1970s, containerization had been realized on more than 10 main routes, and by the end of 1972, there had been over 160 full container vessels with 2,770,000 t loaded in service, which made the transportation capacity of containers to 1,280,000 TEU together with semi-container vessels.

The main characteristics of this period are that containerization was realized progressively; the first container vessels appeared and specialized container terminals were established; 20 and 40 ft containers were dominant in containerization.

(3) The growth period (1971 to the late 1980s)

Due to the high efficiency, low cost of container handling and high benefits, high quality of container transportation as well as convenience of intermodal transportation, container transportation was favored by the shippers, carriers, ports, and departments concerned. During 1971–1989, it developed rapidly, and the international ocean shipping routes had extended from Europe and America to Southeast Asia, Middle East, and other main routes in the world. At the end of 1971, the high-speed container vessel "Kamakura Maru" with loading capacity of 51,139 t, 1950 20 ft containers and navigation speed of 26 knots began its sailing on the Far East-Europe route. In 1972, the Land Sea Transport put the ultra-large super high-speed full container vessels in service, with length of 288 m, loading capacity of 1968 TEU, power of 98,000 kilowatts (120,000 horsepower), and navigation speed of 33 knots. From then on, more large and high-speed container vessels had been put in service successively by the Land Sea Transport and Scanduch Group, which was jointly operated by three shipping companies from Japan, Great Britain, and Germany. Hence, the container transportation had developed into its second era from the first era, which was featured by the loading capacity of 700 TEU and navigation speed of 22–23 knots. In the second era, the loading capacity had been increased by two times and navigation speed by 3–5 knots. The container vessels in this era were mainly high-speed full container vessels with loading capacity of 2000 TEU; the transporting distance had been extended from simply connecting ports on both sides of the ocean to crossing two oceans; the transporting routes had formed the network by branch routes and the land bridge transportation appeared. In this period, not only the developed countries were trying to expand their container fleets, but also the developing countries began to establish their own container fleets and joint operation became prevalent among container shipping companies.

In 1980s, there was new development in container transportation worldwide. Affected by two successive oil crises and the transformation of first container transportation era, the third generation container vessels began to emerge, which aimed at reducing resource consumption and improving transportation efficiency. In 1984, the Evergreen company put two container vessels successively into round-the-world sailing, which initiated the voyage round the world. Since the container marine transportation worldwide had made such great progress, the container terminals had welcomed its growth period.

The main characteristics of this period are that container vessels and container terminals continuously developed and the capacity of container transportation was greatly increased; the modernization of port machinery and the application of computer technology had improved the management skills and tools; the intermodal transportation began to arise.

(4) The popularization period (1990s to now)

Since 1984, the shipping market worldwide had broken away from the depression of oil crises and got back on the path of steady development. At present, the degree of containerization of general cargo transportation has been over 80%. According to statistics, by 1998, there had been over 6800 container vessels of all kinds in the world and the total loading capacity of 5,790,000 TEU. In 1990s, especially after 1994, the full recovery of international economy had an active effect on the shipping market, and in container transportation market, the cargo amount shipped on all routes gained robust growth. According to statistics, by November 1, 1994, the number of container vessels in service had reached 5715, and loading capacity of 4,100,000 TEU. Though the cargos shipped and container vessels involved were focused on three main routes, namely the Far East-North America (with the route length of 11,000 nautical miles), the Far East-Europe Mediterranean, and the North America-Europe Mediterranean (with the route length of 4000 nautical miles), the trend of investment in ultra-large vessels had become increasingly evident among all the major shipping companies. Sufficient amount of cargos were shipped on the three main routes, which were perfect for the large container vessels to sail on, so it was definitely the first choice for the ultra-large container vessels to sail on. In 1996, the Maersk put the "Maersk Queen" with the loading capacity of 6000 TEU in service on the Far East-Europe route. By 2004, the loading capacity of the container vessels of the Maersk in service had reached 8360 TEU. In the next two years, the ultra-large container vessels invested increased in times, and it could be predicted, the scale of the container fleets would expanded continuously. All of these showed the coming of the popularization period of container transportation.

As turned into the mature phase, container transportation has extended to all countries with maritime transportation in the world, and containerization has become the main trend of freight transportation by sea. The integral technology of hardware and software tends to be perfect and has made the intermodal transportation and the "door-to-door" phase come.

The main characteristics of this period are that the container vessels were upsizing and automating, the operating efficiency of terminals improving, the collecting and distributing systems constantly perfecting, management skills and tools modernizing. The electronic data interchange (EDI) systems were generally adopted, with which the dynamic tracking and management of the containers were realized. The development of intermodal transportation of containers was enormously promoted.

The development of container transportation in China took a late beginning with relatively outdated technology but huge potential. In September 1973, the vessel "BH1" loaded with small containers, arrived at Tianjin Port from Kobe, Japan, which symbolized the opening of the first container liner in China. Due to the slow development in container transportation, from 1973 to 1976, only 1188 containers with 3773 t cargos were handled at Tianjin Port.

On December 25, 1981, the No.12 container berth at Tianjin Port was completed and put into operation, meanwhile it formally passed the national acceptance. This berth was 398 m long and the water depth −10 m. The corresponding terminal yards covered an area of 92,200 m^2, configured with two warehouses for road and railway transfer, respectively, three railway sidings, two quay cranes, one rubber-tyred gantry crane, two straddle carriers, ten big forklifts and 21 small forklifts, eight trailers, and 11 Maffey trailers. In 1982, the No.21 container berth at the third basin of Tianjin Port was also completed and put into operation. In December 1985, another three container berths at the fourth basin of Tianjin Port formally passed the national acceptance and began to operate as the key project of the state's sixth five-year plan. Hence, there had been four deep container berths for the Tianjin Port Container Terminal Company, which were capable of berthing 50,000 dwt vessels and the annual handling capacity was 40,000 TEU as designed, ranking the first in China in the year. In 1991, the contract of building the first container quay crane with the maximum load and outreach by the Shanghai Port Machine Factory for Tianjin Port Container Terminal Company (the first container terminal in China) was signed in Shanghai, and in August 1992, the crane was put into service. The outreach of the quay crane was 4 m, the maximum load of 40.5 t, which was internationally advanced in 1990s and met the handling requirements for the fourth generation container vessels.

In 1977, the Ministry of Transportation decided to improve the tenth operating zone of Shanghai Port into semi-container berth, and imported container machineries with 4,560,000 dollars. In December 1980, the remodeled fourth and fifth container berths from the freight terminal were put into operation. Under the efforts of the State Council and all parties, the container liner from Shanghai to Australia was formally open, and the vessel "PXC" loaded with

162 containers started sailing from Shanghai, which symbolized the end of the history that China had no international maritime container shipping route. After that, the Ministry of Transportation first formulated the *Interim provisions on the collection of container port fees for international shipping routes*, the container terminals in China thrived and the intermodal transportation started successively. The first international container train started from Tianjin Port via Erenhot to the Republic of Mongolia. On August 19, 1989, the first sea-railway combined container train started from Dalian Port to the Changchun No. 1 Auto Plant. The New Eurasian Continental Bridge started from Lianyungang made the land transportation of containers to Europe possible.

At the beginning of container transportation in China, the management of container handling was achieved with paper T-cards, which were placed on the yard maps hanging on the wall. Each T-card represented one container, and the name of vessel, type of container, name and amount of goods, consignor and consignee were recorded on it. Initially, with a small amount of vessels and containers, it was adequate for operation, but as the throughput capacity of containers increased rapidly, the use of paper T-cards was outdated. Since the modern electronic computers had become the indispensible tools in management of large amounts of containers in and out in the container terminals worldwide, led by the former Ministry of Transportation, Tianjin Port and Shanghai Port jointly imported the electronic computer technology of container management from NEC, Japan, and sent computer professional technicians and operators to Japan, respectively, to take training and practice on electronic computer management and operation. Meanwhile, the independent development of management information system in China also made good progress.

(1) Independently designing software system of container management.
The Jungonglu and Zhanghuabang container companies of Shanghai Port jointly founded the Container Computer Technology Company, focusing on developing computer software system for terminals and maintenance of hardware. The container management software imported from Japan was not suitable for China, so they self-designed and self-developed the computer software system suitable for Chinese ports and successfully put into operation in terminals.
(2) Computer replacing T-cards in management.
The container management of Chinese ports quickly met the requirements of containers in and out in large amounts, which was a huge step toward the internationally advanced management in port.
(3) Computer management first implemented in container terminals at ports.
The computer management was gradually involved in everywhere in container terminals in China, from the initial management of full containers in the yards, to the accumulation and statistics of days of empty containers in the yards. It was a convenience for the shipping companies to understand the number of their containers remained in the terminal and a reference to allocate and

transport empty containers in time. This system also provided the stowing software of container vessels, which laid a good foundation for the loading plan of container vessels.

With all the solid foundations, the faster and better information management system was required in container terminals in China. In 1987, the project of industrial experiment on intermodal transportation for international containers led by the Ministry of Transportation was started at Shanghai Port. This project was meant to open a demonstration line centered in Shanghai of intermodal transportation for international containers by combining the industrial experiment and technology reform and establish a transportation management system for international containers and a modern information management system conforming to the international rules and situation in China. This project ended up in a huge success, and the results of industrial experiment were promoted widely, which laid the foundation for the standardization of the operation processes in container terminals in China.

The generation of electronic data interchange (EDI) improved the transmission speed of sea freight documentations. On June 14, 1995, the former Ministry of Transportation organized a review on the feasibility study report of the project of "*Establishing a model of EDI system for international container transportation*" and passed it. On August 29, 1995, this project was approved by the State Planning Commission. It was planned that the development of EDI system for international container transportation would be completed by 1997, and models of EDI system established at ports in Shanghai, Tianjin, Qingdao, Ningbo, etc and in China Ocean Shipping (Group) Corporation (COSCO). This EDI system could only achieve the data interchange between shipping companies and container terminals, while the customs, the commodity inspection, the animal and plant quarantine and the health quarantine were not covered. In August 2001, the project of the EDI system for customs clearance on logistics information platform in Shenzhen was started, which realized the connection to the customs system. The import and export enterprises could input the contract information in their own offices through Internet, and the customs would review it automatically via computer and conduct all the clearance procedures involved within 24 h. Since November 1, 2001, the new clearance model had been adopted in Shanghai Customs which made the clearance of the goods in advance and release on arrival. Since March 1, 2002, the paperless customs clearance had been put in trial in eight cities including Shanghai, which has greatly improved the efficiency of customs clearance, facilitated the international trade procedures, reduced the circulation cost of goods, and enhanced the international competitiveness of import and export enterprises.

After recent years' progress in container logistics, the container terminals have gained a leap-forward development. The application of computers in container terminals in China has already reached the level of advanced management of ports in the world. Besides Internet and EDI, other sophisticated technologies are also applied in container terminals. For example, the application of wireless communication technology (TETRA, Mesh, etc.) has made the container terminals the leaders in information transmission. Also, the RFID, GIS, GPS have laid good foundations for automation and intelligence of operation in container terminals.

Besides in the informatization, a lot of resources have been invested in the infrastructures and facilities of the container terminals and huge progresses achieved. For example, the capability of the handling facilities at the apron has been upgraded from one container one time to three 40 ft containers at the same time. Also, with the application of rail-mounted gantry cranes with longer span, the handling quality and efficiency in the yard have been improved greatly. The application of automatic guided vehicles (AGVs) in horizontal movement has realized the development of automation and intelligence in the container terminals.

It has been over 40 years since the USA first adopted oil vessels and general cargo vessels into container vessels to conduct maritime container transportation along its coasts and won huge benefits. Due to the outstanding superiority of container transportation compared with the traditional way, container terminals have been constructed in large amount throughout the world. The container terminals have been playing an extremely important role in enhancing the competitiveness of the ports, owing to the accelerated circulation of vehicles and vessels, increased transporting speed and reduced transporting cost.

The main functions of container terminals are:

(1) Providing terminals for container transportation system;
(2) Providing containers storage areas, as buffering zones for the change of transportation mode;
(3) Working as connections and hubs of water transportation and land transportation for containers.

Usually, there are two stages for the cargos transported by sea (see Table 1.1), namely dynamic stage (transported in the vessels) and static stage (handled and stacked in terminals). Table 1.1 has shown the percentage of time required and the percentage of labor required in different stages of traditional transportation for general cargos.

Compared with general cargos, the containers with goods packed in still need to be changed, distributed, stacked, and stored in the terminals. It can be concluded from Table 1.1 that in 35% of the time, the cargos transported by sea are in static stage, while they will consume 80% of the labor. So it is the key to reduce the labor consumption and the static time in the terminals. The emergence of container transportation has

Table 1.1 The percentage of time in days and the percentage of labor in different stages of traditional transportation for general cargos

Stage	Percentage of time in days (%)	Percentage of labor (%)
Dynamic stage	65	20
Static stage	35	80

greatly improved the productivity in the terminals, reduced the labor invested and costs in transportation, and effectively shortened the time of cargos handling involved in the terminals.

1.3 Layout and Facilities of the Container Terminals

The layout of the container terminals is to determine the number of berths, the width of the apron, the size and configuration of the yards, the performance parameters, and number of handling and transporting facilities, which mainly depend on the vessel forms and loading capacity of the vessels as well as their arrival density. The yards and the combination of different horizontal transporting facilities form different technical systems of handling, which also affect the layout of the terminals. For a general container terminal, there are three parts (see Fig. 1.2), including the handling area at the apron (near to the berths), the yards, and the gatehouses.

The highly mechanized and highly efficient mode of mass production in container terminals needs to integrate the vessels and the terminals as a whole to ensure the efficient operation of the highly strict flow production line and fully perform the three main functions of the container terminals. The essential infrastructures in the container terminals include the berths, the apron, the container yards, the container freight station, the control tower, the gates, the maintenance workshop, etc.

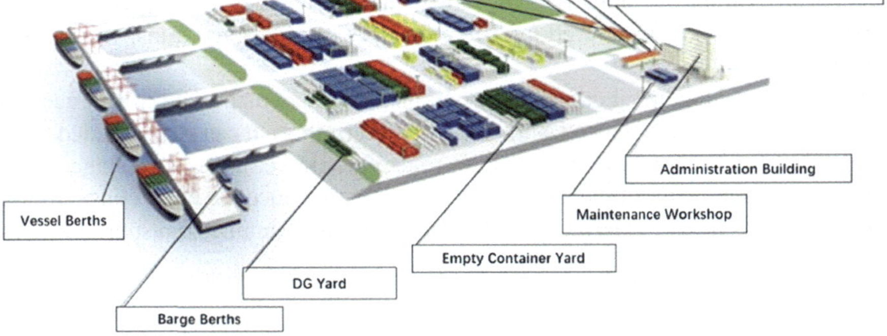

Fig. 1.2 The layout of a container terminal

(1) Berth

The berth is the area within the terminal constituted by the quays for the vessels docking and the corresponding waters. The length and water depth of the berth depend on the port type, terminal type, and docking vessel's form and size. With the development of large container vessels, the length and water depth of the berths in the container terminals are also extended. At present, the berths in the specialized terminals for full container vessels are usually 300 m long and depth of water −11 m.

The mooring facilities for vessels docking, including bollards and fenders (rubber pier), comprise the quay. The effective length of the quay required is usually 1.2 times of the vessel length when the vessel is berthing and unberthing.

(2) Apron

The apron is the area between the berth quays and the container yards (the flood control walls). Since there are quay cranes (QCs) equipped in the apron, which is also the main location for the import and export containers to be reloaded, the width of the apron usually depends on the spans of the QCs and the handling processes. There are normally three parts:

① The distance between the berth quays to the first track (on the sea side) of the QCs generally ranges from 2 to 3 m;
② The distance between the tracks (from the sea side to the land side) of the QCs generally ranges from 15 to 30 m;
③ The distance between the second (on the land side) track of the QCs and the container yards (the flood control walls) generally ranges from 10 to 25 m.

From the arrangements mentioned above, it can be concluded that the width of the apron in the container terminals generally ranges from 30 to 60 m. Besides the QCs and their tracks equipped in the apron, there are also high-voltage and low-voltage electrical boxes, interfaces for marine telephones, cable trench for QCs, water supplies, lighthouses, etc. The apron should be always unblocked and wide enough to stack the hatchway covers of the vessels and pass the horizontal transporting facilities simultaneously.

(3) Container Yard (CY)

The container yard refers to the area to store the containers to be loaded in the vessels according to the pre-made stowage plans as well as the containers to be unloaded from the vessels according to the delivery schedules before the vessels enter the port. The areas of the CYs could be different, depending on the loading capacity of the vessels entering the port and the berthing rate. Simply speaking, the container yard is the place within the terminal to store containers, including marshalling yards and back-up yards.

① Marshalling yard
The marshalling yards are located between the apron and the back-up yards, which are designed to store containers to improve the loading and unloading efficiency of the vessels. The main functions are to pre-store the export containers to be loaded in the vessels before the vessels' arrival and to temporarily store the import containers when the vessels are unloaded. The total area of the marshalling yards takes a relatively large portion of the container yards and the size depends on the handling techniques adopted and the number of stacking tiers.

② Back-up yard
The back-up yards refer to areas within the container yards except the marshalling yards, which are designed to store empty and full containers. The back-up yards include transshipment container yards, import full container yards, empty container yards, reefer container yards, dangerous goods container yards, etc.

Actually, there are no well-defined boundaries between the marshalling yards and the back-up yards, but only geographically. In practice, people usually store the export containers in the front of the CYs, the transshipment containers in the middle, and the import containers, the reefer containers, dangerous good containers, empty containers in the back. With the perfect computer systems, especially the radio data transport (RDT) facilities, it will be more flexible to plan the functions of the CYs under real-time control, even to mix-store the full and empty containers or just store the full containers in the front but the empty ones in the back.

(4) Container freight station (CFS)
The CFSs are designed to pack and unpack the less than container load (LCL) cargos as well as to store, safeguard, collect, and deliver them, usually referred to as warehouses. Different from the conventional warehouses, CFSs' main functions are packing and unpacking instead of storing.

The CFSs are usually located at the rear of the terminals, on the side near to the external roads or railways, which could let the vehicles directly enter the CFSs instead of passing the container yards.

In recent years, with the development of the container transportation and the refining market segmentation driven by the competition, the functions division has changed a lot. At some huge container ports, as the throughput increases, shipping companies began to outsource their empty containers to the external yards for professional management. Meanwhile, the external CFSs started to appear, professionally unpacking and delivering import containers and packing export containers. Those external yards and external CFSs are usually referred to as depots.

(5) Control tower
The control tower, also referred to as control center, central control office, commanding tower (office), is the commanding and controlling center of all the operations in the container terminal. Its functions are to fully utilize the production resources in the terminal and to supervise, regulate, and command the implementation of all the operation plans. It is usually located on the top of the administration building, where all the operations within the whole terminal can be viewed.

In the control tower, computer systems, wind meters, and meteorological forecast systems are configured, as well as the very high frequency wireless walkie-talkies (VHF), closed circuit television (CCTV), and telescopes to monitor the operations, and telephones and fax machines to communicate with others. It is the center of all the operations in the terminal.

(6) Gate house
The gatehouses are the entrance and exit of the terminal, the connection of containers and cargos in the containers as well as the boundary of responsibility inside and outside of the terminal. As it is the only way for containers to enter and exit the terminal, it is necessary to check the relative documentations of the containers as well as to transfer the responsibility of the container and the cargos, to inspect the apparent conditions like the container no., the seal no., the container body, and the cargos appearance.

The gatehouses are usually located at the rear of the container terminals. Wagon balances are configured according to the needs to ensure the safety of facilities and stowage as well as the control requirements of customs. Computers, IC card machines, releasing railing, automatic container no. identification system are also equipped.

(7) Maintenance shop
The maintenance shop, also referred to as repair shop, is the place to inspect, repair, and maintain the containers and special-purpose facilities for container handling. Its main functions are to ensure the maintenance quality and integrity of the handling facilities, which are very important for improving the terminal efficiency and keeping its superiority in container transportation.

The maintenance shop is usually located at the rear of the terminal or near to the maintenance area, which will not interfere with the operations in the terminal. Traveling cranes, lathes, welding and cutting machines, workbenches, air compressors, repairing galleries, and accessory warehouses are configured.

Besides the main infrastructures in the container terminals mentioned above, administration offices for administrative departments and other ancillary facilities like power supply, communication, canteen, computer rooms, oil depots, water supply and drainage, illumination, and roads are also configured.

1.4 Handling Facilities Commonly Used in the Conventional Container Terminal

In order to improve the handling efficiency and accelerate the reshuffle of the vessels, containers, and other resources, efficient special-purpose facilities and equipment are adopted in the conventional container terminal, which makes the loading and unloading operations mechanized. The entire mechanization system in the conventional container terminal includes handling facilities at the quay, horizontal moving facilities and handling facilities in the yards.

1.4.1 Handling Facilities at the Quay

The standardization of containers and customization of container vessels have provided excellent conditions for the efficient mechanization of handling processes in the terminals. At conventional container ports, loading the containers into the vessels and unloading the containers off the vessels are realized by the quay cranes (QCs) (see Fig. 1.3). The QC is a special-purpose equipment for the container ports, with enormous structures, huge dead weight, and high expense.

Fig. 1.3 Quayside container cranes

The QC consists of the portal with traveling mechanism, booms, stays to support the booms and other components. The booms include seaside boom, landside boom, and middle boom, which is used to connect the seaside boom and landside boom. The main function of the booms is to support the weight of the trolley with hoisting mechanism, while the hoisting mechanism is applied to support the weight of the container and the spreader. The seaside boom is designed with luffing mechanism so as to avoid collision with superstructures on the vessel when the QC is traveling.

According to the appearance of the gantry, the QCs can be classified as the A-shaped and the H-shaped. Due to the huge throughput in the terminals, the H-shaped are mostly adopted. For the form of the boom, the QCs can also be classified as the luffing, the shuttle-type and the foldable, while the luffing QCs are widely applied in the terminals. For the number of trolleys, there are QCs with single trolley and double trolleys. For the number of spreaders, there are QCs with single spreader, double spreaders, and triple spreaders.

Since there are cells configured in the cabin of the container vessels, the positioning of the containers in the cabin is very convenient when the QCs are operating and no manual assistance is necessary. The cabin operations for the general cargos have been eliminated. According to the operation experience in the container terminals worldwide, one berth is generally configured with one to three QCs.

The operation processes of the QCs are listed below:

① Before the vessel is docking, the QCs will travel to an appropriate position in order not to interfere with the vessel's safe docking.

② After the vessel has docked, the QCs will travel to the specific operation hold position.

③ The trolley will be moved to just over the container to be unloaded and the spreader be lowered. The operation sequence of vessel unloading is usually from seaside to landside, tier by tier from top to the bottom, and the sequence of vessel loading is vice versa.

④ The container will be locked with the torque devices on the spreader (by fixing the four corners onto the torque devices) and lifted.

⑤ The trolley will move to the landside along the cantilever with the container and put it onto the horizontal moving facility in the apron.

⑥ The torque devices will be released and the spreader separated from the container.

⑦ The spreader will be retrieved and the trolley move to the seaside along the cantilever to continue with the next operation.

The relative parameters of the QC can be determined as follows.

(1) Loading capacity

Loading capacity is an index of the QC's capability and usually determined according to the rated capacity and the weight of the spreader.

$$Q = Q_t + W$$

where Q is the loading capacity of a QC; Q_t is the rated capacity; W is the weight of the spreader.

The rated capacity refers to the maximum gross weight of a container that a QC can lift. For example, the maximum gross weight of the ISO 1A, 1AX, 1AA 40 ft container is 30.5 t. The loading capacity of a QC is the rated capacity plus the weight of the spreader. When determining the loading capacity, the following operation conditions should be considered.

① The requirement of lifting the hatchway covers. Generally, the hatchway cover's weight is less than 28 t, but special ones could reach 35.6 t with the size of 14 m × 14 m.
② The requirement of lifting the non-standard containers. The maximum gross weight of the non-standard containers could reach 38 t or even higher.
③ The possibility of lifting two 20 ft containers simultaneously. The maximum gross weight of two 20 ft containers is 40.6 t.
④ The requirement of lifting other bulk and heavy cargos.

(2) Main geometric parameters
The geometric parameters of a QC are determined according to the vessel forms and container types to be handled, the operation conditions at the ports and operation modes in the container yards.

① Load-lifting height

The load-lifting height of a QC includes the lifting height above the track and lifting height below the track, which depends on the container vessel forms, draught, tidal range, number of tiers of containers stacked on deck, standard height of the port and tilting of the vessels. When determining the load-lifting height of a QC handling a Panamax container vessel, it must be ensured that the lightly loaded vessel at high water level with three tiers of containers could pass and four tiers of containers could be stacked; the QC could reach the bottom tier of containers within the cabin when the vessel is fully loaded at low water level. The limit of the lateral tilt angle is 3° outward when operating.

② Outreach at seaside

The outreach at seaside refers to a QC's maximum horizontal distance to the seaside between the centerline of the seaside track and the centerline of the spreader, which mainly depends on the width of the vessel berthing at the port and the allowable maximum height of containers stacking on deck. When the vessel is tilting 3° outward, the QC should be able to lift the top containers above the deck on the outermost side of the vessels

③ Outreach at landside

The outreach at landside is the maximum horizontal distance to the landside between the centerline of the landside track and the centerline of the spreader. When determining the inreach, two factors should be taken into account. One is to temporarily stack containers as a buffer when the moving facilities like straddle carriers, trailers, etc, cannot be timely in position;

the other one is to place hatchway covers, which means the distances for different modes of power supply should be considered.

④ Span

The span is the horizontal distance between the centerlines of two tracks of the QC, which will affect the stability of the crane. In order to ensure the stability of the QC and efficiently dispatching the containers at quays, at least three lanes should be configured within the span. By comparing different kinds of moving facilities, it is found that a single transfer line with straddle carriers needs the largest width. If three transfer lines with straddle carriers are to be considered, then the span should be 16 m. At present, the spans of large cranes are usually 30 m, which will enable the configuration of six transfer lines.

1.4.2 *Horizontal Moving Devices*

1. Tractor

The container tractors, which are not configured with carrying units, have to be combined with trailers to transport containers within the terminals or on the roads outside (Fig. 1.4).

① By the modes of the cabs, the tractors include the cabover type and bonnet type. The cabover type of tractors (see Fig. 1.5a) can provide good view in operation and small turning radius due to the short cab, short wheelbase, and short body. However, since the engine is configured just beneath the driver's seat, it is not comfortable to drive due to the impact of vibration. The bonnet type of tractors (see Fig. 1.5b) can provide a relatively comfortable environment in operation,

Fig. 1.4 Container tractor and trailer

Fig. 1.5 The container tractors by the modes of cabs. **a** the cabover type **b** the bonnet type

more safety for the drivers in a crash and convenient maintenance since the engine and the front wheels are configured in front of the cab. However, due to the long cab and long body, bigger turning radius is needed. At present, the bonnet type of tractors with short body and small turning radius are increasingly applied.

② By the functions, the tractors include the line-haul tractors and the depot tractors

The line-haul tractors are featured with high power, high speed, and strong climbing capacity, while the depot tractors with great traction power and configuration with hoisting devices.

2. Trailer

The trailers are unpowered platforms to carry containers. With the development in port transportation, trailers become more and more specialized and standardized, and different types with various functions have appeared.

There are three kinds of combinations for tractors and trailers: semi-trailer, independent trailer, and double trailer.

The semi-trailer is the tractor pulling the turntable trailer loaded with container, see Fig. 1.6a. As shown in the figure, the weight of the container is jointly supported by the tractor and trailer, resulting in a relatively small axial pressure. Since the rear axle bears part of the container's weight, greater traction power can be obtained. This kind of tractor-trailer combination is short in full-length, convenient for reversing

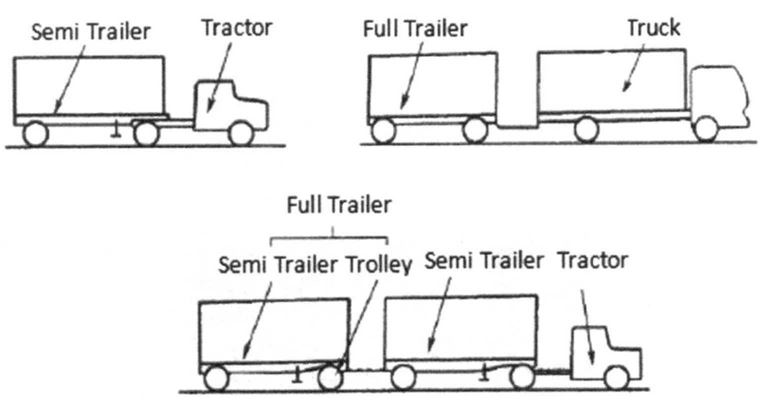

Fig. 1.6 Combinations of container tractors and trailers. **a** semi trailer; **b** independent trailer; **c** double trailer

and turning with high safety and reliability. Outriggers are configured at the bottom of the front of the trailer, easy for swapping trailer transport.

The independent trailer is to connect the tractor and trailer with traction strut. The tractor can serve as a general truck, and the trailer can also be independently supported by the outriggers, see Fig. 1.6b. The independent trailer is a common combination of tractor and trailer just behind the semi trailer, and more difficult to operate than the semi trailer.

The double trailer is to combine another full trailer just behind the independent trailer, and the tractor actually pulls two trailers, see Fig. 1.6c. When driving forward with high speed in such combination, the rear trailer will swing and it is difficult to operate when reversing, which impedes its application.

3. Straddle Carriers

The straddle carriers are specialized container handling facilities with multi-functions of moving, stacking, and reloading, usually applied in container terminals and container transshipment yards to horizontally transport and stack containers as well as to load and unload the container semi trailers, see Fig. 1.7.

At container ports, the straddle carriers can finish the following operations:

① Handling and moving between the operation points of container handling devices and yards;
② Handling and moving between the marshalling yards and back-up yards;
③ Handling and moving between the back-up yards and CFSs;
④ Reloading the trailers.

The straddle carriers have their unique advantages. With one straddle carrier, various operations (including picking up, moving, stacking, loading and unloading other vehicles) can be finished, resulting in reduced amount of machines in the terminals and simple management. The straddle carriers are more flexible to pick up containers without precise positioning of QC, which has improved the efficiency of QCs. They also have high maneuverability, either moving or stacking, and can be applied in combination as mobile machinery.

In spite of many unique advantages, the straddle carriers are considered as equipment with high failure rate, in certain countries as high as 30–40%, resulting in increased maintenance fees. With technology development as well as appropriate operation and management, the straddle carriers are successfully applied at some ports. For example, the straddle carriers are commonly adopted at container ports in Japan. At container ports in Belgium, a mode of 24 h-operation and 6 h-maintenance is applied, which has reduced the failure rate to 5–10%. The straddle carriers are seldom applied in China.

The straddle carriers are specialized container handling devices with high price. In order to reduce the number of straddle carriers and cut down the equipment investments and handling costs, the handling from the quays to the yards with straddle carriers has been replaced by the yard trucks, and the straddle carriers are only in charge of stacking operations in the yards.

Fig. 1.7 A straddle carrier

Since one QC has to be configured with four straddle carriers or more to finish reloading and stacking operations, one to handle the trucks in and out, and another one as backup, which means at least six straddle carriers have to be assigned to a QC. Besides, the high wheel loads of the straddle carriers result in high construction fees. Since the straddle carriers must align with the QC when loading containers but not necessary when unloading containers, the straddle carriers systems are suitable to container terminals with a large amount of import containers and a small amount of export containers.

1.4.3 Handling Facilities in the Yards

1. Gantry cranes

The gantry crane system was first adopted at Amsterdam Port in the Netherland. The gantry cranes are facilities applied in the yards to stack containers and handle transporting vehicles. There are two kinds of gantry cranes in the yard, rubber-tyred gantry cranes and rail-mounted gantry cranes. Handling processes with gantry cranes are first unloading the containers from the vessel and transferring them to the yards with yard trailers or other facilities, and then applying rubber-tyred gantry cranes or rail-mounted gantry cranes to stack the containers in the yards or reload the containers onto land vehicles (container trucks or rail wagons).

The rubber-tyred gantry (RTG) cranes commonly adopted in the conventional container terminals are specialized facilities to handle, move, and stack containers, see Fig. 1.8.

Take the RTG crane as an example, the gantry consists of frames and girders supported on the rubber tyres. The trolley configured with spreader travels along the rails on the bridge beams and finishes stacking and handling operations together with the trailer.

The RTG cranes are featured with high flexibility and universality. They are capable of traveling forward and backward as well as from one block to another via turning the wheels by 90° with the turning devices.

The main parameters of a RTG crane are the loading capacity, span, load-lifting height, wheel loads, operation speed, etc.

Fig. 1.8 Rubber-tyred gantry cranes

(1) The loading capacity of a RTG crane is determined according to the rated capacity and the weight of the spreader. The rated capacity usually depends on the maximum gross weight of the container to be lifted.
(2) The span of a RTG crane is the distance between the centerlines of the wheels on both sides. The size depends on the number of container rows to span and the width of the trailer passing lane. According to the layout of the container yards, a RTG usually spans six container rows and one trailer passing lane.
(3) The load-lifting height refers to the vertical distance from the spreader to the ground, which depends on the number of container tiers to be stacked. If four tiers of containers are stored in one bay in the yard, the fifth tier is needed for the spreader to move over the containers, which means the spreader should be at least five-tier high from the ground. Currently, the load-lifting height of the RTG crane is usually around 11 or 12 m.
(4) The wheel loads of a RTG crane include maximum operation wheel load and maximum non-operation wheel load. The number of wheels is determined according to the design requirements of the wheel loads in the yards, while the gantry traveling mechanism is either four-wheel-drive or two-wheel-drive.

    The maximum operation wheel load refers to the maximum pressure each wheel bears when lifting the rated load with the operation speed of 16 m/s. The maximum non-operation wheel load refers to the maximum pressure each wheel bears when lifting no container under non-operation speed with the wind blowing perpendicular to the girder of the crane.

    The maximum wheel load is the reference of determining the loading capacity of the ground when the cranes are traveling.
(5) The operation speed is usually determined according to the handling cycle. Low operation speed will affect the operation schedule in the yards, while high operation speed will result in huge swing of the containers and reduced safety.

In order to avoid the collision between the RTG cranes as well as that between the RTG crane and the containers, manual deviation correction system and anti-collision devices are configured on the cranes. For the safety measures, over-load protections, diesel engine over-speed protections, signal devices showing over-heated water and low oil pressure, wind speed indicator, typhoon anchors, emergent stop buttons, limit switches and signal indicators of different mechanisms, etc., are configured. Besides, differential global positioning system (DGPS), electrical control and management system (ECMS), Remote Crane Management System (RCMS) as well as gantry lifting devices to facilitate turning and reduce tyres wear can be selectively installed.

2. Reach stacker

The reach stackers are specialized facilities currently popular in the container terminal yards, see Fig. 1.9. Since the traveling direction is perpendicular to the operating direction, relatively wide space is required. However, they can stack containers relatively high and operate multiple bays of containers, and they are generally welcome due to their high flexibility. The reach stackers are capable of stacking three to four tiers of full containers, or seven to nine tiers of empty containers, which results in an

Fig. 1.9 A reach stacker

efficient utilization of the stacking areas. At present, the reach stackers are mainly applied only as ancillary facilities in the container yards, but they are specialized container handling devices with great potential.

The reach stackers are capable of moving, stacking and handling as well as operating crossing different blocks. They can also handle special cargos by equipping with hooks and grabs. However, they can only operate in the yards with smaller blocks and more passing lanes. The operating efficiency of a single reach stacker is relatively low, which can be improved by applying several reach stackers in combination. The wheel loads are relatively high, resulting in serious wear of wheels and grounds. By comparing the advantages and disadvantages of the reach stackers, they are presently applied as ancillary facilities.

3. Container forktruck

The container forktrucks (see Fig. 1.10) are common handling facilities with multiple functions, mainly applied to handle, stack and move containers in a short distance as well as handle vehicles at the general ports with relatively low operating amount. They are also applied as ancillary facilities in the container yards. The spreader is fixed on the top in front of the mask, connecting with the containers via twist locks and lifting from the top. They can also lift and move containers by inserting the forks into the sockets at the bottom of the containers. The requirements for the operation performance are listed below:

Fig. 1.10 A container forktruck

① The lifting capacity must meet the requirements of handling all kinds of containers;
② The lifting height must meet the requirements of stacking tiers;
③ The loading center should be located at the half of the container width, namely the distance between the front wall of the forks and the gravity center of the cargos should be 1220 mm;
④ In order to meet the requirements of handling containers, specialized spreaders on the top should be equipped besides the standard forks.
⑤ In order to facilitate the positioning with the containers, the forks should be capable of moving or swinging sideway.

The forktrucks can move the containers in the following two ways.

(a) Moving with spreaders: moving the containers with specialized spreaders on the top.
(b) Moving with forks: moving the containers by inserting the forks into the sockets at the bottom of the containers, which is usually applied to 20 ft containers or empty containers.

By the structures and forms, container forktrucks can be divided into front container forktrucks and side container forktrucks. The front container forktrucks are commonly used at present. The side container forktrucks are alike general side forktrucks. When handling a container, the mask and forks are stretched out sideway, and then withdrawn after taking the container. Comparing with the front container forktrucks, the lateral size of the side container forktrucks when moving containers is much smaller, the requirement for the width of the passing lanes is relatively low (around 4 m). When moving containers, the loading center is located between the front wheels and rear wheels, resulting in good stability in traveling and uniform distribution of wheel loads. However, the structures and operations of side container forktrucks are complicated, leading to bad view and low handling efficiency.

Even though the container forktrucks have the advantages of good universality, low expense, and easy operation, their disadvantages are more. The container forktrucks are not suitable to terminals with large throughput capacity due to the low efficiency. When handling containers, the positioning is relatively difficult. When moving containers, the wheel loads are relatively high and wide passing lanes are required. As a result, the container forktrucks are applied as ancillary facilities to complete handling operation in combination with other machines in the terminals.

1.5 Handling Facilities in the Automated Container Terminals (ACT)

In last section, the handling facilities in the conventional container terminals are introduced. As the container shipping volume is increasing constantly and container vessels becoming larger, the port operators are focusing on how to keep the container handling stable and efficient, energy saving and environment friendly with low cost. The appearance of the automated container terminals has satisfied the requirements mentioned above. The handling facilities in the ACTs include facilities at quay, horizontal moving facilities and handling facilities in the yard.

The handling facilities at quay include quay cranes (QC), double trolleys quay cranes, multipurpose gantry cranes, etc. The horizontal moving facilities include automated guided vehicles (AGV), lift-AGVs, etc. The handling facilities in the yard include rail-mounted gantry (RMG) cranes, automated rubber-tyred gantry(ARTG) cranes, etc.

1.5.1 Handling Facilities at Quay in the ACTs

The layout of ACT is similar to the conventional terminal; the operations of loading and unloading vessels are all accomplished by the QCs. Besides the QCs in the

Fig. 1.11 Double trolleys quay cranes

conventional terminals, some specialized QCs like automated double trolleys quay cranes (see Fig. 1.11) are also applied.

Compared with the conventional QCs, double trolleys quay cranes are equipped with two self-propelled trolleys and handling operations can be finished in relays at transit platform. The handling operations at seaside and landside are accomplished, respectively, by two trolleys to improve efficiency. The main trolley (at seaside) handles containers in the vessels, and the auxiliary trolley (at landside) handles containers on the high tracks, which is referred to as "handling in relays." The operating efficiency of the double 40 ft and double trolleys quay cranes developed and manufactured by ZMPC for Long Beach Container Terminal (LBCT), USA can reach 103 TEU/h.

1. Structural Analysis of Automated Double Trolley Quay Cranes

(1) The QCs are equipped with two self-propelled trolleys. The two trolleys travel on their own tracks and do not interfere with each other. The main trolley can travel on the tracks laid on the boom, and the tracks of the auxiliary trolley are laid on the linking beams of the gantry. The two trolleys can be operated separately. The driver of the main trolley is responsible for positioning when lifting the container, while the auxiliary trolley can be operated without drivers. Monitoring room is configured under the linking beams, and the operators can monitor and control the operations of the two trolleys and AGVs.

(2) Transit platform with the space of two containers is configured under the seaside beam at landside. The transit platform is the main component to fulfill the

functions of the double trolleys quay crane as well as the connection of the main trolley and the auxiliary trolley, where the relays between the two trolleys are accomplished. The main trolley lifts and moves the container in the vessel to the transit platform, and the auxiliary trolley transfers the container from the transit platform onto the AGV. Since the operations of unloading containers from the vessels are accomplished by two trolleys in relays, the handling efficiency has been greatly improved.

(3) Since self-propelled trolleys are adopted and the moving loads are relatively high, single box beam with high bending resistance and high twisting resistance as the boom can solve the problems caused by lifting double containers with high outreach (61 m). But it cannot be ignored that the trolley tracks are directly laid on the webs of the box beams, when the trolleys are traveling on the beams, their wheel loads will cause relatively high local pressure. The webs of the single box beam are usually thin-walled, so necessary measures should be adopted to prevent the buckling of the webs.

(4) Advanced electronic information technologies are applied to achieve anti-collision and anti-swing of the two trolleys as well as automation controls like automatic positioning and identification. The latest advanced electronic anti-swing technology in the world is adopted to the main trolley, and the mechanical anti-swing measures adopted to the auxiliary trolley. On the four surfaces of the upper frame on the spreader, eight steel wire ropes form the wrapping system in the most optimized way to avoid swing of the spreader in each direction.

(5) AGVs undertake the transporting on the ground, which has solved the problems that the ground transporting vehicles and the cranes do not match, causing low operating efficiency of the double trolleys quay cranes. Hence, when the main trolley unloads the containers from the vessel and puts them onto the transit platform in the shortest route, the auxiliary trolley takes the containers immediately and loads them onto the AGVs and then to the yards.

(6) In the structural details, since the gantry trolleys have to travel on the inner sides, the tracks are laid on the track girders connecting to the lateral linking beams sideway via lateral outriggers instead of on the top of the lateral linking beams. When the gantry trolleys are operating, large horizontal side forces can be caused on the webs of the lateral linking beams, where the concentrated wheel loads are applied. Hence, at the connections of the lateral linking beams and the lateral outriggers, appropriate structures should be adopted to transfer horizontal forces.

(7) The whole crane has excellent stiffness. The double-stayed structures are adopted in the front and rear booms to improve the structural stiffness and reduce the sway of the crane to the minimum when the two trolleys brake simultaneously.

(8) Leading automation and control technologies are applied to the crane as well as safety measures and faults indicating devices, which have also accomplished the automated control of the AGVs on the ground while the trolleys are handling containers.

2. The technical parameters of the double trolleys quay cranes

The technical parameters of the double trolleys quay cranes include loading capacity, outreach, span, and inreach.

(1) Loading capacity
Since double 20 ft container spreader is configured on the QCs, the loading capacity is designed as 65 t under the spreader, according to ISO regulations, domestic and overseas practice experience as well as operation requirements of lifting containers of different sizes (including 45, 48, 53 ft, etc.), over-weighted containers, and hatchway covers.
(2) Outreach at seaside
Comparing with designing and manufacturing cranes capable of handling vessels with 16–18 rows of containers stacking on deck and extending the outreach in the future, directly designing and manufacturing cranes capable of handling vessels with 22 rows of containers stacking on deck is with lower costs and no interference with the handling operations. To satisfy the requirements of handling 18,000 TEU container vessels and development in a long run, the crane should be designed capable of handling vessels with 23 rows of containers stacking on deck, with the outreach at seaside as 70 m.
(3) Span
As the container vessels and QCs are becoming increasingly larger in the world, the stability of the cranes with larger outreaches should be considered when determining the span. Meanwhile, the situation of concentrative operations of multiple facilities when handling large container vessels should also be considered. By taking into account the possibility of concentrative operations of multiple facilities as well as the stability of the cranes, the span of the QCs should be set as 35 m.
(4) Outreach at landside
In order to ensure concentrative operations of multiple facilities at the ports, two lanes are added behind the landside tracks of the QCs. Meanwhile, considering the requirements of lifting hatchway covers, the outreach at landside of the QCs should be set as 20 m.

1.5.2 Horizontal Moving Facilities in ACTs

The horizontal moving facilities adopted in ACTs are mainly AGVs and lift-AGVs.

1. Automatic Guided Vehicle (AGV)

The first ACT in the world (ECT) adopted AGVs capable of autonomous navigation and positioning as the horizontal moving facilities. Such ACTs have flexible layout, with yards of various sizes and shapes and back-up yards configured perpendicular to the coastline. The first generation of AGVs could only travel with fixed routes,

Fig. 1.12 The AGV and the AGV mate

but at CTA in Hamburg, they had been optimized, capable of traveling with variable routes and turning anywhere without obstacles.

AGVs can be powered with internal combustion engine, pure barratries, or hybrid power. One AGV can take two 20 ft containers or one 40 ft container at one time with the positioning accuracy up to ±5 cm. With the help of ultrasonic detecting device and other anti-collision devices, they are capable of detecting obstacles within the brake distance to avoid collisions. The remote management and control of AGVs and ASCs in the container yards by the central controlling office has achieved full automation.

Since AGVs cannot be loaded or unloaded by themselves, there is a possibility that they may wait or be waited in the operating interaction areas. In order to improve the operating efficiency and solve the "handshake" problems of AGVs and yard cranes, ZMPC has developed the AGV mates (see Fig. 1.12) applied at the ends of operation areas in the yards.

2. Lift-AGV

In order to improve the operating efficiency of the AGV system and decouple AGVs in the head areas of the container blocks, the horizontal moving techniques with lift-AGVs have been developed in recent years. Different from the traditional AGVs, the lift-AGVs are capable of lifting the container automatically. Combined with the lift-AGV mates (see Fig. 1.13), the lift-AGVs are decoupled in the head areas of the container blocks. When the lift-AGV loaded with containers enters the head area, the containers could be unloaded to the mate platform with the automatic jacking devices of the lift-AGV, and also the empty lift-AGV could also be loaded with containers on the mate platform in the same way.

Fig. 1.13 A lift-AGV and the lift-AGV mate

1.5.3 Handling Facilities in the Yards in ACT

Since the ECT in Rotterdam in Europe was put in operation in 1993, the ACTs have gained great progress and development. The semi-automated terminals with automatic operation systems in the yards have been applied widely in the world, and the technologies, equipment, processes, and controlling systems have become very mature. There are mainly two kinds of handling facilities in the yards, namely automatic rail-mounted gantry cranes (ARMGs) and automatic rubber-tyred gantry cranes (ARTGs)

1. ARMG

The ARMGs can be operated automatically without drivers. In the controlling office, multiple ARMGs can be remotely controlled simultaneously. They have been applied at the Times Port in GB and the PPT in Singapore. The ARMGs can also be operated in a semi-automatic way with divers.

At present, the automatic operation mode with remotely controlled ARMGs (see Fig. 1.14) has been established in China. Compared with the traditional RMGs and RTGs, the new ARMGs have advantages in positioning and detecting technology, safety protection technology, and man-machine interactive system, providing efficient and stable security for the operations. Since the ARMGs are stable in technology and handling efficiency, it is assumed that the ARMGs are suitable handling facilities in the yards in the ACTs.

Besides the parameters of traditional gantry cranes, the ARMGs have their unique spans and configuration of over-hanging arms.

① From the angle of operation configuration, the larger the span of an ARMG, the higher the utilization rate of the yard. However, from the angle of mechanical

Fig. 1.14 ARMGs in the yards

design, the larger the span of an ARMG, the higher the requirements on strength and stiffness of the metal structures of the whole crane, especially on the dynamic stiffness, which makes the structural design and manufacturing processes more complicated and difficult.

It is generally assumed that all-fixed-gantry-legs design is suitable for ARMGs with span smaller than 35 m, however, if the span reaches 35 m or above, such design is liable to rail gnawing. As the span increases, the length of the boom will also increase, which makes the boom liable to deform under loadings, leading to rail gnawing due to outward displacement of the legs. If such design must be adopted, the vertical stiffness of the boom should also be increased, causing increased weight of the whole crane and wheel loads as well as the cost. So in order to avoid rail gnawing with large spans, the design of fixed gantry legs on one side and hinged legs on the other is adopted on the ARMGs. However, such design will result in low dynamic stiffness in the trolley traveling direction. In order to solve the problem, the stiffness of the boom should be increased, also causing increased weight of the whole crane and wheel loads as well as the cost of the track base.

Additionally, the larger the span of an ARMG is, the more serious the deformation of the metal structures of the whole crane will be. As the traveling speed of the gantry increases, the difference of resistance to traveling on the gantry legs of both sides will also increase, leading to different movements of the gantry legs and higher liability to rail gnawing. So as to solve such problem, electrical control equipment should be added to ensure the same traveling speed of the gantry traveling mechanisms on both sides, which will make the manufacture expense of the whole crane increase. Meanwhile, the larger the span, the longer the trolley has to travel, leading to decrease of the operating efficiency.

So taking the geological conditions of the terminal, handling efficiency and terminal layouts into account, the span usually ranges from 35 to 40 m.

② At present, ARMGs are configured with over-hanging arms or none, and the ARMGs with over-hanging arms include the single-armed on one side and the double-armed on both sides.

When the ARMGs are configured with over-hanging arms, gantry structures are usually adopted considering the pass of the containers, resulting in increased distance between the track centers of the trolley (usually around 16 m) as well as low structural stiffness, high weight, and high wheel loads. However, for the ARMGs without over-hanging arms, the operation passing lanes are configured inside the tracks and the pass of the containers is not a consideration, such design results in simple structures and good stiffness.

The weight of ARMG with over-hanging arms is much higher than that without over-hanging arms. With the same span, the weight will double and the wheel loads also increase.

For the ARMGs with over-hanging arms, since the operations take place outside the tracks, it is good for the organization of the operating vehicles in the terminal as well as operation safety. However, it also results in lower utilization rate of the yards.

2. ARTGs

The automation degree of ARTGs is increasing constantly with improving operation performance and flexibility (see Fig. 1.15). The ARTGs are configured with container stacks detecting systems, anti-collision systems, stacking guiding systems, trailer position detecting systems, automatic position indicating systems, and so on to ensure the collaborative operations with the AGVs.

1.6 Key Information of Containers and Data Structure

Containers are large goods holders with certain strength, stiffness, and specifications for circulation use. When transporting goods with containers, the goods can be loaded directly at the warehouse of the shipper and unloaded at the warehouse of the consignee, without reloading them when the vehicles or vessels are changed during the journey.

As goods holders, the containers are different from general holders since they have to follow some special requirements. The ISO has specific standards on container sizes, terms, testing methods as well as regulations on technical features such as structures, performance, etc. The standardization of containers has promoted the container circulation throughout the world and played a vital role to the rational exchange of the international goods.

In order to avoid mistakes during transportation, some attribute values should be set to the containers before the journey, including information about the container, status of the goods inside as well as transportation information.

Fig. 1.15 An ARTG

1.6.1 Information About the Container

(1) **Container no.** is the only identification for the container in theory, just like the ID no., which must be unique. The container no. includes two parts, namely holder code and sequence number. The container no. generally has 11 characters, for example like "YMCU2008570," the first four characters are holder code. According to the ISO regulations, the holder code should be represented with four capital Latin letters, the first three letters can be specified by the holder and the fourth letter must be "U." The last seven of the container no. is sequence number and represented with Arabic numerals. If the sequence number is less than 7 characters, the other characters should be filled with "0" before the effective characters, such as "0531842."

(2) **Size** refers to the length of the container, with feet as the unit. Containers in container terminals are 20, 40, and 45 ft. Usually, a standard 20 ft container is referred to as a TEU. Containers with other sizes, such as 10, 12, 30, 48, 53 ft, etc, are usually handled in general cargo terminals. The Fig. 1.16 has shown the most common 20 and 40 ft containers.

(3) **Container height**, namely the height of the container, has the sizes of 8′6″ and 9′6″ (see Fig. 1.17). Usually, the containers with the height of 8′6″ are

Fig. 1.16 20 ft and 40 ft containers

Fig. 1.17 Containers with the heights of 8′6″ and 9′6″

referred to as flat containers, and the containers with the height of 9′6″ referred to as high cubics. In the container terminals, the specific sizes in height are not concerned and the containers are only defined as flat or high cubic, among which the flat containers (8′6″) are the most.

(4) **Container types**, namely the descriptions on container types, which determine the goods types inside. Main container types handled in the container terminals include general purpose containers, bulk containers, flat rack containers, open-top containers, reefer containers, tank containers, etc, most of which are the general containers. The container type codes are shown in Table 1.2.

(5) **Container holders**, also known as container operators, refer to entity units that the containers belong to. Such information is quite important for the management of empty containers since they are usually stacked according to their holders in the terminals.

Table 1.2 Container types

Container types	Container type codes
General purpose	GP
General high cubic	GH (HC, HQ)
Bulk	BU
Garmentainer	HT
Open top	OT
Reefer	RF
Reefer high cubic	RH
Tank	TK
Flat racks	FR

1.6.2 Information About the Goods

(1) Net weight of the container is the gross weight of the goods inside, regardless of the weight of the container, and the unit is ton.

(2) Gross weight of the container is the weight of the goods inside of the export container plus the weight of the container, and the unit is ton. This is another important attribute of export containers. When the export containers are loaded into the vessels, the full containers must not be stacked on the empty ones, and the full containers at different tonnage levels should also be stowed separately, or stacked under empty ones, which means the containers in the yards should also be stacked with the same rules. So the gross weight of the container is closely related to location selection in the yards, stowage in the vessels, loading sequence, etc.

(3) Special goods requirements specify whether the container is over-limited or packed with dangerous goods. All the containers with records of special goods requirements should be specially treated when unloaded from the vessels and charged with extra handling fees. If a container is over-limited, this field of attribute will indicate the types of over-limit, including over-height, over-length, over-width, over-height and over-width, over-length and over-width, over-height and over-length, over-height and over-length and over-width. All the over-limited containers must be handled by the experienced drivers with steel wire ropes bound in advance instead of handled directly with quay cranes. The dangerous good containers can be classified into 9 grades according to the dangers of the goods inside and the specific criteria are list in Table 1.3.

As described in the table, the first grade is explosive, the second is gas, the third is flammable fluid, the fourth is flammable solid, self-combustible substance and substances in water releasing flammable gas, the fifth is oxidizing substances and organic peroxides, the sixth is toxic and infectious substances, the seventh is radioactive substances, the eighth is corrosives, the ninth is miscellaneous dangerous goods.

Table 1.3 Grading criteria of dangerous goods

Grade	Descriptions
D1.1	Substances and articles which have a mass explosion hazard
D1.2	Substances and articles which have a projection hazard but not a mass explosion hazard
D1.3	Substances and articles which have a fire hazard and either a minor blast hazard or a minor projection hazard or both, but not a mass explosion hazard
D1.4	Substances and articles which present no significant hazard
D1.5	Very insensitive substances which have a mass explosion hazard
D1.6	Extremely insensitive articles which do not have a mass explosion hazard
D2.1	Flammable gases
D2.2	Non-flammable gases
D2.3	Toxic gases
D3	Flammable liquids
D4.1	Flammable solids
D4.2	Substances liable to spontaneous combustion
D4.3	Substances which, in contact with water, emit flammable gases
D5.1	Oxidizing substances (agents) by yielding oxygen increase the risk and intensity of fire
D5.2	Organic peroxides—most will burn rapidly and are sensitive to impact or friction
D6.1	Toxic substances
D6.2	Infectious substances
D7	Radioactive substances
D8	Corrosives
D9	Miscellaneous dangerous substances and articles (Marine pollutants which are not of an otherwise dangerous nature are listed in Class 9)

(4) Goods descriptions are the descriptions about the goods types inside the containers. This attribute is not quite important to the terminals as they need to handle the containers instead of the goods inside.

1.6.3 Shipping Information

(1) Discharging port is the port where the container will be unloaded. The discharging port is a very important attribute of a container. Generally, in one voyage of a vessel, there will be several discharging ports. In order to avoid reshuffle when unloading containers at intermediate ports, containers of different discharging ports should be stowed separately in the vessel, which means they should also be stacked separately in the yards. So the attribute of discharging port is closely related to location selection in the yards, stowage in the vessels, loading sequence, etc. Once the discharging port is input with

mistakes, the container will be unloaded at the wrong port, which will bring a huge loss to the shipper.

(2) Destination port is the port at which the container finally arrives. Generally, shipping companies all have their fixed shipping routes. If the destination port of the cargos is inconsistent with the port at which the vessel calls, the cargos will be discharged to the nearest port of the destination, and then delivered to the destination port the clients request by other means of transportation. Therefore, when booking space with the shipping company, it is required to claim a specific destination port, in case the cargos are delivered to the wrong place. In particular, when trading with inland areas, inland city shall not be chosen as the discharging port. Instead, the port close to the city where vessel transportation can be arranged should be selected as the discharging port and inland city only as the destination port.

(3) Transshipment port is the port through which containers are transferred. Transshipment refers to the process in which a container loaded at one port is brought into another port in the vessel and unloaded to the yards, and then loaded into another vessel and shipped to the third port.

(4) Logistics status refers to the transition status of a container and whether it is packed or not. The transition status includes import, export, transshipment, and transit. According to logistics status, the containers can be divided as import full containers, import empty containers, export full containers, export empty containers, import transshipment containers, export transshipment containers, international transshipment containers, domestic trade containers, transit containers, etc.

The full containers refer to containers packed with goods and empty containers are containers without goods. Empty containers are dispatched by the shipping company. Transshipment containers refer to containers unloaded at one port before the destination and loaded into another vessel. Domestic trade containers refer to containers transported domestically. Transit containers refer to the containers that enter the port with the vessel but not unloaded and continue to be transported to other ports. The logistics status is usually a combination of attributes mentioned above.

(5) Bill of lading no. is the no. of the bill of lading. The bill of lading is used to take delivery of goods. A shipment of goods will only be allocated with one bill of lading and one bill of lading no., which are issued by the shipping company that transports the goods.

(6) Seal no. is usually behind the container no. marked with customs inspection. A seal is a device similar to a lock applied by specific people after the goods are loaded into a container and the door is properly closed. Once the seal is properly locked, it cannot be opened unless it is violently broken (i.e., cut open), and the damaged seal cannot be reused. Each seal has a unique identification no.. As long as the container has a complete appearance, the container door is properly closed and the seal is normally locked, it can be proved that the container has not been opened without permission during the transportation. The condition inside the container shall be supervised by the person who packed the container.

1.6.4 *Position Information*

1.6.4.1 Position in the Vessel

The stowage position of a container in the vessel is also referred to as slot number. A slot for a 20 ft container is a standard slot, and two adjacent standard slots can be combined as a slot for a 40 ft container. The stowage position of a container can be represented with six Arabic numerals, including bay, row, and tier in three dimensions. The first two are bay no., the middle two are row no., and the last two are tier no.

(1) In one longitudinal line from bow to stern (usually on the same plane of main deck) of a container vessel, the slots arranged longitudinally with a 20 ft container as the modules are referred to as bays. Its numbering rule is to be numbered from the bow to the stern with odd numbers (i.e., 01, 03, 05 …) in the same longitudinal line, with a 20 ft container as the module, see Fig. 1.18. The even number between two odd numbers represents the slot of a 40 ft container (the bay between 01 and 03 is represented by 02).

(2) The slots arranged laterally from starboard to portside of a container vessel are referred to as rows. The row no. is represented by even numbers on the starboard and odd numbers on the portside within the main deck transverse plane, with the middle slot as the cardinal number. If the total number of rows within the whole transverse plane is odd, then the middle slot is represented with 00, the starboard with 01, 03, and the portside with 02, 04.… If the total number of rows within the whole transverse plane is even, the 00 in the middle will be canceled, and then the portside be represented with even numbers and the starboard with odd numbers.

(3) With the main deck as the horizontal plane, the cabin and the deck are strictly separated, and the slots arranged vertically are referred to as tiers. The tier no. is represented with even numbers (with 02, 04, 06 … from the bottom up in the cabin and with 82, 84, 86 … from the bottom up on deck), with the height of a container as the module. If odd numbers (e.g., 01, 03, 05, …) appear, it means the container at this tier is only half the height of a standard container.

Figure 1.19 has shown a simple chart of a bay in the vessel, with clear row

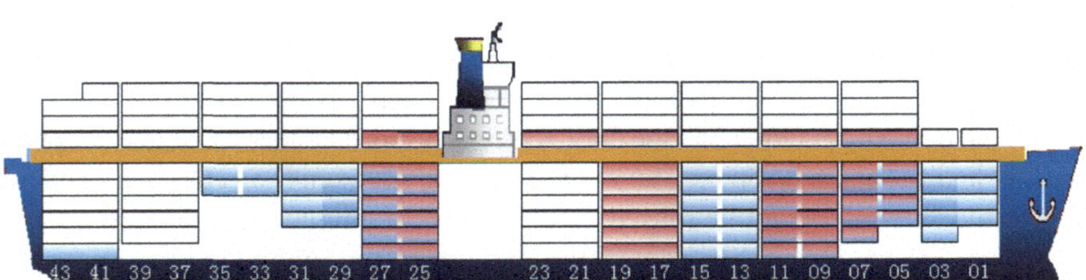

Fig. 1.18 Bays distribution on the vessel

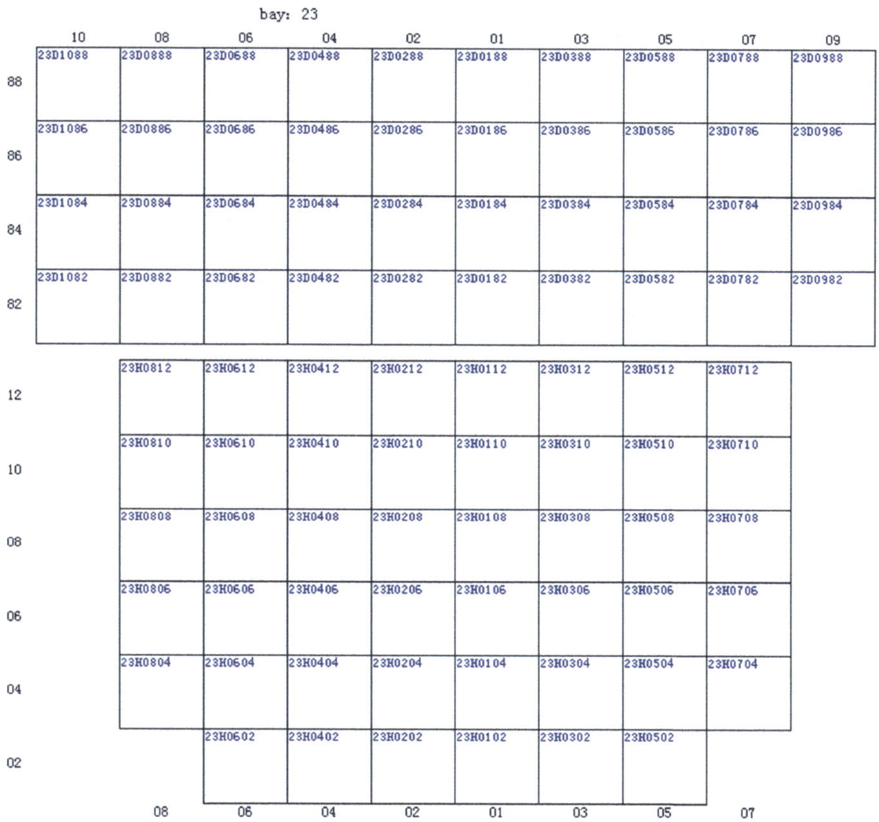

Fig. 1.19 A bay in the vessel

numbers and tier numbers.

(4) The slot no. has indicated a specific position in the vessel, including information of bay, row, and tier. If a container is placed in the slot with the no. of 070082, it means this is a 20 ft container placed at bay 07, in the middle row, at the first tier on deck. With slot no., the complete three-dimensional space in the vessel has been represented.

1.6.4.2 Location in the Yards

(1) Introduction of container blocks

Container blocks in the yards are areas used to temporarily store all kinds of containers, such as export containers to be loaded into the vessels, import containers unloaded from the vessels, transshipment containers, etc. Proper arrangement of blocks and container locations can reduce container reshuffle rate and waiting time of quay cranes and yard cranes, solve the bottleneck problems in the operation, and increase the loading and unloading speed at the apron. In addition, it can maximize the utilization rate of the yards and the passing capacity of the terminal, reduce the cost of production, maximize the handling capacity of the terminal and further improve its industrial competitiveness.

(2) Classification of blocks

Different types of containers have different stacking requirements in the yard blocks. For example, dangerous goods containers should be monitored strictly during the process of stacking, with the requirement of spray. In order to maintain the temperature in the reefer containers, it is necessary to have power interfaces available in the blocks. Therefore, according to the classification of container types, container blocks can be generally divided into the following types:

① By logistics status: import container blocks, export container blocks, and transshipment container blocks;
② By container status: full container blocks and empty container blocks;
③ By handling facilities: reach stacker blocks, gantry crane blocks, forktruck blocks;
④ By container types: general container blocks and special container blocks (including temperature controlling blocks, dangerous goods container blocks, over-limited container blocks, damaged container blocks, etc.)

It should be noted that all the block locations are relatively fixed except that the dangerous good container blocks and the temperature controlling blocks should be provided with devices like spray and power interfaces. For other container blocks such as import container blocks and export container blocks, the container clerks in the terminal may adjust the proportion of each container block appropriately according to the quantity of import and export containers, the actual storage situation in the terminal, the arrival situation of the vessel, and the container use situation of the shipping companies.

(3) Definitions of block, row, bay and tier

① Block

The block definition is the plane frame defining the container storage area. Any container terminal or depot operating container transportation business must plan the storage area of the container blocks, dividing it into several container blocks and number them according to the operation requirements. The no. of the container block is hereinafter referred to as the "block no.," usually composed of 2–3 letters and numerals, such as: A1, A2, A3… B1, B2, B3…, etc.

Each block is also divided into rows and bays, which are numbered with "row no." and "bay no.," respectively. Generally, the block handled by the gantry crane is defined as the standard blocks with six rows, but the number of bays is determined by the length of the block. The block should not be too long; otherwise, the efficiency of import and export throughput will be reduced due to the limitations of container truck lanes. The block plane is divided into several cells by the rows and bays, referred to as "stack serials." Stack serials are a series of stacks perpendicular to the ground (as shown in Fig. 1.20). The height of stack serials is the allowable stacking height of the block, counted with number of containers.

② Bay

Bays refer to sections formed by dividing a block in the length direction with a minimum unit length of a 20 ft container (Fig. 1.21).

In most terminal systems, there is a limit to the block length, namely, an upper limit to the number of bays in the block. The bay no. is represented by two characters, and the odd number is used to indicate the bay no. (small bay) used for stacking the 20 ft containers, that is, the bay no. starts from 01, 03, 05, and until 65. The combination of adjacent odd bays constitutes an even bay (large bay), such as 02, 06, 10. The even-numbered bays are used for stacking 40 ft containers, in which the bay 02 is formed by combining bay 01 and bay 03, and the bay 06 is formed by combining bay 05 and

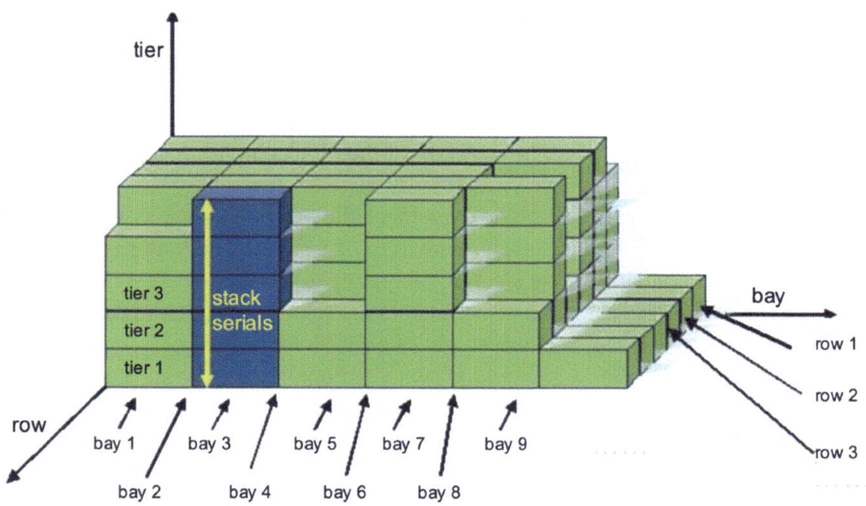

Fig. 1.20 Schematic diagram of a block

Fig. 1.21 Schematic diagram of a container yard

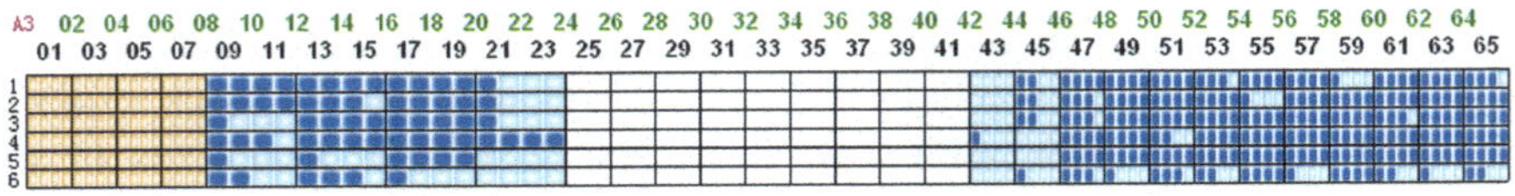

Fig. 1.22 Schematic plan of a container block

bay 07… Therefore, when a 40 ft container is placed at the bay 02, the corresponding bay 01 and bay 03 cannot be placed with 20 ft containers.

The following figure shows the bays distribution of a block, namely 01, 03, 05… Since the bay 02 is the combination of 01 and 03 and the bay 04 is the combination of 03 and 05, another 40 ft container cannot be placed at bay 04 after a 40 ft container is placed at bay 02 (Fig. 1.22).

③ Row

The row no. refers to the serial number in the direction perpendicular to the bays within the horizontal plane of a block, which is represented by an Arabic numeral, usually with 1, 2, 3…. The number of rows for stacking containers in each block is 5–6. Corresponding to each bay in the container block, there is a section diagram consisting of rows and tiers. The left one of the figures below is the actual stacking situation of the yard. It can be seen from the figure that there are 5 rows of containers stacked in the block. The right one is the sectional view of the block, from which the stacking status of each container at each bay of the block can be clearly seen.

④ Tier

Tier refers to the height of the containers allowed to be stacked, represented by an Arabic numeral, usually with 1, 2, 3, 4… from the bottom to the top. In general, the allowable stacking height of full containers is 4 tiers, while the stacking height of empty containers is relatively high. As can be seen from Fig. 1.23, the stacking height of each block is 4 tiers.

Each cell in the container block where a container can be stacked is called a "stack." It is the minimum unit that constitutes a container yard. A stack no. includes information of block, bay, row, and tier. Most block no. consist of one letter plus one numeral, therefore, the stack no. is usually composed of six letters and numerals. For example, the stack "A31022" means the 2nd tier the 2nd row and the 10th bay in the block A3.

CCLU2390872 GUOTA /E03 20 GP Flat Container EF 26730 CNSHA 21H0706 ☆A34914	SCZU7829227 GUOTA /E03 20 GP Flat Container EF 24405 CNSHA 23H0208 ☆A34924	CCLU3165027 GUOTA /E03 20 GP Flat Container EF 25250 CNSHA 23H0408 ☆A34934	CCLU3247125 GUOTA /E03 20 GP Flat Container EF 26250 CNSHA 21H0608 ☆A34944	CAXU6811098 GUOTA /E03 20 GP Flat Container EF 27602 CNSHA 23H0706 ☆A34954	GUOTA /E03 ☆A34964
CCLU2541217 GUOTA /E03 20 GP Flat Container EF 26250 CNSHA 21H0408 ☆A34913	TGHU3576851 GUOTA /E03 20 GP Flat Container EF 24116 CNSHA 21H0108 ☆A34923	CCLU2763240 GUOTA /E03 20 GP Flat Container EF 6610 CNSHA 23H0208 ☆A34933	CCLU3154357 GUOTA /E03 20 GP Flat Container EF 26250 CNSHA 21H0608 ☆A34943	CCLU2632678 GUOTA /E03 20 GP Flat Container EF 23728 CNSHA 23H0108 ☆A34953	GUOTA /E03 ☆A34963
GESU2735410 GUOTA /E03 20 GP Flat Container EF 24770 CNSHA 23H0308 ☆A34912	FSCU3453268 GUOTA /E03 20 GP Flat Container EF 24116 CNSHA 21H0508 ☆A34922	CCLU2846622 GUOTA /E03 20 GP Flat Container EF 26250 CNSHA 21H0308 ☆A34932	TGHU2708212 GUOTA /E03 20 GP Flat Container EF 21250 CNSHA 21H0610 ☆A34942	CCLU2456075 GUOTA /E03 20 GP Flat Container EF 23728 CNSHA 23H0508 ☆A34952	GUOTA /E03 ☆A34962
CCLU2546245 GUOTA /E03 20 GP Flat Container EF 20250 CNSHA 23H0610 ☆A34911	CCLU2596605 GUOTA /E03 20 GP Flat Container EF 6249 CNSHA 21H0812 ☆A34921	CCLU2489232 GUOTA /E03 20 GP Flat Container EF 22590 CNSHA 23H0708 ☆A34931	SCZU7124026 GUOTA /E03 20 GP Flat Container EF 16250 CNSHA 21H0612 ☆A34941	CCLU2059078 GUOTA /E03 20 GP Flat Container EF 23728 CNSHA 21H0708 ☆A34951	TGHU3647127 GUOTA /E03 20 GP Flat Container EF 12300 CNSHA 23H0812 ☆A34961

Fig. 1.23 The sectional view of a bay

Chapter 2
Operation Management in the Container Terminal

2.1 Introduction of Operations of Container Transportation

Container transportation means to transport different goods via containers with road transportation, rail transportation, inland waterway transportation, marine transportation, and air transportation. Meanwhile, extra protection is also provided to the goods in transportation. It is easy to transfer from one mode of transportation to another during the whole transportation process.

The process of the container transportation chain is from the consignor to the consignee—from the manufacturer or producer in the export country to the buyer in the import country. The consignor means the exporter, and the consignee means the importer.

The transportation of container goods has changed fundamentally in the whole process of circulation as well as in the transportation organization. The transportation laws and regulations of different countries, as well as the economic and geographical conditions on specific routes, determine the different modes of distribution and circulation processes of transportation of container goods. The general process of container transportation is shown in Fig. 2.1.

The inland container transfer (or distribution) center refers to the inland freight transfer site for the transshipment, unpacking and consolidation of containers and their goods under the supervision of the customs and other competent authorities of the state.

The feeder port refers to the terminal mainly berthing vessels with many ports of call, incapable of loading and unloading large vessels, and without trunk transportation, which is configured at the coastal ports and inland for the purpose of container collection and distribution. Its function is to collect the containers nearby and transport them to the import and export hub port.

In the process of transportation of container goods, there are two kinds of circulation forms: full container load (FCL) and less than container load (LCL).

N. Zhao et al., *Digital Management of Container Terminal Operations*,
https://doi.org/10.1007/978-981-15-2937-5_2

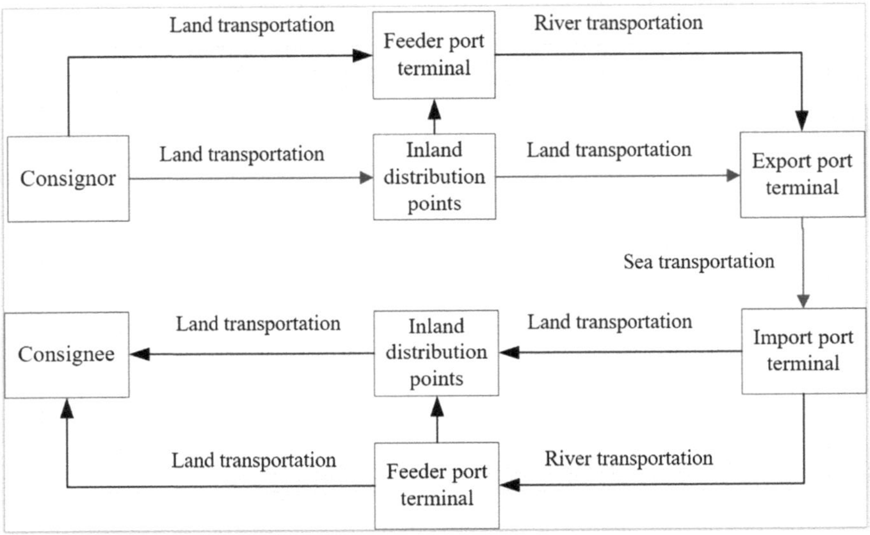

Fig. 2.1 General process of container transportation

1. The circulation procedures of FCL

Generally speaking, the circulation of FCL includes the following procedures.

(1) The consignor arranges the containers and packs the goods into the containers in his own factory or warehouse.
(2) The container goods are transported to the container terminal by inland transportation or inland river transportation.
(3) The handover is dealt with in the container terminal, and the container goods are temporarily stored in the yard according to the yard plans, waiting for loading into the vessel.
(4) The container goods are loaded into the vessel according to the loading plans.
(5) The container goods are transported to the discharging port by water transportation.
(6) The container goods are unloaded from the vessel according to the unloading plans.
(7) The container goods are temporarily stored in the yard according to the yard plans.
(8) The container goods are transported to the consignee's factory or warehouse by inland transportation.
(9) The consignee unpacks the containers in his own factory or warehouse.
(10) The empty containers are returned.

Three transportation systems are available for inland transportation from the consignor to the container terminal yard and from the container terminal yard to the consignee.

(1) Transportation by the consignor himself means the consignor is responsible for the configuration of empty containers and the transportation of packed containers and conducts the handover with the shipping company after transporting the containers to the container terminal yard.
(2) Transportation by the carrier means that the arrangement of empty containers and the (inland) transportation of the packed containers as well as the payment of freight are taken over by the shipping company. The carrier's liability begins in the consignor's factory or warehouse.
(3) Mixed transportation means that the shipping company is responsible for the configuration of empty containers and supervision, while the consignor is in charge of the transportation of the packed containers as well as the payment of freight. When the carrier is responsible for the transportation, the cost of inland transportation shall be part of the total cost.

2. The circulation procedures of LCL

Generally speaking, the circulation of LCL includes the following procedures.

(1) The consignor is responsible for the transportation of the goods to the CFS in the container terminal.
(2) The CFS is responsible for the container configuration and packing.
(3) The CFS transports the containers packed with goods to the container terminal.
(4) The containers are temporarily stored in the yard according to the yard plans, waiting for loading into the vessel.
(5) The container goods are loaded into the vessel according to the loading plans.
(6) The container goods are transported to the discharging port by water transportation.
(7) The container goods are unloaded from the vessel according to the unloading plans.
(8) The container goods are temporarily stored in the yard according to the yard plans.
(9) The container goods are transported to the CFS.
(10) The CFS unpacks the containers and delivers the goods.
(11) The empty containers are returned.

In the process of transportation, the shipping space is first arranged according to the voyage scheduled by the shipping company. The consignor can book the shipping space, make the documents, and transport the goods to the port all by himself, or ask a professional logistics company to do the relevant work. When the booking is completed, the shipping company will arrange transportation of the empty containers to the location of the consignor for packing. Upon completion of the packing, the consignor or his agent fills out the documents. The containers are transported to the export port by road, railway or inland waterway and stored in the terminal yard. After the vessel arrives at the port, the containers are loaded into the vessel according to the corresponding stowage plan and transported. When the containers arrive at the import port, they are unloaded from the vessel and stored in the yard, waiting for customs clearance and other import procedures. After that, the containers are delivered from the yard and transported to the consignee by inland means of transportation.

FCL and LCL are organized in four modes:

(1) The LCL is packed at the shipping origin, transported to the destination, and delivered to the consignee in FCL.
(2) The LCL is packed at the shipping origin, transported to the destination, unpacked and distributed to the consignee in LCL.
(3) The FCL is packed at the shipping origin, transported to the destination, and delivered to the consignee in FCL.
(4) The FCL is packed at the shipping origin, transported to the destination, unpacked and distributed to the consignee in LCL.

2.2 Main Documents in the Container Terminal

2.2.1 Document System in the Container Terminal

Operations in container terminals are an important section in container transportation. As service enterprises, the container terminals have a lot of customers, including shipping companies, shipping agents, freight forwarders, customs clearance companies, truck fleets, railway transportation departments, etc, and a lot of business documents involved, including container load plan (CLP), bill of lading (B/L), sea waybill, dock receipt (DR), equipment interchange receipt (EIR), manifest, pre-BAPLIE, BAPLIE, delivery order, cargo receipt, etc.

There are various types of documents in container terminals, which can be classified from different perspectives and requirements:

According to the documents in the import and export processes, they can be divided into import business documents and export business documents.

According to the documents preparation and circulation, they can be divided into documents from the external, internal circulation documents, and documents to the external.

According to the subjects of the documents, they can be divided into vessel operation documents and container operation documents.

2.2.2 The Functions of the Documents in the Container Terminal

Business documents run through the whole operation processes in the container terminal. Different documents have different contents, but they all play important roles in the processes of container terminal operations.

1. Documents record the basic information of the goods and are the main references for the container terminal to understand the goods and arrange them.
2. Documents are the references for the handover of goods and business, as well as the transfer and definition of responsibilities.
3. Documents are the references of operating charges.
4. Documents can be used to reflect the process of freight activities, as well as the status of each business section.
5. Documents are used for information storage and business tracking.
6. If there is an accident or dispute in the course of operation, the documents can be used as the references for responsibility definition.

2.2.3 Main Documents in the Container Terminal

(1) Bill of Lading

The bill of lading is the evidence of the contract of carriage of cargos by sea and the receipt of a shipment of cargos including carrier taking over, shipment and carrier guarantee; it is the main business document that international freight agent handles.

The functions of the bill of lading:

① The bill of lading is the reference of the contract of carriage of cargos concluded between the carrier and the shipper. The clauses and contents in the bill of lading usually claim the rights and duties of the carrier and the shipper, respectively, and are the references for handling disputes between the parties in the course of transportation.

② The bill of lading acts as a receipt of cargos. The carrier issues a bill of lading to the shipper after taking over the cargos or loading the cargos on board, indicating that the carrier has received the cargos specified in the bill of lading and is responsible for keeping them.

③ The bill of lading is a document of title, that is, the certificate of property right. By transferring the bill of lading, the ownership of the cargos is transferred.

Classification of the bills of lading:

① According to the titles of the bills of lading, they can be divided into Straight B/L, Order B/L, Bearer B/L.

② According to whether the cargos are loaded into the vessel, they can be divided into Received for Shipment B/L and Shipped on Board B/L.
③ According to the modes of shipment, they can be divided into Direct B/L, Transshipment B/L, Through B/L, and Charter Party B/L.
④ According to the presence of notation on the bill of lading, they can be divided into Clean B/L and Unclean B/L.
⑤ According to issuing time and requirements for use, they can be divided into Anti-dated B/L, Advanced B/L, and Stale B/L (Fig. 2.2).

(2) Sea Waybill
A sea waybill refers to a non-negotiable document issued by the carrier to the shipper to indicate that the cargos have been received and would be delivered by the carrier to the consignee with this document in the destination port of shipment.
Compared with the bill of lading, the sea waybill has the following characteristics:

① It is easier and more prompt for the consignee to take delivery of cargos when using the sea waybills.
② The sea waybill has the functions of receipt of cargos and evidence of marine transportation contract, but it is not the document of title, which cannot be negotiated. This is the main difference between the sea waybill and the bill of lading.
③ When the goods are delivered, the consignee takes delivery of the goods only with the delivery notice and identity certificate instead of the sea waybill, and the carrier releases the goods with the consignee's notice of arrival and consignment as well as identity documents.
④ The sea waybills are mainly used by transnational corporations in their freight operations in different regions, including short-distance shipping, long-term, stable cargo transportation between mutual trust customers, and cargo transportation of relatively low value or no risk of capital and so on.

(3) Dock Receipt
The dock receipt (DR) is the document issued by the container freight station, the container terminal yard, or the inland transfer station entrusted by the container carrier, after receiving FCL or LCL. It is a specialized export document for international container transportation.
The DR is a comprehensive document, which integrates the documents such as booking form, customs declaration, mate's receipt, freight notice, which improves the efficiency and circulation speed of container cargo consignment.
The functions of the DR:

(1) After the shipping company or the shipping agency confirms the booking and affix the document seal with the customs clearance qualification on the DR, the DR will be given to the shipper or his agent, which means the contract of transportation starts to be implemented.

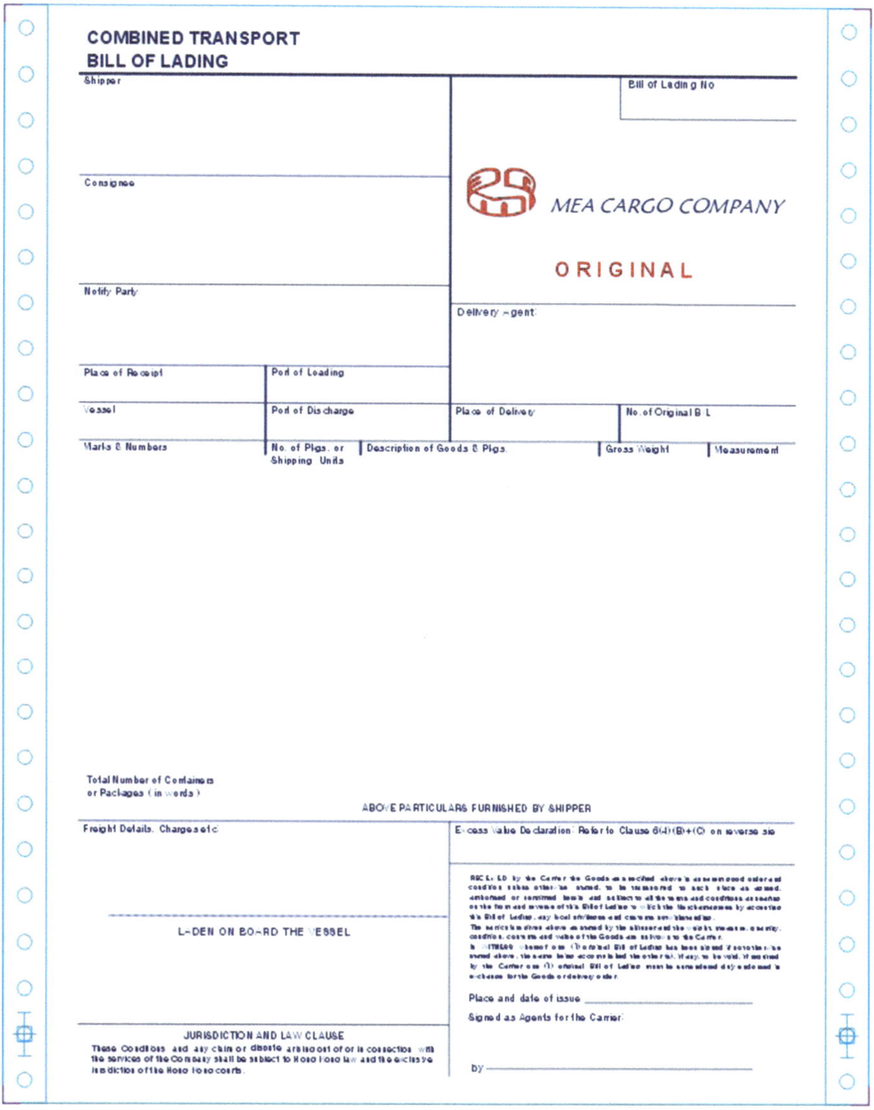

Fig. 2.2 A sample of the bill of lading

(2) The DR is one of the declaration documents of the export goods.
(3) The DR is the document to exchange for the shipping bill of lading.
(4) The DR is the evidence that the carrier has received the consignment and is responsible for it.
(5) The DR is the reference for the shipping companies and the ports to organize loading and unloading, tallying and stowage.
(6) The DR is the reference of freight settlement.

(7) If stipulated in the letter of credit, the DR can be used as a document of exchange settlement with the bank (Fig. 2.3).

(4) Cargo Receipt
The cargo receipt is a document issued by the container terminal yard or container freight station to prove that the delivery of cargos has been completed by

▽

Shipper （发货人）

D/R No.（编号）

Consignee （收货人）

场站收据
DOCK RECEIPT

第六联

Notify Party （通知人）

Received by the Carrier the Total number of containers or other packeges or units stated below to be transported subject to the temrs and conditions of the Carrier's regular form of Bill of Lading (for Combined Transport or port to Port Shipment) which shall be deemed to be incorporated herein.

Date （日期）：

Pre carriage by （前程运输） Place of Receipt （收货地点）

Ocean vessel （船名） Voy. No.（航次） Port of Loading （装货港）

场站章

Port of Discharge （卸货港） Place of Delivery （交货地点） Final Destination for Merchant's Reference （目的地）

Particulars Furnished by Merchants

Container No. （集装箱号）	Seal No. （封志号） Marks & Nos. （标记与号码）	No. of containers or P'kgs. （箱数或件数）	Kind op Packages: Description of Goods （包装种类与货名）	Gross Weight 毛重（公斤）	Measurement 尺码（立方米）
TOTAL NUMBER OF CONTAINERS OR PACKAGES(IN WORDS) 集装箱数或件数合计（大写）					

Container No. （箱号） Seal No. （封志号） Pkgs. （件数） Container No. （箱号） Seal No. （封志号） Pkgs. （件数）

Received （实收） By Terminal clerk （场站员签字）

FREIGHT & CHARGES	Prepaid at （预付地点）	Payable at （到付地点）	Place of Issue（签发地点）
	Total Prepaid （预付总额）	No. of Original B(s)/L （正本提单份数）	BOOKING （订舱确认） APPROVED BY

Service Type on Receiving ☐-CY, ☐-CFS, ☐-DOOR	Service Type on Delivery ☐-CY, ☐-CFS, ☐-DOOR	Reeter Temperature Required. （冷藏温度）	°F	°C

TYPE OF GOODS （种类）	☐Ordinary, （普通） ☐Reefer, （冷藏） ☐Dangerous, （危险品） ☐Auto. （裸装车辆） ☐Liquid, （液体） ☐Live Animal, （活动物） ☐Bulk （散货） ☐______	危险品	Glass: Property: IMDG Code Page: UN NO.

Fig. 2.3 A sample of a DR

both parties and to indicate the status of the delivery when the cargos are delivered to the consignee. It also certifies that the carrier's liability for the carriage of the cargos has been terminated.

The second form of the cargo receipt is the delivery order, which is the document with which the consignee or his agent can claim the goods from the container terminal or container freight station. The cargo receipt has a big difference in nature from the bill of lading; it is not a document of title and cannot be negotiated; instead, it is just a document that the shipping company or its agent instructs the container terminal or the container freight station to deliver the cargos to the designated owner.

(5) Equipment Interchange Receipt

The equipment interchange receipt (EIR) is the certificate that the container holder or its renter entrusts the container terminal yard to hand over the container and related equipment to the container user.

The main contents of the EIR include vessel, voyage no., container user, bill of lading no., container no., container type, size, place of delivery, status, purpose of gate-in and gate-out, damage record, etc.

The main function of the EIR is to track and manage the containers and define responsibilities.

The EIRs are divided into two types: IN and OUT (Fig. 2.4).

Since March 20, 2018, Shanghai Port has carried out the pilot work of electronic international container EIR at Shanghai water transport ports, and the paper EIR has been not in use.

2.3 Common Handling Techniques in Conventional Container Terminals

Compared with the general cargo terminal, the most prominent advantage of the container terminal is the high efficiency of operations. The handling techniques and processes are the guarantee of high efficiency of terminal operation. The proper selection of handling techniques and processes is the precondition of smooth operation in the terminal.

The so-called handling techniques refer to the methods and procedures for loading, unloading, and moving cargos in the terminal. According to certain operating procedures and port conditions, the loading, unloading, and moving tasks are completed in the principle of rationality and economy for different cargos, means of transportation, and handling facilities.

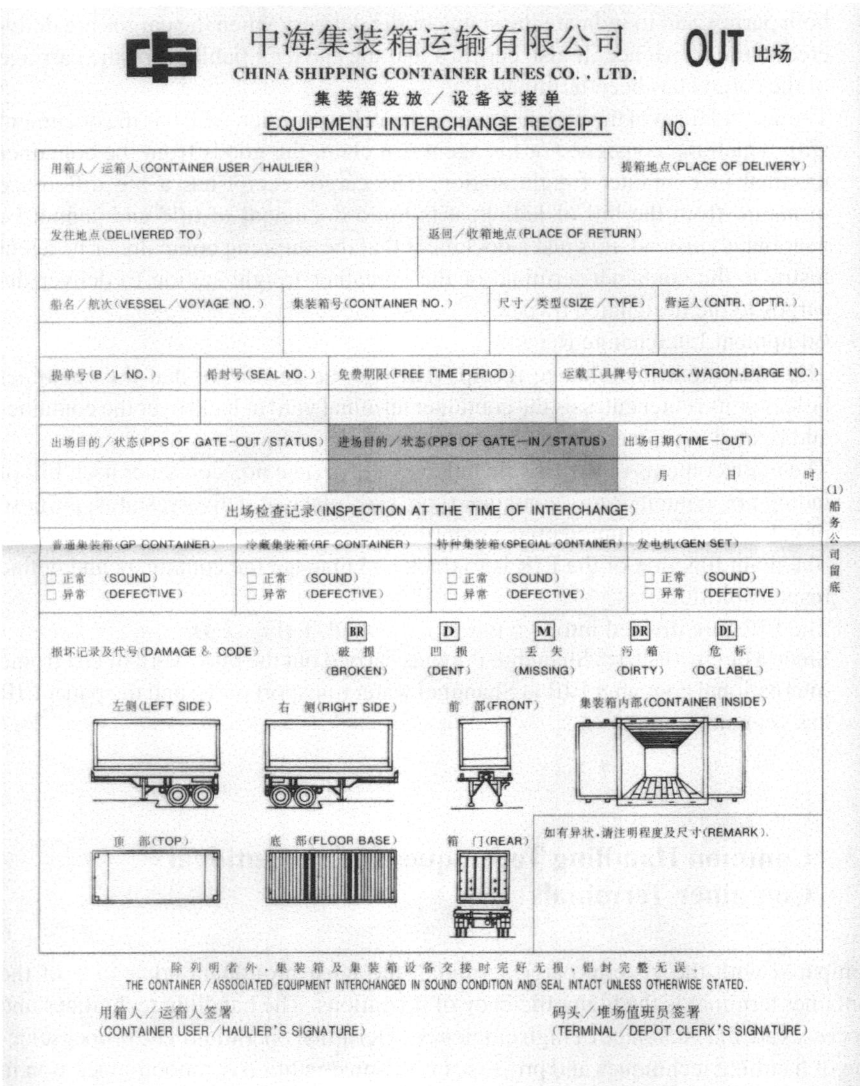
中海集装箱运输有限公司
CHINA SHIPPING CONTAINER LINES CO., LTD.
OUT 出场
集装箱发放／设备交接单
EQUIPMENT INTERCHANGE RECEIPT
NO.

用箱人／运箱人(CONTAINER USER／HAULIER)			提箱地点(PLACE OF DELIVERY)
发往地点(DELIVERED TO)		返回／收箱地点(PLACE OF RETURN)	
船名／航次(VESSEL／VOYAGE NO.)	集装箱号(CONTAINER NO.)	尺寸／类型(SIZE／TYPE)	营运人(CNTR. OPTR.)
提单号(B／L NO.)	铅封号(SEAL NO.)	免费期限(FREE TIME PERIOD)	运载工具牌号(TRUCK, WAGON, BARGE NO.)
出场目的／状态(PPS OF GATE-OUT／STATUS)	进场目的／状态(PPS OF GATE-IN／STATUS)	出场日期(TIME-OUT) 月 日 时	

出场检查记录(INSPECTION AT THE TIME OF INTERCHANGE)

普通集装箱(GP CONTAINER)	冷藏集装箱(RF CONTAINER)	特种集装箱(SPECIAL CONTAINER)	发电机(GEN SET)
□ 正常 (SOUND) □ 异常 (DEFECTIVE)	□ 正常 (SOUND) □ 异常 (DEFECTIVE)	□ 正常 (SOUND) □ 异常 (DEFECTIVE)	□ 正常 (SOUND) □ 异常 (DEFECTIVE)

损坏记录及代号(DAMAGE & CODE)
BR 破损 (BROKEN)
D 凹损 (DENT)
M 丢失 (MISSING)
DR 污箱 (DIRTY)
DL 危标 (DG LABEL)

如有异状，请注明程度及尺寸(REMARK).

(1) 船务公司留底

除列明者外，集装箱及集装箱设备交接时完好无损，铅封完整无误。
THE CONTAINER／ASSOCIATED EQUIPMENT INTERCHANGED IN SOUND CONDITION AND SEAL INTACT UNLESS OTHERWISE STATED.

用箱人／运箱人签署
(CONTAINER USER／HAULIER'S SIGNATURE)

码头／堆场值班员签署
(TERMINAL／DEPOT CLERK'S SIGNATURE)

Fig. 2.4 A sample of an EIR

2.3.1 Handling System of Rubber-Tyred Gantry Crane Technology

The process scenario of rubber-tyred gantry (RTG) crane technology can result in high efficiency in yard utilization and is commonly adopted in China, especially in

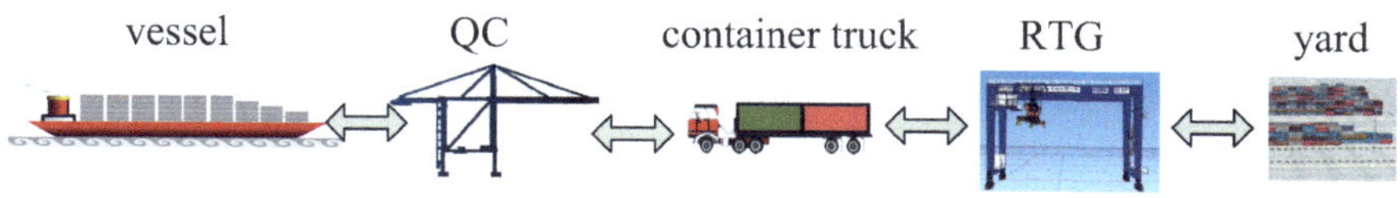

Fig. 2.5 Technological processes of RTG

those conventional container terminals of large scale. The following contents of this book are all based on such technological method.

In the system of rubber-tyred gantry crane technology, the loading and unloading operations at the apron are undertaken by the quay cranes (QCs).

The rubber-tyred gantry crane is responsible for loading, unloading, and stacking operations in the yard. The horizontal movement from the apron to the yard and between the blocks in the yard is completed by container trucks. The rubber-tyred gantry cranes can generally span 6 rows of containers, plus 1 row of truck passing lane. The stacking height is of 3–5 tiers of containers. The rubber-tyred gantry cranes are equipped with steering devices, which can achieve movement of the cranes from one block to another for operation (Fig. 2.5).

(1) Technological processes

$$\text{vessel} \rightarrow \text{QC} \rightarrow \text{container truck} \rightarrow \text{RTG} \rightarrow \text{yard}$$

For the import containers, the specific technological processes are that the container is unloaded from the vessel by the QC at the apron, loaded onto the container truck waiting under the QC, moved horizontally to the yard, and then handled by the RTG to the location in the yard. For container delivery, the container is loaded by the RTG onto the external container truck and transported away from the terminal. The technological processes of the export container have the reverse features.

(2) Advantages of RTG technology

① The yard is highly utilized. The RTG can stack up to 30 containers at one bay, and the clearance between the containers is small, so that the storage space can be effectively used.
② The cost of yard pavement is relatively low.
③ The facility is easy to operate, and the technicians only need the operation training of medium skill level.
④ Compared with the straddle carrier system, the damage to the containers is less.
⑤ The RTG turns with 90° steering and fixed axle steering, taking up small passage area.
⑥ Compared with the rail-mounted gantry crane, the RTG can move from one block to another without the limitation of the tracks.

⑦ The automatic control of the running tracks can be realized by the straight-line automatic control devices and controlled by the computer, easy to achieve automatic handling of containers.

(3) Disadvantages of RTG technology

① Compared with the straddle carrier system, the system has not adequate flexibility. Although cross-blocks operation can be realized, it takes a long time to travel.
② Since the span of the RTG is large and the stacking tier is high, it is difficult to deliver containers, resulting in high reshuffle rate.
③ The RTG has to be configured with container trailers to undertake horizontal movement, which increases the operation steps.
④ The initial investment is also high. Each QC has to be configured with more than 4 RTGs, and the high cost of RTGs increases the fixed cost of the terminal.

The RTG technology system is applicable to the terminals with small land area, and most conventional container terminals in China have adopted this system.

2.3.2 Handling System of Rail-Mounted Gantry Crane Technology

Compared with the system of the RTG technology, the span of the yard facility is larger and the stacking capacity is higher in the system of rail-mounted gantry crane (RMG) technology. The RMGs can stack 4–5 tiers of containers and can span 14 or more rows of containers (Figs. 2.6 and 2.7).

(1) Technological processes
The handling processes of RMG usually have the following two types:

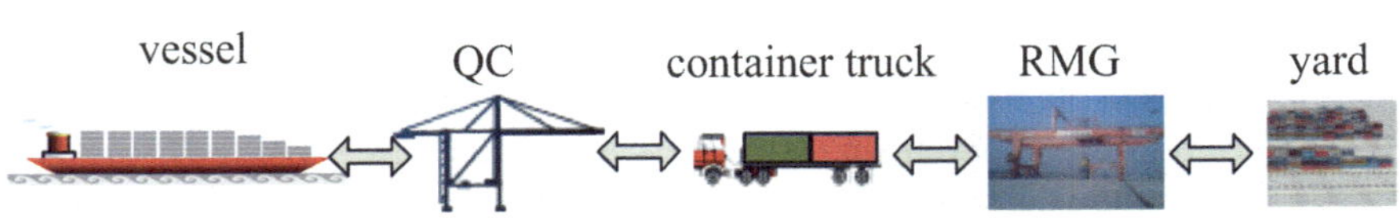

Fig. 2.6 Technological processes of RMG (1)

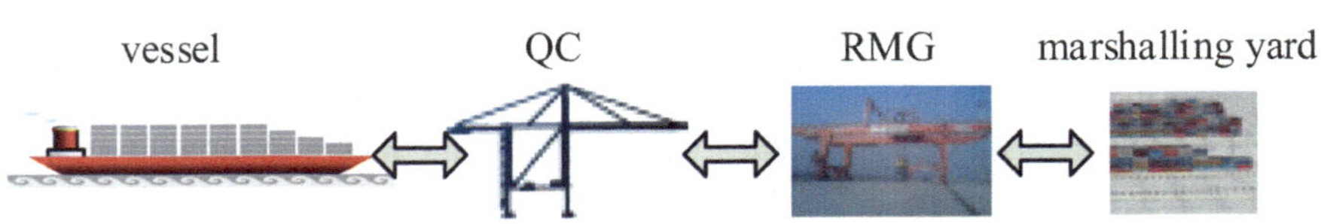

Fig. 2.7 Technological processes of RMG (2)

vessel → QC → container truck → RMG → yard

vessel → QC → RMG → marshalling yard

The first type is applicable to the backup yard in the terminal. The processes are that the container is unloaded from the vessel by the QC, moved by the container truck horizontally to the backup yard, and then handled by the RMG to the location in the yard. The second type is applicable to the marshaling yard in the terminal. After the container is unloaded from the vessel by the QC, the QC is directly connected with the RMG to complete the operation in the marshaling yard.

(2) Advantages of RMG technology

① The utilization rate of the yard is high.
② Simple mechanical structure results in high reliability of operation and maintenance.
③ The maintenance and management costs as well as the operating costs are low.
④ The RMG is driven by electricity, resulting in low energy consumption.
⑤ The RMG moves on the tracks, which facilitates the computer control and automatic handling of containers.

(3) Disadvantages of RMG technology

① The RMG can only move on the tracks, resulting in low flexibility and limited operation range.
② Large span makes the delivery and transposition of the containers difficult.
③ Investment is relatively high.

The RMG technology system is applicable to the water terminals with limited land area, large container throughput and high degree of computer control.

2.3.3 Handling System of Trailer Technology

The trailer system is a handling technology first adopted by the SeaLand Service Company, the USA, and so it is also called the "SeaLand mode." The loading and unloading are undertaken by the QC at the apron. The main feature of the system is that the container is placed on the trailer for the entire duration in the yard (Fig. 2.8).

(1) Technological processes

vessel → QC → trailer → yard

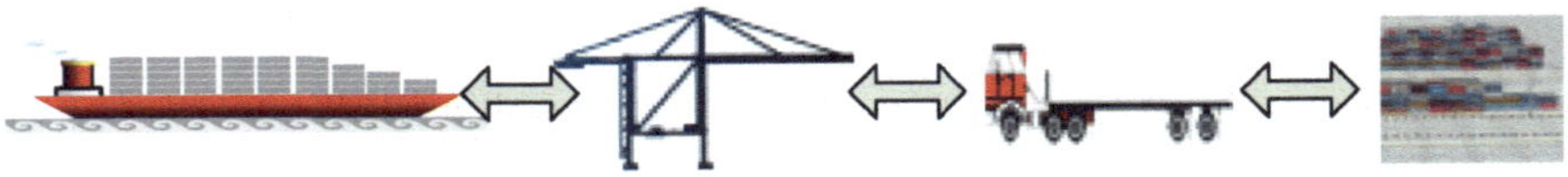

Fig. 2.8 Technological processes of trailer system

The import container is directly loaded by the QC onto the trailer, and the trailer with container is trailed by the tractor to the yard for parking. When leaving the terminal, the tractor directly trails the trailer loaded with container out from the yard and delivers the container to the customer's door. The export container loaded on the trailer is trailed by the tractor from the outside into the terminal and parked in the yard. When loading, the tractor trails the trailer loaded with container from the yard to the apron, and the container is loaded into the vessel by the QC.

(2) Advantages of trailer technology

① The operation number of containers in the terminal has been reduced, resulting in high handling efficiency and low container damage rate.
② The trailer can be directly used for land transportation, suitable for door-to-door transportation.
③ The low wheel pressure of the trailer has relatively low requirements on the bearding capacity of the yard, reducing pavement investment.
④ The operating organization is simple, and technical requirements on handling technicians and management personnel are relatively low.
⑤ It is not necessary to configure expensive handling facility in the yard.

(3) Disadvantages of trailer technology

① In order to make it convenient to park the trailer and perform the trailing operation, a large area is required, resulting in low utilization rate of the yard.
② The demand for trailer as well as the investment is large, the operation could be interrupted due to the lack of the trailers in the peak of traffic volume.
③ It is not easy to realize automation.
④ Large terminals with such systems have long hauling distances and are prone to traffic congestion in the terminals during rush hours.

The handling system of trailer technology is applicable to container terminals with small container throughput, large space and more FCL, and less LCL. There are very few terminals in the world that adopt such handling technology system.

2.4 Common Handling Techniques in Automated/Semiautomated Container Terminals

Compared with the conventional container terminals, the most prominent advantage of the automated/semiautomated container terminals is the remotely controlled operations, which have overcome the limitation of severe weather on manual operations, thus significantly improved the efficiency of operations. There is not much difference between automated/semiautomated container terminals and conventional container terminals in terms of overall layout, but there are big differences in handling techniques. The quay cranes (QC) are generally applied for vessels' loading and unloading operations. For the operations in the yard, most of the conventional container terminals generally select the rubber-tyred gantry cranes (RTGs) and a small number select rail-mounted gantry cranes (RMGs), while the automated terminals basically select RMGs.

At present, the automated container terminals are divided into four categories according to different horizontal moving facilities, namely fully automated terminals with AGVs or three-dimensional (3D) distribution system for horizontal movement, and semiautomated terminals with straddle carriers or container trucks for horizontal movement.

2.4.1 Fully Automated Terminals with AGVs for Horizontal Movement

It is the AGVs with autonomous navigation and positioning functions that are adopted by the first automated terminals in the world (ECT) as horizontal movement facility. The layout of such type of automated terminal is more flexible, and AGVs are applicable to the terminal with different sizes and geometric shapes of the yards, and the backup yards configured perpendicular to the coastline of the terminal. The first generation of AGVs could only drive along fixed routes, which has been optimized by the CTA in Hamburg, where the driving is more flexible and not restricted by the working routes. The AGVs are capable of turning and driving anywhere without obstacles.

In order to improve the operating efficiency, the QCs are generally configured with double trolleys. The working range of the main trolley is between the vessel and the transit platform, while the ancillary trolley is responsible for the connection of the transit platform and AGVs. In the yard, usually two ARMGs are configured for one block (Fig. 2.9).

AGVs can be powered by an internal combustion engine, batteries, or hybrid power. One AGV can carry two 20 ft containers or one 40 ft container at a time with the positioning accuracy up to ± 5 cm. Equipped with ultrasonic detecting device and other anti-collision devices, the AGVs are capable of detecting obstacles within the brake distance to avoid collisions. The AGVs and ASCs in the container yard

Fig. 2.9 Simulation of the yard layout perpendicular to the shoreline

are remotely managed and controlled by the central controlling office, realizing the complete automation.

Since AGVs cannot be loaded or unloaded by themselves, there is a possibility that they may wait for each other in the operation interaction areas with the ARMGs. In order to improve the operating efficiency and solve the "handshake" problems between AGVs and ARMGs, ZMPC has developed the AGV mate, working in the operation area at the end of the yard. Typical terminals for AGV mate applications are ECT in the Netherlands, CTA in Hamburg, Germany, Euromax in the Netherlands.

2.4.2 Semiautomated Terminals with Straddle Carriers for Horizontal Movement

Similar to the AGVs, the straddle carriers also have flexible traveling paths and are suitable for terminals of different sizes and shapes. Meanwhile, the straddle carriers are capable of self-loading and self-unloading, resulting in freely transshipment of containers between different blocks. For such handling techniques, the storage yards are generally arranged perpendicular to the coastline of the terminal and two ARMGs are configured in each block, respectively, responsible for straddle carriers and container trucks operations at both ends (Figs. 2.10 and 2.11).

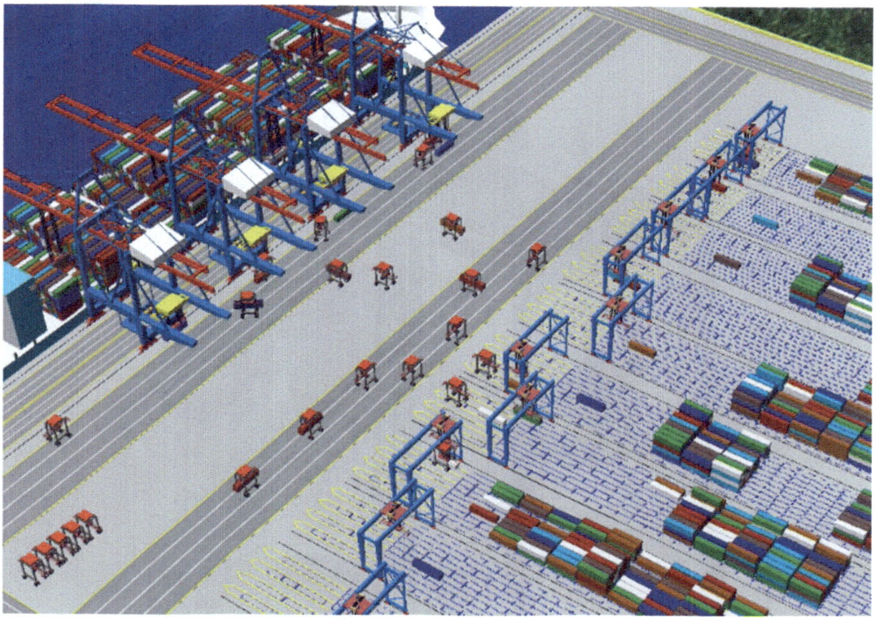

Fig. 2.10 Simulation of automated terminal with straddle carriers

The straddle carrier can be powered by an internal combustion engine, hybrid power, or electricity. Typical terminals applying straddle carriers are Haniin terminal in South Korea, the TTI terminal in Spain and Virginia Terminal in the USA.

2.4.3 Fully Automated Terminals with 3D Distribution System for Horizontal Movement

This technique is developed independently by ZPMC and applicable for small and medium-sized terminals of regular shape (usually rectangle) with 1 or 2 berths. The 3D distribution system is composed of lower bridge structures, lifting trolley, lower bridge shuttles, and ground shuttles (Fig. 2.12).

The lifting trolley is responsible for the container transfer between the lower bridge and the ground shuttle. Since the container stacking direction in the vessel is different from that in the yard, the lifting trolley is capable of $\pm 90^{\circ}$ turning. The lower bridge shuttle is responsible for the container transfer between the QC and the lifting trolley. The ground shuttle is capable of transporting the containers from the shuttle on the lower bridge to any designated position in the yard.

Since all the facilities run on the tracks, the scheduling and controlling system is simple and reliable, which can reduce the traffic congestion caused by horizontal

Fig. 2.11 A straddle carrier

movement. The lifting trolleys and shuttles running on the tracks are all electricity-driven without internal combustion engine. The RMGs in the yard do not have to travel long distance, which is energy-saving and environment-friendly. Modern means such as magnetic label are adapted to the facilities running on the tracks to achieve precise positioning with the accuracy of 5 mm as well as effective obstacle avoidance and emergent operation stop (Figs. 2.13 and 2.14).

The typical example applying such technique is the demonstration line that the ZPMC built in its Changxing base in Shanghai, but there has been no practical application yet.

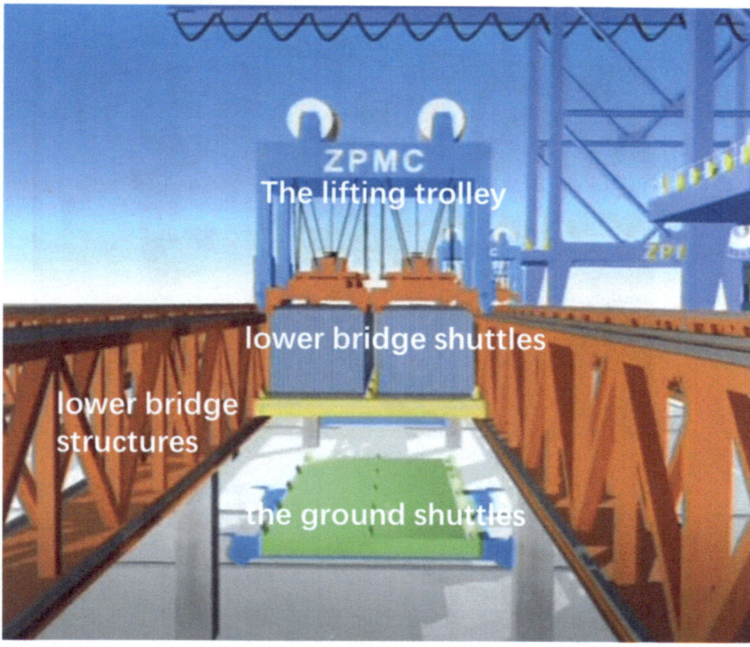

Fig. 2.12 Composition of the 3D distribution system

Fig. 2.13 Simulation of automated terminal with 3D distribution system

Fig. 2.14 A demonstration line of the automated terminal applying the 3D distribution system

Fig. 2.15 Haniin and Busan terminals in South Korea

Table 2.1 Comparison of handling technique features of each generation of container terminals

Technical features	ECT Delta, DDE, DDW	CTA, Hamburger, Germany	Euromax, Rotterdam, the Netherland	Tobishima, Japan	Virginia International Terminal, USA	Pasir Panjang, Singapore
QC	Semiautomated with single trolley	Semiautomated with double trolleys	Semiautomated with double trolleys	Semiautomated with single trolley	Semiautomated with double trolleys	Semiautomated with single trolley
Operation area of the horizontal vehicles at quays	Within the span of the QC	Within the outreach of the QC at landside	Within the outreach of the QC at landside	Within the outreach of the QC at landside/Within the span of the QC	Within the outreach of the QC at landside	Within the span of the QC
Horizontal movement and driven power	AGV/internal combustion engine	AGV/internal combustion engine	AGV/internal combustion engine	AGV/internal combustion engine	Vehicles on tracks/electricity	Tractor and trailer combination/internal combustion engine
Yard arrangement	Perpendicular to the coastline	Perpendicular to the coastline	Perpendicular to the coastline	Parallel to the coastline	Perpendicular to the coastline	Parallel to the coastline
Operation area of the horizontal vehicles in the yard	At the landside of the yard	At the landside of the yard	At the landside of the yard	Inside the yard	Inside the yard	Inside the yard
Horizontal movement distance	Short	Short	Short	Long	Long	Long
Yard cranes/driven power	ARMGs, two for each block/electricity	ARMGs, two for each block, capable of passing through/electricity	ARMGs, two for each block/electricity	ARTGs/internal combustion engine	ARMGs, two for each block/electricity	Automatic overhead bridge/ARMGs/electricity
Stacking tier and width	4 tiers(1 over 4), 6 rows	5 tiers(1 over 5), 10 rows	5 tiers(1 over 5), 10 rows	–	–	8 tiers (1 over 8) OHBC: 5 rows on each side; RMG: 12 rows

(continued)

Table 2.1 (continued)

Technical features	ECT Delta, DDE, DDW	CTA, Hamburger, Germany	Euromax, Rotterdam, the Netherland	Tobishima, Japan	Virginia International Terminal, USA	Pasir Panjang, Singapore
Rehandling in the yard	When a container is unloaded from a container truck and sent directly to the other end of the block, it needs to be rehandled by a yard crane adjacent to the seaside	There is no need for rehandling, since the yard cranes of the same block have different track center distances, they can run through each other.	When a container is unloaded from a container truck and sent directly to the other end of the block, it needs to be rehandled by a yard crane adjacent to the front end	–	–	–
Operation area of the container trucks in the yard	Landside handling point	Landside handling point	Landside handling point	Inside the yard	Inside the yard	Inside the yard
Ratio of QC:AGV:ARMG	1:8.8:6	1:6:3	1:8:4.8	–	–	–
Initial investment	High	Higher	High	High	High	High
Impact of faults to operations	The failure of one crane has a great impact on the other's operation in the same block	The impact is small; the mutual influence has been effectively avoided.	The failure of one crane has a great impact on the other's operation in the same block	The failure of one crane will lead to the complete shutdown of the operations in the block.	The failure of one crane has a great impact on the other's operation in the same block	When a facility fails, it can be removed from the operation area without delay
Requirements on automatic handling technology in the yard	High	Higher	High	High	High	High
Technical level	The first generation	The second generation	The third generation	The third generation	The third generation	The first generation

2.4.4 *Semiautomated Terminals with Container Trucks for Horizontal Movement*

Besides the three types of automated container terminals mentioned above, some of the existing terminals adopt the remotely controlled RMGs in the yard, but the horizontal movement is still undertaken by the container trucks. These terminals are mainly concentrated in South Korea and Taiwan, such as Haniin terminal and Busan new port in South Korea, Taipei Port, and Kaohsiung Yangming Port in Taiwan (Fig. 2.15).

Since these terminals have only achieved semiautomation in the yard, the automation degree of the whole terminal is low. With the rapid growth of port throughput and the trend of large-scale container vessels, especially the 18,000 TEU vessel of 3E class is about to be put into use, the port operators have begun to focus on the stability and high efficiency of container handling, energy conservation, and environment protection as well as reducing operating costs. Improving the automation degree of the container terminal is the inevitable trend for reducing the human costs, improving the through capacity, reducing energy consumption, and improving the image of the ports.

2.4.5 *Comparison of Handling Technique Features of Typical Automated Terminals*

The development of international automated container terminals can be roughly divided into three stages. According to the technical level of facility automation as well as management and controlling system, they can also be referred to as automated container terminals of three generations. The comparison of handling technique features of each generation of container terminals is shown in Table 2.1.

2.5 General Introduction on Import and Export Operations in the Container Terminal

The container terminal plays a central role in the whole process of container transportation, where all the transporting processes converge. In addition, the container terminal controls the speed and efficiency of international transportation and intermodal transportation to a large extent. The container terminal provides many services to shipping company, shippers, and consignees. Generally, the services offered by the container terminal can be divided into three stages of activities.

(1) The arrival of containers;
(2) The storage of containers;
(3) The departure of containers.

2.5.1 Export Operations in the Container Terminal

The processes of the export operations in the container terminal are shown in Fig. 2.16.

"Receiving" means that the containers are transported to the container terminal with vehicles by road, rail or with barges, and the process is the arrival of containers. "Storage" means the temporary stacking of containers in the yard or warehouse until the vessel arrives at the port for loading, which is the storage of containers. "Loading" means loading the containers into a vessel and carrying them out of the port, and the process is the departure of containers.

For export containers, some general information about the containers is sent by the shipping company to the terminal before the containers are shipped. The information is input into the terminal's information system so that terminal planners can prepare a container collection plan. Then, the detailed information about the containers and goods inside is also sent by the shipping company to the terminal and input into the information system. The terminal can make a thorough yard plan for temporary storage when the containers are waiting for the arrival of the vessel after receiving, according to the loading and stowage plan.

Containers are usually transported to the terminal for shipment by road, rail, or inland waterway a few days before the vessel arrives in a pre-planned sequence. When a container arrives with the documents, it can be immediately checked and updated against the container record already in the terminal information system, inspected for damage, the seals on the doors and weighed. An equipment interchange receipt (EIR) is prepared which records the arrival time of the container and specific information about the container and its condition. The EIR is provided to the driver as a receipt (to confirm that the container is handed over to the terminal for custody) and a control document so that the container can be delivered to the terminal in a controllable manner. The EIR is issued for both the packed containers that are ready to be shipped to the storage area of the terminal (ready for loading), and the empty containers either to the CFS (ready for packing) or to the storage area (waiting for further instructions from the container holder). Containers transported by rail or inland waterways also follow corresponding procedures of receiving documentation, inspecting, and handling.

For the goods that need to be packed at the CFS of the terminal, they need to arrive at the port within a certain time before the arrival of the vessel, so as to ensure that there is time for grouping the goods with other goods in the same container. The

Fig. 2.16 Processes of the export operations in the container terminal

shipping company prepares the packing list and delivers it to the CFS, and the CFS personnel load the goods into empty containers according to the packing list. Customs clearance and other procedures are usually dealt when the goods are packed at CFS. When packing is completed, the containers' doors are sealed. The information about the containers and the goods inside is updated in the terminal information system, and the packed containers are transported to the yard waiting for the vessel's arrival at the port and further loading operations.

We can summarize the activities of the first part as the "receiving" of containers, and the second part as the storage of containers. Upon completion of the receiving procedures, the containers (from the place of reception or the CFS) are transported to the designated locations in the storage blocks in the terminal. The container location in the yard has been prearranged according to the vessel to be loaded, discharging port, size, and weight. The reefer containers, dangerous goods containers, or over-limited containers shall be placed in designated special container blocks. When unloaded into the yard, the specific location of each container is recorded in the corresponding container record in the terminal information system so that the container can be found quickly and accurately at the time of being loaded into the vessel.

The final stage is the loading operation, in which the export containers are moved from the yard to the apron in a carefully planned sequence. There, the containers are inspected again and loaded into the vessel's planned loading position. It is necessary to tally and check whether the loaded containers are correct and complete. Containers for loading may also be those that have just been transported to the terminal, which may be moved to the vessel directly without storage in the yard. Even so, the relevant information should be input into the information system comprehensively in each process related to the containers. After the last container is loaded on board and the containers loaded on deck have been fastened, the stowing information is updated and corrected, ready to be sent to the next port of call of the vessel, after which the vessel can depart. In this way, the third stage of the export container's activities is completed and the containers leave the terminal.

2.5.2 Import Operations in the Container Terminal

The processes of the import operations in the container terminal are shown in Fig. 2.17.

Fig. 2.17 Processes of the import operations in the container terminal

"Unloading" is the process that the containers are unloaded from the vessel to the terminal, which is the arrival of containers. "Storage" means the temporary stacking of containers in the yard or warehouse until the containers are delivered, which is the storage of containers. "Delivery" means the delivery of the containers by road vehicles, rail, or barge, which is the departure of containers.

Although the activity sequence of import and export containers is different and the flow direction through the terminal is completely opposite, they still have a lot in common, especially in the way of container information recording, which is all input into the management information system (MIS).

For the import containers, information about the containers, including the stowing positions in the vessel, is transmitted to the terminal in advance by the shipping company, so that the terminal planners can prepare a detailed unloading plan and a yard plan and establish records for the containers to be unloaded. The vessel shall be unloaded in accordance with the plan as closely and efficiently as possible to minimize the vessel's duration at the port. The containers are hoisted successively and placed beside the vessel for inspections of the body state and door seals. Any damage must be recorded and reported to the shipping representative. The containers are recounted, and the terminal information system is updated to record the arrival time of the containers at the port.

After the containers are unloaded from the vessel, they will be moved to the yard. Like the export containers, the import containers will be sent to the pre-designated locations in the yard according to the size, nature of goods, destination and mode of transportation. The container records in the terminal information system are updated again, and the stacking locations of the containers are recorded, so that the containers can be found immediately when they need to be shipped out of the terminal or to CFS. In addition, the containers may need to be temporarily moved to a specific area for inspections of customs or quarantine department before they are returned to the yard for delivery. All the movement information is planned and recorded in the information system so that the exact location of each container at each moment can be tracked to avoid the loss of the containers.

The import containers will leave the terminal in different ways. Some are delivered to the consignee's location by road, rail, or inland waterway, while some are shipped to the CFS for unpacking. The shipping company will inform the terminal which containers should be transferred to the CFS for unpacking and which containers be delivered intactly. Instructions about moving certain empty containers which are temporarily stacked and to be packed at the shipper's location will also be issued by the shipping company as well as notice to the consignee (or its agent) that when the goods will be shipped to the port, and when to go to the terminal or the CFS for the goods delivery. According to the terms in the contract, the shipping company can also entrust a local transportation company to make the necessary transporting arrangements to deliver the containers to the consignee by road, rail, or inland waterways.

Within the time specified by the terminal, the truck driver who comes to pick up the container will arrive at the terminal, hand over the authorization document for container delivery, receive the EIR as the receipt, and get the instruction to pick up the container from the yard. When the truck is leaving the terminal, it is necessary to check again whether the container is damaged, the seal intact, and fill in the EIR. The containers will be transported to the inland destinations after the container records in the terminal information system are updated, and the security and other procedures are completed. Of course, the documentation, inspections, and handling procedures are much the same for container delivery by rail or inland waterways. This is the final step of the import container operations.

For goods that need to be unpacked at the CFS, the delivery processes are slightly different. The containers to be unpacked at the CFS are moved from the yard to the unpacking area of the CFS according to the feasible schedule established by the terminal planner, and the container records are updated to indicate that the containers have left the yard. At the CFS, the container conditions are checked and the containers are opened under the supervision of the customs and the goods are unloaded from the containers. The goods are checked against the packing list provided by the shipping company and then delivered to the warehouse, while the empty containers are returned from the CFS to the empty container blocks, and the information system of the terminal is updated accordingly. Meanwhile, the preparation and management of goods documents should be completed so that the consignee or his agent can pick up the goods. When the goods leave the terminal, the CFS records are updated to indicate the goods' departure.

The activities that the container must complete are introduced above, which have effectively played the central role in the international transportation chain. Although the specific processes of import and export container activities are different, they all follow the same principles (the processes of the arrival of containers, the storage of containers, and the departure of containers), regardless of the container moving directions.

Chapter 3
Management of the Vessel Unloading Operations in the Container Terminal

3.1 General Introduction of Vessel Unloading Processes

The loading and unloading operations in the terminal are interlinked. To fully understand the vessel unloading operations in the container terminal, it is necessary to study the information flow in each step as a whole, analyze the relationship between each step, and further clarify the logic relationship in operations involved in each module.

The import and export operations in the container terminal can be divided into four types: (1) vessel unloading operations, (2) container delivery operations, (3) container collection operations, and (4) vessel loading operations. The first two types are the two stages of import operations, and the last two are the two stages of export operations, as shown in Fig. 3.1.

The contents in this chapter focus on the vessel unloading operations in the container terminal, and the processes can be roughly divided into three steps as shown in Fig. 3.2.

① Information collection and processing, which is the first step of preparation work for all the operations, as well as the data foundation of the information system's normal operation.
② Resource planning and scheduling, which refers to resource allocation and process arrangement of loading and unloading operations in the terminal, including planning and scheduling of berths, blocks, quay cranes, yard cranes, internal container trucks, etc.
③ The actual operation, which refers to the actual loading and unloading operations carried out by the on-site operators according to the scheduling arrangement from the central controlling office.

The original version of this chapter was revised: Figures have been corrected. The correction to this chapter is available at https://doi.org/10.1007/978-981-15-2937-5_7

N. Zhao et al., *Digital Management of Container Terminal Operations*,
https://doi.org/10.1007/978-981-15-2937-5_3

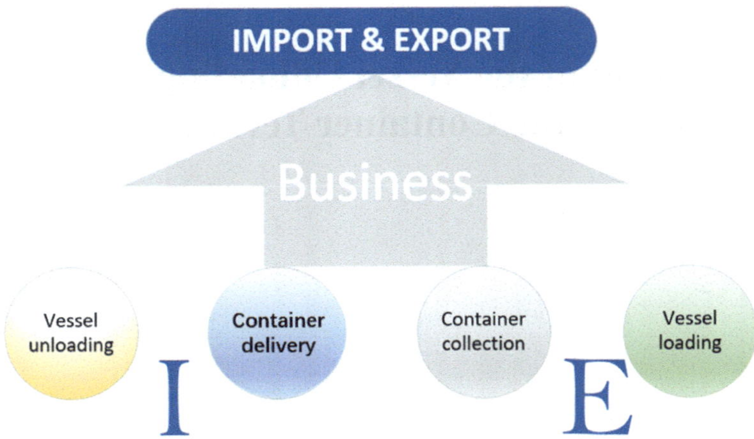

Fig. 3.1 Main operations in the container terminal

Fig. 3.2 Three major steps in vessel unloading operations

The import containers are shipped, unloaded from the vessel, and finally landed into the yard in the terminal, during which the preparation and control of the operations generally go through the steps shown in Fig. 3.3.

The figure above shows the operation processes of vessel unloading operations of the import containers in the terminal. The whole processes have covered the operations in several departments of the terminal, from information collection, to planning, controlling, and on-site operations. The following is a detailed introduction of the operation processes.

(1) **Information input of BAPLIE and manifest**
In the vessel unloading operations of the import containers, all the container data and the corresponding manifest data of the containers to be unloaded at the port shall be input into the system. The former is usually called import BAPLIE, while the latter is called import manifest. The import BAPLIE in specific words, that is, the stowage plan of the vessel, records the specific

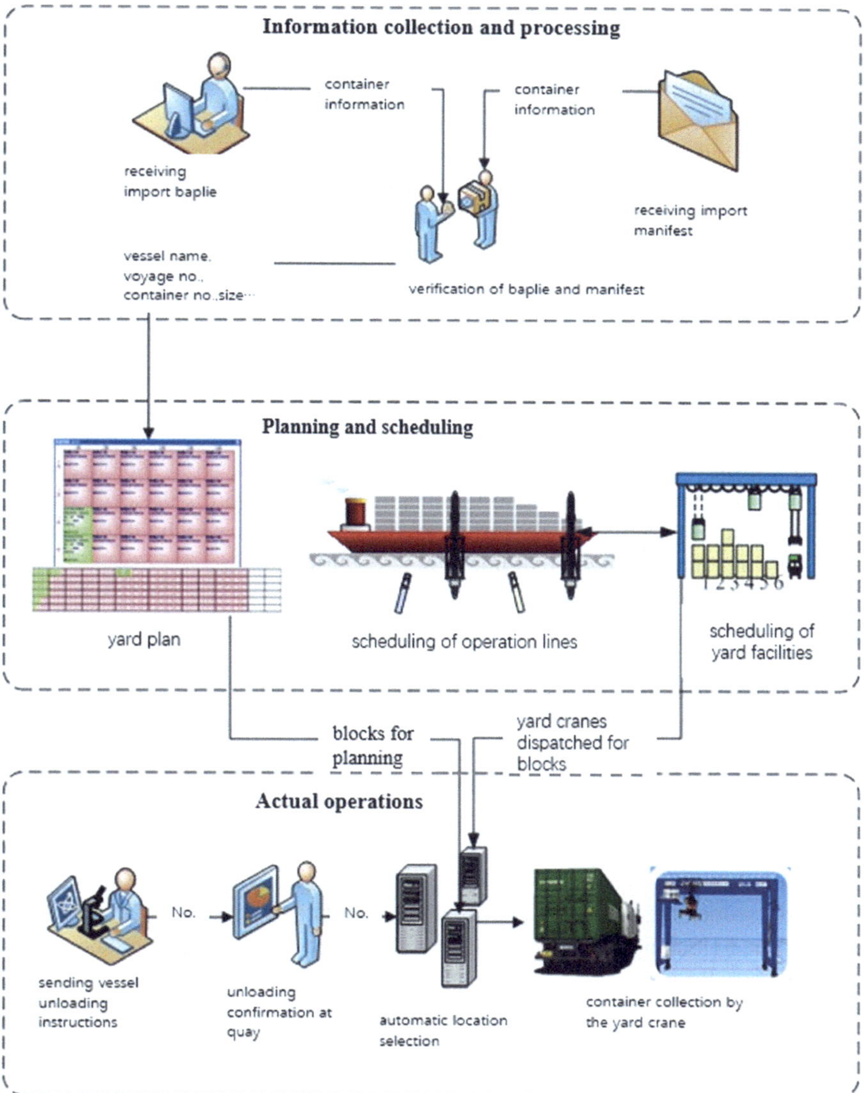

Fig. 3.3 Vessel unloading operation processes

stowing positions of the containers at a certain voyage as well as the information of stowing requirements. The import manifest records the goods information of all containers at a certain voyage, including the bill of lading no., shipping mark, the grade of the dangerous goods, etc. The information center of the terminal receives the BAPLIE, manifests information sent by the shipping company or the agent company, and inputs them into the computer management system of the port so as to establish the data of the import containers via some means, which can be generally conducted manually or with EDI.

(2) **Information verification of BAPLIE and manifest**
The BAPLIE is the reference of unloading operations, and the manifest is the reference of checking the data of short-unloading, over-unloading as well as the reference of container delivery acceptance by the consignee or CFS unpacking, which have reflected various data of the import containers. Mutual verification of the container data in the BAPLIE and manifest can help to find the errors in manual input or EDI transmission, so as to effectively improve the data quality of the import containers and ensure the validity of the import data.

(3) **Vessel plan**
The monthly plan, short-term plan, and 24 h plan for vessels are prepared according to vessel information in the system database input and checked by the information center of the terminal. The vessel plan is the first working procedure of the whole operations in the terminal, by which all the original data of routes, container volume, etc., are mastered.

(4) **Confirmation of vessel's berthing**
The berthing time and the starting time of the vessel unloading operations are registered after the vessel has docked at the port.

(5) **Storage plan of the import containers**
Before the actual vessel unloading operations start, the yard plan of the import containers must be made, that is, the storage plan of the containers unloaded from the vessel, indicating the specific locations of the containers in the yard when unloaded. The storage plan can specify the exact location for each container, or more coarsely, to a bay or a row, or even more coarsely to only an area (with bays or rows). The practice has shown that too specific plan will reduce the unloading speed, which is not conducive to the improvement of the unloading efficiency. Only a small number of special containers, such as special purpose containers, tank containers, reefer containers, dangerous goods containers, etc., need to be designated to relatively specific storage locations.

(6) **Scheduling of operation lines**
The operation line scheduling is the process of implementation and adjustment of the handling operation plans. Before the vessel's docking, the vessel control scheduler has mastered container information at each bay, such as container type, size, status, and dangerous goods, according to day and night plan for vessels and stowage BAPLIE of the vessel in the database, then reasonably allocates the handling facilities at quay (like QCs), to enable the QCs' concurrent operations to the maximum and avoid shut down of the QCs caused by human factors.

(7) **Scheduling of yard cranes and internal container trucks**
The yard cranes are dispatched to designated positions according to the yard plan of the import containers, and the current positions of the yard cranes and

the internal container trucks are configured for the QCs, which is achieved in the TPS control module.

(8) **Vessel unloading instructions**
The vessel control scheduler sends the unloading instructions of several containers to be unloaded at the current bay so that the tallyman at the quay can see the containers to be unloaded on the handheld device, and the operators at the quay can select to confirm the unloading. There are generally less than 30 instructions on each operation line since the number of rows that can be displayed on the handheld device is limited, and too many are difficult to find.

(9) **Container unloading by the QCs to the container trucks**
The QC driver on the corresponding operation line receives the unloading instructions and operates the QC to start the unloading operation, completing the first vertical movement of the containers and unloading the containers onto the container trucks. The processes of vessel unloading are flexible, and the unloading sequence of the containers at the same bay in the vessel can be arbitrarily changed, as long as it is convenient and reasonable to operate without strictly following the unloading plan.

(10) **Confirmation of the container unloading**
This process is as follows: the driver of the QC unloads the container from the vessel onto the container truck. The operator at the quay checks the container no. of the unloaded container (generally the last four characters), selects the corresponding container no. on the handheld device, and keeps the container truck no. as the system default, which would not be wrong under normal circumstances. If there is an error, the correct truck must be selected on the handheld device or notify the central controlling office to dispatch the correct truck. As can be seen from the above, the task of the operator with the handheld device is to record the specific container and the specific container truck carrying the container into the database.

(11) **Intelligent location selection for the unloaded containers in the yard**
The algorithm of the intelligent location selection in the yard is performed by the database server of the container terminal operation system. Its function is to dynamically monitor the container which has already been unloaded onto the container truck but not yet allocated a specific location in the yard. Once such target is found, a reasonable location in the yard will be selected by the algorithm according to the data flow information, and the calculation result shall be written into the container unloading task in the yard.

(12) **Establishment of the container unloading task in the yard**
The server establishes the container unloading task in the yard according to the algorithm of the intelligent location selection and sends it to the TPS terminal on the container truck and the on-board terminal of the yard crane.

(13) **Container truck moving container to the yard**
After receiving the task of container unloading on the TPS terminal, the container truck driver will then move the container to the corresponding block designated by the task.

(14) **Confirmation of the container unloading into the yard**
The yard crane driver receives the container unloading task on the on-board terminal, unloads the container from the container truck into the yard, and confirms the operation. After the confirmation, the unloading task will be closed, and the container truck will be released, which means the completion of the unloading task of this container.

3.2 Information Collection and Processing in Vessel Unloading Operations

3.2.1 Import BAPLIE

The BAPLIE records the specific location information of the containers loaded in the vessel of certain voyage as well as container information and loading and unloading requirements. The BAPLIE is the reference of the vessel loading and unloading operations.

3.2.1.1 The Contents and Functions of the Import BAPLIE

The contents of the import BAPLIE are mainly divided into two parts, container information and stowage positions.

1. **The main composition of the container information of the BAPLIE**

(1) **Container no.** is the unique identification of the container. In theory, it is just like the ID no. that cannot be repeated. The container no. includes two parts, namely holder code and sequence number. The container no. generally has 11 characters, for example, like "YMCU2008570," the first four characters are holder code. According to the ISO regulations, the holder code should be represented with four capital Latin letters, the first three can be specified by the holder, and the forth must be "U." The last seven of the container no. is sequence number and represented with Arabic numerals. If the sequence number is less than seven characters, the other characters should be filled with "0" before the effective characters, such as "0531842."

(2) **Size** refers to the length of the container, with feet as the unit. Containers mainly handled in the container terminal are 20, 40, and 45 ft. Containers with other sizes, such as 10, 12, 30, 48, and 53ft, are usually handled in the general cargo terminal.

(3) **Container height**, namely the height of the container, has the sizes of 8′6″and 9′6.″ In the container terminal, the specific sizes in height are not concerned, and the containers are only defined as flat and high cubic, among which the flat containers (8′6″) are the most.

(4) **Container types**, namely the descriptions on container types, which determine the goods types inside. Main container types handled in the container terminal include general purpose containers (GP), bulk containers (BU), (flat rack containers (FR), open-top containers (OT), reefer containers (RF), tank containers (TK), etc., most of which are GP.

(5) **Special goods requirements** specify whether a container is over-limited or loaded with dangerous goods. All the containers with records of special goods requirements should be specially treated when unloaded from the vessels and charged with extra special handling fees. If a container is over-limited, this field of attribute will indicate the types of over-limit, including over-height, over-length, over-width, over-height and over-width, over-length and over-width, over-height and over-length, over-height and over-length and over-width. All the over-limited containers must be handled by the experienced drivers with steel wire ropes bound in advance instead of handled directly with quay cranes. The dangerous goods containers can be classified into nine grades according to the dangers of the goods inside, and the specific criteria are listed in Table 3.1.

(6) **Container holders**, also known as container operators, refer to the companies that the containers belong to. Such information is quite important for the management of empty containers.

(7) **Loading port** is the port where the container is loaded.

(8) **Discharging port** is the port where the container will be unloaded.

(9) **Transfer port** is the port through which the container will be transferred. Transshipment refers to the process in which a container loaded at one port is brought into another port in the vessel and unloaded to the yard and then loaded onto another vessel and shipped to the third port.

(10) **Status** refers to the transition status of a container and whether it is packed or not. According to status, the containers can be divided as import full containers (IF), import empty containers (IE), import transshipment containers, export

Table 3.1 Cargo codes (excerpts)

No.	Cargo code	Cargo name
1	01	Coal and products
2	011	Coke
3	012	Raw coal
4	0121	Anthracite
5	0122	Bitumite
6	0123	Lignite
7	013	Washed coal
8	015	Coal products
9	019	Other coal and products not listed
10	022	Gasoline
11	023	Kerosene

transshipment containers, international transshipment containers, domestic trade containers, transit containers, etc.

2. **Stowing positions of the containers in the vessel**

The stowing positions of the containers in the vessel are referred to as slots, including "bay," "row," and "tier" in three dimensions.

(1) Bay no. is the longitudinal stowing position of the container in the container vessel. Its numbering rule is to number from the bow to the stern with odd numbers (i.e., 01, 03, 05 …) in the same horizontal line, with a 20 ft container as the modulus. An even number between two odd numbers represents the slot of a 40 ft container or a 45 ft container. For example, bay 02 is the combination of bay 01 and bay 03.

(2) Row no. is the lateral stowing position of the container in the container vessel. There are two notation methods. One is to number from the starboard to the portside as 01, 02, …. The other is to number from the middle to both sides, the starboard is represented with odd numbers as 01, 03, 05, …, and portside is represented with even numbers as 02, 04, 06, …. The most common is the second method.

(3) Tier no. is the vertical stowing position of the container in the container vessel. The numbering methods include numbering on deck and numbering in the cabin. The tier no. on deck from the bottom up is 82, 84, 86… while in the cabin from the bottom up is 02, 04, 06….

There are six characters for the slot no. The first two are bay no., the middle two are row no., and the last two are tier no. If a container is placed in the slot with the no. of 040202, it means this container is placed at the fourth bay, the second row from the portside, and the first tier in the cabin.

3.2.1.2 Format and Conversion of the EDI Platform File of the Import BAPLIE

With the continuous expansion of container transportation, the loading and unloading volume also grows rapidly. If such a great deal of container information in the BAPLIE is input manually after fax, it will not only greatly reduce the speed of information acquisition but also waste the operating time and more likely to cause incorrect input, thus causing hidden troubles for smooth operation of loading and unloading. Therefore, most major ports have applied electronic data interchange (EDI) technology at present, with which the designated data can be uniformly coded by the EDI center, so that data transmission can be electronic, and the port can obtain information through decoding.

When data are exchanged electronically, the carrier of the data is the message. United Nations electronic data interchange in administrative, commercial, and transportation (UN/EDIFACT) is the general international standard of EDI. In order to be in line with international standards and to promote the application of EDI, the Ministry of Communications of China has further improved the message standards on the basis of the *demonstration project of the electronic information transmission and operation system in the international container transportation* as well as negotiation and agreements with all the parties involved in the data exchange. The format of BAPLIE, import manifest, handling report, over-unloading/short-unloading/damage report, and stacking report in this version is consistent with the format of the platform document issued in the *project of the transportation EDI information network (phase I)*.

In order to facilitate the understanding for the beginners, the import BAPLIE of a certain voyage of the vessel ASJ5 is taken as an example to show the original EDI file and the converted BAPLIE information, so as to deepen the perceptual understanding for the beginners.

As shown in Fig. 3.4, the operator in the information department of the container terminal will obtain EDI files through email or FTP or other data transmission means. The original EDI file is actually a text file suffixed with ".edi," as the file "ba093244.edi" shown in Fig. 3.4.

The EDI file of a BAPLIE can be opened with notepad tool, and the following script is the import BAPLIE code of vessel ASJ5, 0927.

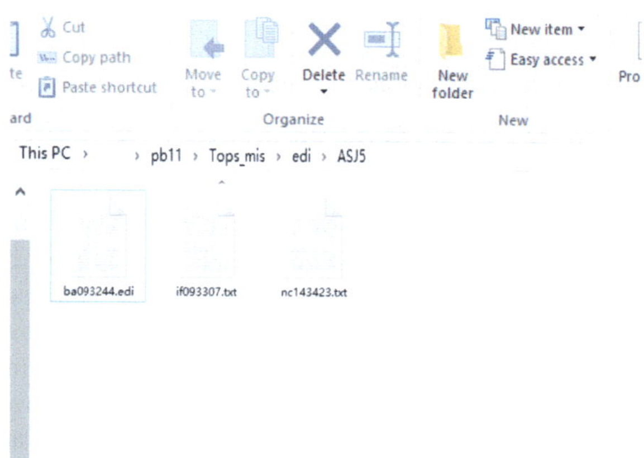

Fig. 3.4 EDI file of a BAPLIE

```
00:BAPLIE:BAYPLAN:9:CUSTHY:GS2:200907060932:CNTSN:CNTSN'
10:ASJ5:安盛集 5:CN:0927::石湖-天津:20090706:20090707:CNTSN:天津:CNSTG:汕头'
11:QZAS:泉州安盛船务有限公司'
50:CLHU3544210:20GP:F:0030282:::::0:0:0:0:0:21000:0:QZAT:安通'
51:ATLQZTJ1700860'
52:CNQZJ:泉州:CNTSN:天津'
50:TEXU3986863:20GP:F:0170204:::::0:0:0:0:0:21000:0:QZAT:安通'
51:ATLQZTJ0800197'
52:CNQZJ:泉州:CNTSN:天津'
50:GESU3576210:20GP:F:0110306:::::0:0:0:0:0:21000:0:QZAT:安通'
51:ATLQZTJ1700866'
52:CNQZJ:泉州:CNTSN:天津'
50:GESU3363200:20GP:F:0090004:::::0:0:0:0:0:21000:0:QZAT:安通'
51:ATLQZTJ1700858'
52:CNQZJ:泉州:CNTSN:天津'
----------------------------------------------------------------------------
50:ATLU0005642:20GP:F:0010104:::::0:0:0:0:0:21000:0:QZAT:安通'
51:ATLQZTJ101848'
52:CNQZJ:泉州:CNTSN:天津'
50:GESU3575913:20GP:F:0110106:::::0:0:0:0:0:21000:0:QZAT:安通'
51:ATLQZTJ0700109'
52:CNQZJ:泉州:CNTSN:天津'
50:XINU1402849:20GP:F:0230304:::::0:0:0:0:0:21000:0:QZAT:安通'
51:ATLQZTJ1800423B'
52:CNQZJ:泉州:CNTSN:天津'
50:UETU2005683:20GP:F:0190406:::::0:0:0:0:0:21000:0:QZAT:安通'
51:ATLQZTJ1100585'
52:CNQZJ:泉州:CNTSN:天津'
50:CLHU3548767:20GP:F:0050004:::::0:0:0:0:0:21000:0:QZAT:安通'
51:ATLQZTJ1100600'
52:CNQZJ:泉州:CNTSN:天津'
50:TGHU3024913:20GP:F:0170006:::::0:0:0:0:0:21000:0:QZAT:安通'
51:ATLQZTJ1700876'
52:CNQZJ:泉州:CNTSN:天津'
50:XINU1428782:20GP:F:0230006:::::0:0:0:0:0:21000:0:QZAT:安通'
51:ATLQZTJ0800197'
52:CNQZJ:泉州:CNTSN:天津'
99:649'
```

The EDI file shown above is in coding. However, the data needed by the terminal operators are well-classified data tables and graphical stacking structure diagrams. Therefore, there must be a designated module in the operation system of the container terminal to complete the decoding of EDI file, as shown in Fig. 3.5.

Obviously, by using EDI to obtain BAPLIE information, the terminal operators' workload can be greatly reduced, and the information accuracy improved. Therefore, there must be a standard for the information encoding and decoding, and this standard is the "*format of the EDI platform file of baplie (BAPLIE)*" formulated by the Ministry of Communications of China.

The specific conversion standard of the EDI platform file of the BAPLIE is listed below:

sender, receiver:
import – vessel agent of the discharging port → terminal, tally, port supervision.
export – pre-stowage of the loading port → tally, port supervision → vessel agent of the loading port.

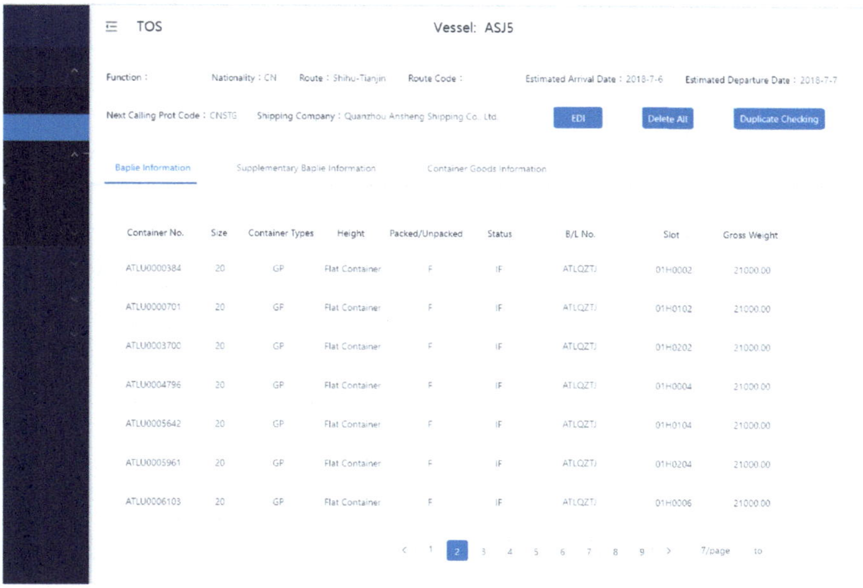

Fig. 3.5 EDI import system of the import BAPLIE

Function: This message provides relevant information about containers and goods loaded in a vessel of one voyage and the specific bay positions of containers in the vessel, which are important data of the vessel's loading and unloading operations at the next calling port as well as the reference for the port to arrange loading and unloading operations.

Corresponding documents: import BAPLIE, export BAPLIE.

Record 00	HEAD RECORD			M
No.	Field name	Format	Annotation	Mark
1	RECORD ID	9(2)	00	M
2	MESSAGE TYPE	X(6)	BAPLIE	M
3	FILE DESCRIPTION	X(35)	BAYPLAN	M
4	FILE FUNCTION	X(2)	9 = original 2 = added 3 = deleted 4 = changed	M
5	SENDER CODE	X(13)		M
6	RECEIVER CODE	X(13)		M
7	FILE CREATE TIME	9(12)	CCYYMMDDHHMM System time	M
8	SENDER PORT CODE	X(5)		C
9	RECEIVER PORT CODE	X(5)		C

(continued)

(continued)

Record 00	HEAD RECORD			M
No.	Field name	Format	Annotation	Mark
10	AGREEMENT ID	X(14)		C

Record 01	OTHER RECEIVERS			CC
No.	Field name	Format	Annotation	Mark
1	RECORD ID	9(2)	01	M
2	RECEIVER CODE	X(13)		M
3	RECEIVER CODE (1-8)	X(13)		C

Record 10	VESSEL			M
No.	Field name	Format	Annotation	Mark
1	RECORD ID	9(2)	10	M
2	VESSELCODE	X(9)		C/M
3	VESSEL	X(35)		M/C
4	NATIONALITY CODE	X(2)		C
5	VOYAGE	X(6)		M
6	TRADE CODE	X(10)		C
7	TRADE	X(35)		C
8	ETD ARRIVED DATE	9(8)	CCYYMMDD	C
9	SAILING DATE	9(8)	CCYYMMDD	M
10	DEPART PORT CODE	X(5)		M/C
11	DEPART PORT	X(35)		C/M
12	NEXT CALLING PORT CODE	X(5)		M/C
13	NEXT CALLING PORT	X(35)		C/M

Record 11	LINES INFORMATION			M
No.	Field name	Format	Annotation	Mark
1	RECORD ID	9(2)	11	M
2	SHIPPING LINE CODE	X(13)		C/M
3	SHIPPING LINE	X(35)		M/C

Record 50	CONTAINER INFORMATION			M
No.	Field name	Format	Annotation	Mark
1	RECORD ID	9(2)	50	M
2	CTN. NO.	X(12)		M
3	CTN. SIZE & TYPE	X(4)		M
4	CTN. STATUS	X(1)	E = Empty F = Full L = LCL	M
5	Stowage Location	9(7)	Bay: 00, Row: 00, Tier: 00	M
6	TEMPERATURE I	X(1)	C = Celsius F = Fahrenheit	C
7	TEMPERATURE SETTING	X(5)	See note	C
8	MIN. TEMPERATURE	X(5)		C
9	MAX. TEMPERATURE	X(5)		C
10	Over-Length Front	9(4)	cm	C
11	Over-Length Back	9(4)	cm	C
12	Over-Width Left	9(4)	cm	C
13	Over-Width Right	9(4)	cm	C
14	Over-Height	9(4)	cm	C
15	GROSS Weight	9(5).9	kg	M
16	TARE Weight	9(5).9	kg	C
17	CTN. OPERATOR CODE	X(13)		M/C
18	CTN. OPERATOR	X(35)		C/M

Note except for the plus (+) minus (−) sign and decimal point, only three digits are allowed in temperature

Record 51	B/L NO. INFORMATION			C
No.	Field name	Format	Annotation	Mark
1	RECORD ID	9(2)	51	M
2	B/L NO.	X(20)		M

Note This record can be used to describe the bill of lading no. of the critical goods in LCL

Record 52	PLACE INFORMATION			M
No.	Field name	Format	Annotation	Mark
1	RECORD ID	9(2)	52	M
2	LOAD PORT CODE	X(5)		M/C
3	LOAD PORT	X(35)		C/M
4	DISCHARGE PORT CODE	X(5)		M/C
5	DISCHARGE PORT	X(35)		C/M
6	PLACE CODE OF DELIVERY	X(5)		C

(continued)

(continued)

Record 52	PLACE INFORMATION			M
No.	Field name	Format	Annotation	Mark
7	PLACE OF DELIVERY	X(70)		C
8	TRANSFER PORT CODE	X(5)		C
9	TRANSFER PORT	X(35)		C

Record 53	OPT. DISH. PORT INFORMATION			C
No.	Field name	Format	Annotation	Mark
1	RECORD ID	9(2)	53	M
2	OPT. DISCH PORT-1 CODE	X(5)		M/C
3	OPT. DISCH PORT-1	X(35)		C/M
4	OPT. DISCH PORT-2 CODE	X(5)		C
5	OPT. DISCH PORT-2	X(35)		C
6	OPT. DISCH PORT-3 CODE	X(5)		C
7	OPT. DISCH PORT-3	X(35)		C
8	OPT. DISCH PORT-4 CODE	X(5)		C
9	OPT. DISCH PORT-4	X(35)		C
10	OPT. DISCH PORT-5 CODE	X(5)		C
11	OPT. DISCH PORT-5	X(35)		C

Record 54	DANGEROUS CARGO INFORMATION			C
No.	Field name	Format	Annotation	Mark
1	RECORD ID	9(2)	54	M
2	CLASS	X(5)		M
3	PAGE	X(7)		C
4	UNDG NO.	9(4)		M
5	LABEL	X(32)		C
6	FLASH POINT	X(5)	Celsius	C
7	EMS NO.	X(6)		C
8	MFAG NO.	X(4)		C

Note This record can be used to describe the dangerous goods information of the critical goods in LCL

Record 55	CARGO INFORMATION			C
No.	Field name	Format	Annotation	Mark
1	RECORD ID	9(2)	55	M
2	CARGO CODE	X(3)	Customs HS code	C
3	CARGO DESCRIPTION (1-5)	X(35)		C

Note this record may or may not be used for LCL cargos. Record 55 may be repeated up to nine times (for cargo description)

Record 99	TRAILER RECORD			M
No.	Field name	Format	Annotation	Mark
1	RECORD ID	9(2)	99	M
2	RECORD TOTAL OF FILE	9(6)	Including: head record, trailer record	M

Record structure:

00 HEAD RECORD	M1
01 OTHER RECEIVERS	C1
10 VESSEL	M1
11 LINES INFORMATION	M1
50 CONTAINER INFORMATION	M9999
51 B/L NO. INFORMATION	C1
52 PLACE INFORMATION	M1
53 OPT. DISH. PORT INFORMATION	C1
54 DANGEROUS CARGO INFORMATION	C1
55 CARGO INFORMATION	C9
99 TRAILER RECORD	M1

The platform file is a unified user data format compiled according to the actual application and corresponding paper documents. The design of the platform file has taken into account the UN/EDIFACT message standard and the data items required by the corresponding documents. Here are some instructions for the platform files.

① The platform file records and fields may vary in length. Each record ends with an apostrophe (') (or enter), and fields in the record are delimited by a colon (:).

Release character (?): if the content of a field in a platform file contains a delimiter (:) or a terminator ('), a release character (?) should be placed before the delimiter (:) or the terminator ('). The release character is not counted in the length of the string.

② In the platform file, the prerequisite records are represented by "M," and the conditional records are represented by "C," which are, respectively, marked in the upper right corner of the data structure for each record. When a conditional record is not selected, the entire record may be omitted.

③ Each field in the record is marked with its no., field name, format, annotation, and mark.

No. indicates the ordinal number of the field in the record. For example, five represents the fifth field of a record.

When a field group needs to be repeated, its subordinate field number is expressed as m-n, where *m* is the ordinal number of the field group in the record, and *n* is the ordinal number of the subordinate field in the field group.

Field name describes the English and Chinese name of the field (or field group) and the number of repetitions, such as (1-3) for up to three repetitions.

Format describes the type and maximum length of the field. For example,

X(2) represents a character with a maximum length of two.
9(2) represents a numeric integer, up to two digits.
9(5) nine represents a numeric value, up to five integers and one decimal. The positive or negative sign of the value does not hold space.
9(12) means if the date type is represented, the value is indicated in the annotation (CCYYMMDDHHMM), with no sign or decimal point.
Annotation is a note to the field and a description of how the code is used.
Mark: *M* indicates that this field must be used in the record.

C indicates that the field is optional in the record, and if not selected, the delimiter cannot be omitted.
M/C means *M* is the first choice, followed by *C*.
C/M means *C* is the first choice, followed by *M*.

For the code and name of the same field, if the code is M/C and the name is C/M, one must be selected, and the code is the first choice. When both the code and the name appear, the code prevails. If the code is C/M and the name is M/C, one must be selected, and the name is the first choice. When both the code and the name appear, the name prevails.

④ Record structure

It reflects the order in which various types of records appear in the platform file, the combination of records and the maximum number of cycles, and indicates that this cycle is required with M and optional with C.

⑤ Other rules

(1) In order to comply with the application syntax rules of the UN/EDIFACT, it should be noted for the numeric fields that there should be at least one number before and after the decimal point symbol, that is, 0.5 and 2.0 are allowed instead of .5 and 2. Nor is the use of a ternary delimiter allowed, i.e., 2500000 instead of 2,500,000.
(2) The container no. shall be numbered with a four-digit prefix and a seven-digit numeral. When the prefix is less than four digits, it shall be supplemented with a space on the right, for example, ABC 1234567. When the check bit is with two bits, it can be numbered with eight-bit numeral. If the number is insufficient, it should be left aligned.

(3) Besides "9" (original file), the file functions also include "2" (adding containers), "3" (deleting containers), "4" (data change), and so on, according to the agreement between users.

3.2.1.3 The General Display Modes of the Import BAPLIE

According to the EDI original file mentioned above and the information imported into the system, the BAPLIE seems to be a group of words without any graphic content. However, like the warehouse plan and the block plan, the BAPLIE also has graphical display mode, and the information is consistent with the display mode of EDI and text information table mentioned above. Therefore, we summarize the common display modes of BAPLIE as follows.

1. **Display in EDI coding**

See Sect. 3.2.1.2.

2. **Display in word message**

The display mode of text information is in the form of data table. This mode is not as visualized as the graphical mode, but it is easy for searching and editing, so it is often used to query and modify BAPLIE information as a very common BAPLIE display mode, as shown in Fig. 3.6.

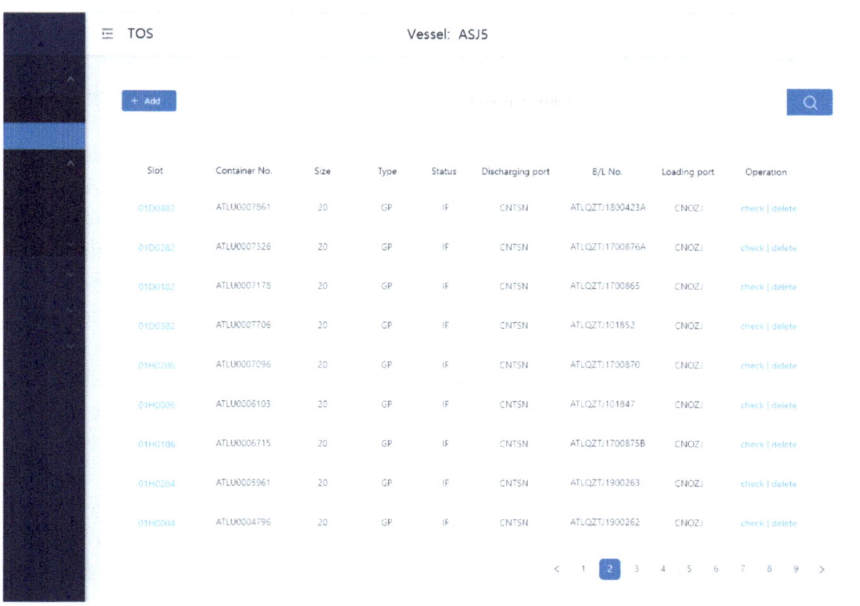

Fig. 3.6 Input and modification of import BAPLIE

It can be seen that the text information display mode of BAPLIE is easy to operate, and it is often used when batch processing or modifying BAPLIE information is needed.

3. **Display in general overview of BAPLIE**

The general overview of the import BAPLIE refers to that the import containers and transit containers at each bay of the whole vessel are graphically displayed to show their relative stacking positions based on the definition of vessel form. In this way, all the bay profiles of the vessel can be displayed on the same page, so as to facilitate reading and form the overall concept. In addition, the general overview of the import BAPLIE can also be used to assist the arrangement of vessel unloading storage plan, and it can also be used as a query way in the actual vessel unloading process. Therefore, the display mode in general overview of the import BAPLIE is concise, clear, and arresting.

Figure 3.7 shows the general overview of the import BAPLIE for vessel ASJ5, 0927. Each container is displayed in different position at different bays according to the stacking position in the vessel. The colors, letters, or icons on the containers represent different meanings, for example, D represents the dangerous good container, R reefer container, F general full container, empty container, high cubic, W cabin reshuffle container, Z transit container, O over-limited container, etc. Obviously, the terminal manager can easily see the positions and approximate number of the special containers in the general overview of the BAPLIE. The display mode in general overview of BAPLIE carries less information than that of the text table but more targeted.

4. **Display in separate bay view**

The display mode in separate bay view means to dynamically display the main information of import and transit containers in a certain vessel at certain voyage in a graphical way based on the internal structure of each bay in the vessel. This display mode is generally applied together with the general overview of the BAPLIE. Although the BAPLIE in general overview can directly display the stacking position of each container and the stacking status of the corresponding bay, the core information of the container, such as the container no., the loading port, the discharging port, size, weight, container type, height, status, special goods requirements and holder, cannot be displayed, which will affect the operation. Therefore, the display mode in separate bay view further refines the contents displayed for each bay. Its feature is that it can display more core container information while giving consideration to the graphic features, so it is more convenient to use. The following Fig. 3.8 shows the bay view of bay 01 of vessel Xinhua 801, 0903 N.

As shown in Fig. 3.8, the position of each container is also displayed according to the internal structure of each bay in the vessel form definition. In addition to directly viewing the text information of the container, the packing status and special nature of

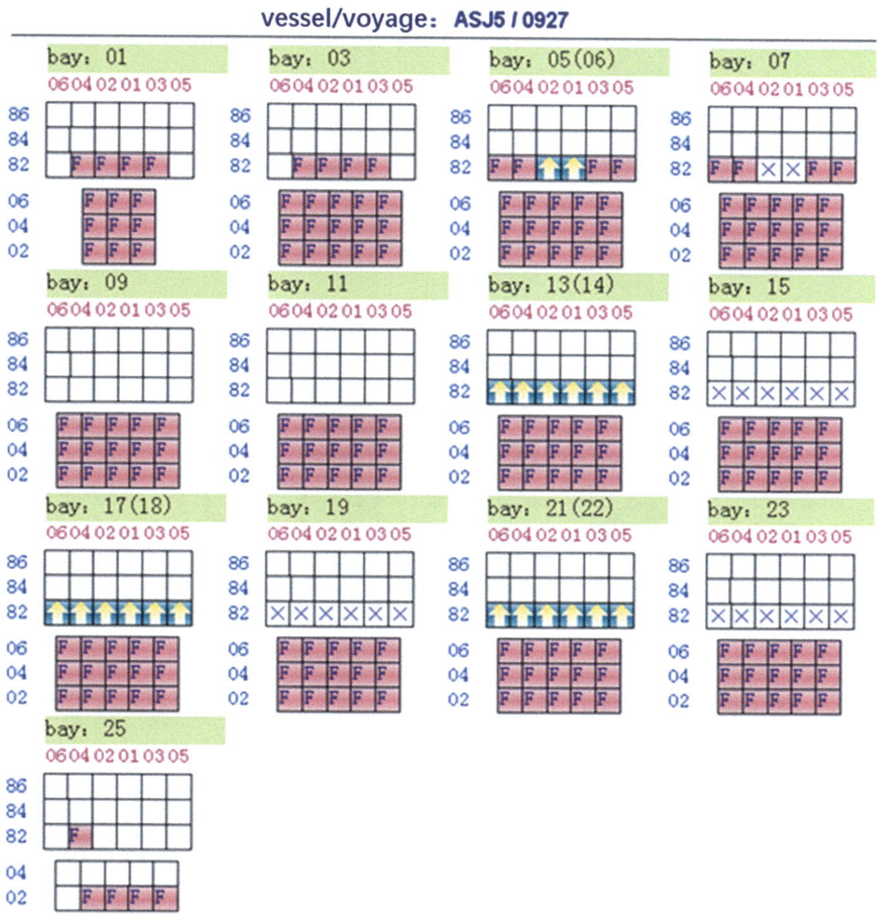

Fig. 3.7 Import BAPLIE displayed in general overview

the container can also be directly seen according to the color, rendering and the icon at the lower right corner of the container in bay view. For example, a container with a white color in the middle and a red color at the corners represents an empty container; a container with a gradient color from top and bottom to the middle represents a full container, and the heavier the container, the darker the color; if the render color is gray-black, that means the container is a transit container not to be unloaded at this port. The icon in the lower right corner represents that the corresponding container has special purpose. For example, ☠ represents dangerous good container, 45 45 ft container, reefer container, which need sockets, high cubic, tank container, open-top container, flat rack container.

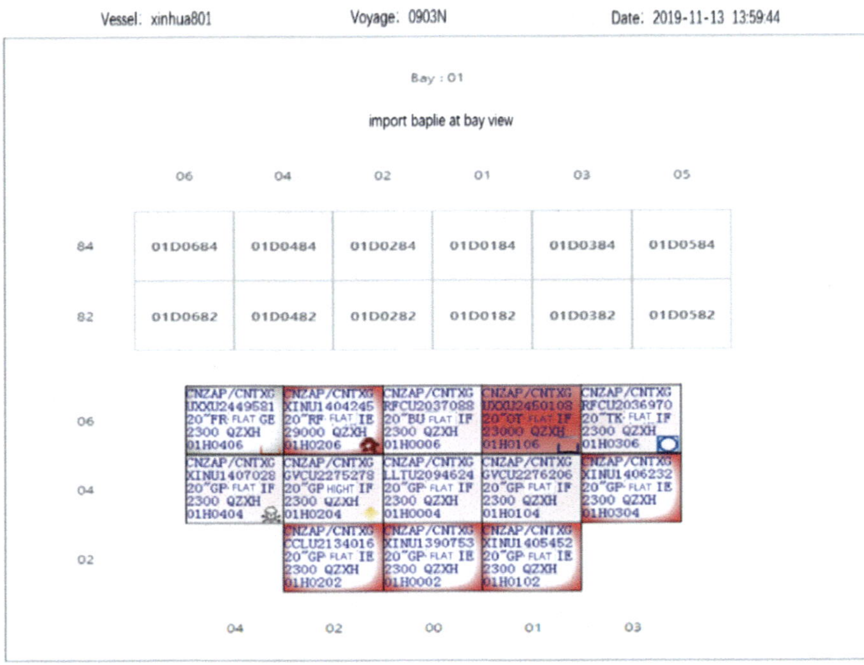

Fig. 3.8 Bay view of the import BAPLIE

5. **Display in vessel side profile**

As the name implies, vessel side profile refers to the side view of container storage positions in the vessel. The general overview and the separate bay view show the container storage situation on the cross-section (at certain bay) of the vessel in lateral direction, while the side profile of the vessel shows the container storage situation in the longitudinal direction, as shown in Fig. 3.9.

The main information in the vessel side profile graph includes:

(1) Cabin structure and stacking height on deck of the vessel;
(2) Location of the engine room;
(3) Berthing direction;
(4) Container stacking height at each bay;
(5) Size of the import containers at each bay;
(6) Size of the stowed containers at each bay.

6. **Display in container volume distribution map**

Volume distribution map refers to the distribution of the number of import and export containers at the side view of the vessel according to the size, deck/cabin, bay position,

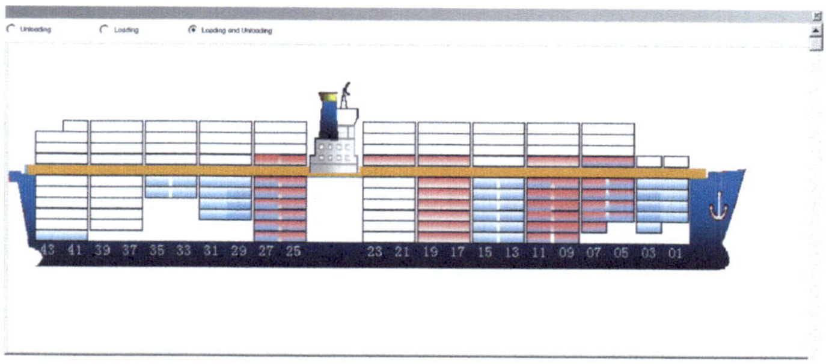

Fig. 3.9 Side profile of the vessel

loading and unloading attributes, and other information. It can be divided into four groups: the number of containers on deck for unloading, the number of containers on deck for loading, the number of containers in the cabin for unloading, and the number of containers in the cabin for loading. As shown in Fig. 3.10, there are eight floors in the volume distribution map. The upper four floors represent the deck operation volume, and the lower four floors represent the cabin operation volume. The grid with pink background (layers 1, 2, 5, and 6) represents the number of containers waiting for unloading operations at a certain bay, while the grid with the blue background (layers 3, 4, 7, and 8) represents the number of containers waiting for loading operations at a certain bay. The longer grid (layers 1, 3, 5, and 7) represents the operation volume at a certain large bay, i.e., the number of 40 ft and 45 ft containers, while the shorter grid (layers 2, 4, 6, and 8) represents the operation volume at a certain small bay, i.e., the number of 20 ft containers. Figure 3.10 shows that 13 20 ft containers are waiting to be unloaded at bay 05, 22 20 ft containers are waiting to be unloaded at bay 07, 39 20 ft containers are waiting to be loaded at bay 27, and so on.

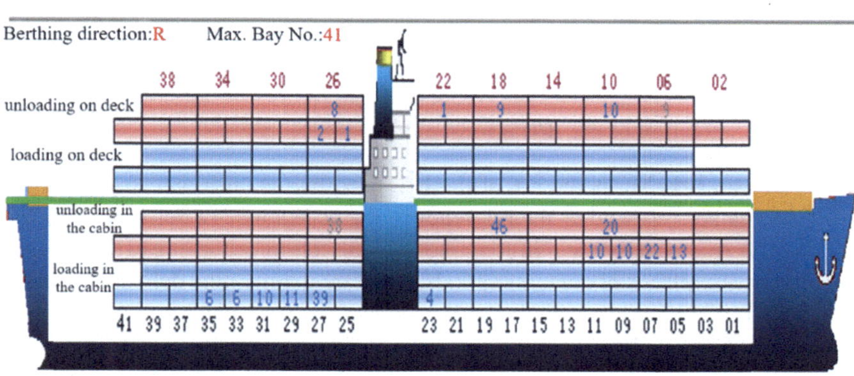

Fig. 3.10 Container volume distribution map

Compared with the vessel side profile, the container volume distribution map does not show the allowable storage structure and the actual storage structure in the vessel side profile.

The main information in the container volume distribution map includes:

(1) The relative horizontal position of each bay in the longitudinal direction of the vessel;
(2) The number of containers at each bay waiting for unloading;
(3) The number of stowed containers at each bay;
(4) Location of the engine room;
(5) Berthing direction.

Although the graphic structure of the container volume distribution map cannot represent the actual structure of the vessel, it is an image of the statistical map, so it is often used more frequently in the actual operation than the side profile.

3.2.2 *Import Manifest*

The import manifest is a summary of the details of the import containers carried by the vessel in accordance with the sequence number of bill of lading (B/L), which mainly involves the information of B/L, details of goods and information of containers. It is an important document for the container terminal to arrange vessel unloading and the original reference to arrange the delivery operation for the consignee.

3.2.2.1 The Contents and Functions of the Import Manifest

1. **The contents of the import manifest**

The contents of the import manifest can be divided into three parts: B/L information, cargo information, and container information.

(1) The composition of the B/L information

① The **B/L No.** is the no. of the B/L used by the customer to pick up the containers, as well as the reference of the acceptance of delivery plan.

② **Terms of delivery**—In container transportation, according to the different delivery places of container goods, there are mainly nine delivery terms for FCL and LCL between the both parties of the vessel and the cargo.

a. Door to door

The door to door delivery means that the carrier accepts the goods in the factory or warehouse of the shipper and is responsible for transporting the goods to the factory or warehouse of the consignee for delivery. In this mode of delivery, the delivery form of goods is FCL.

b. Door to CY
The door to CY delivery means that the carrier accepts the goods in the factory or warehouse of the shipper and is responsible for transporting the goods to the CY of the terminal or its inland CY, and delivers the goods to the consignee in CY. In this mode of delivery, the delivery form of goods is also FCL.
c. Door to CFS
The door to CFS delivery means that the carrier accepts the goods in the factory or warehouse of the shipper and is responsible for transporting the goods to the CFS of the discharging port terminal or its inland CFS, and then delivers the goods to the consignees after unpacking. In this mode of delivery, the carrier generally accepts the goods in the form of FCL and delivers the goods in the form of LCL.
d. CY to door
The CY to door delivery means that the carrier accepts the shipper's goods (in FCL) in the CY of the terminal or its inland CY and is responsible for transporting the goods to the consignee's factory or warehouse for delivery (in FCL).
e. CY to CY
The CY to CY delivery means that the carrier accepts the goods (in FCL) at the CY of the loading port terminal or its inland CY and is responsible for transporting the goods to the CY of the terminal or its inland yard at the discharging port, and delivers them to the consignee in the CY.
f. CY to CFS
The CY to CFS delivery means that the carrier accepts the goods (in FCL) in the CY of the loading port terminal or its inland CY and is responsible for transporting the goods to the CFS at the discharging port or its inland CFS, and generally delivers the goods to the consignee after unpacking.
g. CFS to door
The CFS to door delivery means that the carrier accepts the goods (in LCL) at the CFS of the loading port terminal or the inland CFS and is responsible for transporting the goods to the consignee's factory or warehouse for delivery. In this delivery mode, the carrier generally accepts the goods in the form of LCL and delivers the goods in the form of FCL.
h. CFS to CY
The CFS to CY delivery means that the carrier accepts the goods (in LCL) at the CFS of the loading port terminal or the inland CFS and is responsible for transporting the goods to the CY of the unloading port terminal or the inland CY for delivery. In this delivery mode, the goods are also accepted in the form of LCL and delivered in the form of FCL.

i. CFS to CFS
The CFS to CFS delivery means that the carrier accepts the goods (in LCL) at the CFS of the loading port terminal or the inland CFS and is responsible for transporting the goods to the CFS of the discharging port terminal or the inland CFS, and delivers the goods (in LCL) to the consignee after unpacking. In this delivery mode, the goods are accepted and delivered both in the form of LCL.

③ **Loading port, transfer port, and discharging port**.

(2) The composition of cargo information
The cargo information includes the cargo serial no., cargo code, cargo marks, total number of cargos, total weight, total volume, etc.

① Cargo serial no.
Usually, a B/L corresponds to one type of cargos, but in the actual practice, there is also a B/L containing multiple types of cargos, and the cargo serial no. is used to identify different types of cargos.

② Cargo code
The cargo code is a standard code developed by a country according to the cargo type. Each cargo type has a unique cargo code. Some cargo codes are shown in the following table.

③ The cargo marks are the shipping marks, which are usually composed of a simple geometric figure as well as some letters, numbers, and simple words. The function is to make the cargos easy to be identified by the personnel concerned during handling, transportation, and storage to prevent the wrong shipment. The main contents include consignee code, consignor code, destination port (place) name and number of pieces, batch number.

④ The total number of cargos, total weight, total volume, etc., means the total number of packages, total weight, and total volume of such type of cargos.

(3) The composition of container information
The composition of container information mainly includes container no., size, container height, container type, etc., which is the same as the corresponding content of container information in the import BAPLIE introduced in Sect. 3.2.1.1. In the import manifest, the information of one cargo type usually corresponds to multiple pieces of container information.

2. **The functions of the import manifest**

The import manifest, which is compiled by the shipping company or the shipping agency, has two main functions.

(1) When a vessel is loaded or set sail, the shipping company shall compile the manifest according to the contents of the final B/L confirmed by the client and then send it to the customs in the form of EDI. The most important is the description of the cargos (number of pieces, weight, name of goods, etc.), with which the cargo content in customs declaration must be consistent when the shipper finally refunds the tax, otherwise the tax refund cannot be made.

(2) The import manifest is the cargo data issued by the shipping company with the vessel, which can be seen as an internal B/L of the shipping company. When the vessel arrives at the port of destination, the cargos need to be declared to the customs accordingly. The description of the cargos must be consistent with the B/L held by the consignee, otherwise it is difficult for the consignee to finish the customs clearance and take delivery of the cargos at the port of destination.

3.2.2.2 Main Difference Between Import Manifest and Import BAPLIE and Their Functions

The import manifest and import BAPLIE are vital and essential information documents in the import operations of containers. The information of import BAPLIE is for the containers, one record for one container, while the import manifest is for the cargos, that is, one record for one shipment of cargos. The information of shipping contract and cargos is not included in the BAPLIE, while the information of container position is not included in the manifest.

The import BAPLIE is mainly used for the arrangement of vessel unloading, while the import manifest is mainly used for the handling of containers delivery and providing data for the unpacking at CFS. They reflect the information of the import containers from two perspectives. Comparing the container data in the BAPLIE and manifest can help find the errors caused by manual input or EDI transmission, so as to effectively improve the quality of import container data and ensure the validity.

3.2.2.3 The EDI Platform Format and Conversion of the Import Manifest

The major ports have adopted EDI technology to obtain import manifest information, in order to reduce or avoid errors caused by manual input and improve the speed of information acquisition. The original TXT file together with the converted manifest information is also shown with the example of vessel ASJ5, 0927.

As shown in Fig. 3.11, the operator of the information department in the container terminal gets the TXT file through email or FTP and other data transmission channels. The original TXT file is actually a text file suffixed with ".txt " as the file "if093307.txt" shown in Fig. 3.11.

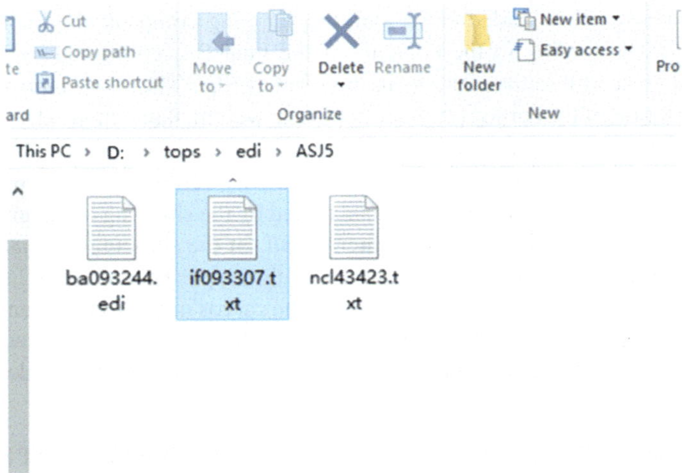

Fig. 3.11 Manifest file

The script of the import manifest for the vessel ASJ5, 0927 is as follows:

```
0:IFCSUM:MAINFEST:9:CUSTHY:GS2:200907060933:CNXIN:CNXIN'
10:000309070116:安盛集 5:CN:0927::石湖-天津:20090707:20090707:CNTSN:天津:CNSTG:汕头'
11:00030673:泉州安通物流有限公司'
12:ATLQZTJ1900259::::::CNQZJ:泉州:DR-DR:20090701:N:'
13:CNTSN:天津:CNTSN:天津::::'
16::磁灶'
17::吴晋发'
41:1::1:PK:PACKAGE:0:0:0'
44:N/A'
47:瓷砖'
51:TGHU3876735:498495:22G1:F:0:21000:2300:0:::::::QZAT:安通'
51:UETU2036344:498336:22G1:F:0:21000:2300:0:::::::QZAT:安通'
51:GESU3577602:498387:22G1:F:0:21000:2300:0:::::::QZAT:安通'
51:GVCU2146373:497801:22G1:F:0:21000:2300:0:::::::QZAT:安通'
12:ATLQZTJ1900262::::::CNQZJ:泉州:DR-DR:20090701:N:'
13:CNTSN:天津:CNTSN:天津::::'
16::水头 0'
17::黄灿林'
41:1::1:PK:PACKAGE:0:0:0'
44:N/A'
47:板材'
51:LLTU2030430:497344:22G1:F:0:21000:2300:0:::::::QZAT:安通'
51:UETU2005615:497524:22G1:F:0:21000:2300:0:::::::QZAT:安通'
51:CLHU3574663:497594:22G1:F:0:21000:2300:0:::::::QZAT:安通'
51:CLHU3815631:497033:22G1:F:0:21000:2300:0:::::::QZAT:安通'
51:ATLU0004796:497583:22G1:F:0:21000:2300:0:::::::QZAT:安通'
12:ATLQZTJ1900260::::::CNQZJ:泉州:DR-DR:20090701:N:'
13:CNTSN:天津:CNTSN:天津::::'
16::磁灶'
17::苏东良'
41:1::1:PK:PACKAGE:0:0:0'
```

```
44:N/A'
47:瓷砖'
--------------------------------------------------------------------------------
12:ATLQZTJ1700876A::::::CNQZJ:泉州:CY-DR:20090701:N:'
13:CNTSN:天津:CNTSN:天津::::'
16::郭少泽'
17::郭少康'
41:1::1:PK:PACKAGE:0:0:0'
44:N/A'
47:板材'
51:ATLU0007326:255937:22G1:F:0:21000:2300:0:::::::QZAT:安通'
51:CNCU2564835:93024:22G1:F:0:21000:2300:0:::::::QZAT:安通'
51:GESU3576102:470096:22G1:F:0:21000:2300:0:::::::QZAT:安通'
99:855'
```

In the container terminal operation system, a specified module is applied to complete the decoding of EDI file, so as to provide the terminal operation management with distinctly categorized data table and graphic stacking map. The decoding process is shown in Fig. 3.12.

The encoding and decoding of manifest information shall comply with the standards of "IFCSUM platform document of manifest" formulated by the Ministry of Communications of China.

The specific conversion standard of the EDI platform file of the manifest is listed below:

sender, receiver:

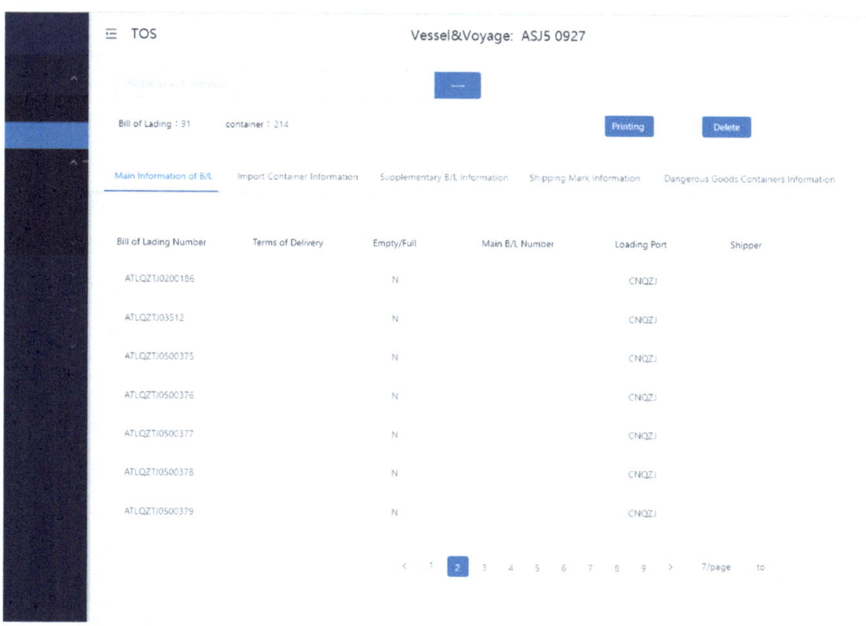

Fig. 3.12 Import manifest fed to the system with EDI

shipping company, shipping agency → terminal, tally, port supervision, etc.
Function: The platform file corresponding to the manifest subset of IFCSUM messages provides information about the cargos for a particular voyage, including details of the carrier, transportation mode, transporting facilities, equipment, and intermodal cargos. The manifest is the proof that the vessel carries the container goods, and it is the necessary document for the vessel to go through the import and export customs formalities, as well as the operation document for the terminal to prepare vessel loading and unloading.
Corresponding documents: import manifest (while the loading list is applied in export operation).

Record 00	HEAD RECORD			M
No.	Field name	Format	Annotation	Mark
1	RECORD ID	9(2)	00	M
2	MESSAGE TYPE	X(6)	IFCSUM	M
3	FILE DESCRIPTION	X(35)	MANIFEST	C
4	FILE FUNCTION	X(2)	9 = original 2 = added 3 = deleted 4 = changed	M
5	SENDER CODE	X(13)		M
6	RECEIVER CODE	X(13)		M
7	FILE CREATE TIME	9(12)	YYMMDDHHMM System time	M
8	SENDER PORT CODE	X(5)		C
9	RECEIVER PORT COD	X(5)		C

Record 01	OTHER RECEIVERS			C
No.	Field name	Format	Annotation	Mark
1	RECORD ID	9(2)	01	M
2	RECEIVER CODE	X(13)		M
3	RECEIVER CODE (1-8)	X(13)		C

Record 10	VSL. & VOY. FIELDS			M
No.	Field name	Format	Annotation	Mark
1	RECORD ID	9(2)	10	M
2	VESSEL CODE	X(9)		C/M
3	VESSEL	X(35)		M/C
4	NATIONALITY CODE	X(2)		C
5	VOYAGE	X(6)		M

(continued)

(continued)

Record 10	VSL. & VOY. FIELDS			M
No.	Field name	Format	Annotation	Mark
6	TRADE CODE	X(10)		C
7	TRADE	X(35)		C
8	ETD ARRIVED DATE	9(8)	CCYYMMDD	C
9	SAILING DATE	9(8)	CCYYMMDD	M
10	DEPART PORT CODE	X(5)		M/C
11	DEPART PORT	X(35)		C/M
12	NEXT CALLING PORT CODE	X(5)		C
13	NEXT CALLING PORT	X(35)		C

Record 11	VSL. & VOY. FIELDS(DETAIL RECORD)			M
No.	Field name	Format	Annotation	Mark
1	RECORD ID	9(2)	11	M
2	SHIPPING LINE CODE	X(13)		C/M
3	SHIPPING LINE	X(35)		M/C

Record 12	FIRST RECORD OF 1 B/L			M
No.	Field name	Format	Annotation	Mark
1	RECORD ID	9(2)	12	M
2	B/LNO.	X(20)		M
3	PRE. VESSEL CODE	X(9)		C
4	PRE. VESSEL	X(35)		C
5	PRE. VOYAGE	X(6)		C
6	PLACE CODE OF RECEIPT	X(5)		C
7	PLACE OF RECEIPT	X70)		C
8	LOAD PORT CODE	X(5)		M/C
9	LOAD PORT	X(35)		C/M
10	DELIVERY TERM	X(9)	i.e., CY-CFS CY-CY	C
11	LOAD DATE	9(8)	CCYYMMDD	C
12	CTN ID	X(1)	Y = B/L of empty container, N = Not B/L of empty container	M
13	M B/L NO.	X(20)	(This is a new item)	M

Record 13	PLACE INFO. OF 1 B/L			M
No.	Field name	Format	Annotation	Mark
1	RECORD ID	9(2)	13	M
2	DISCHARGE PORT CODE	X(5)		M/C
3	DISCHARGE PORT	X(35)		C/M
4	PLACE CODE OF DELIVERY	X(5)		C
5	PLACE OF DELIVERY	X(70)		C
6	TRANSFER PORT CODE	X(5)		C
7	TRANSFER PORT	X(35)		C
8	PLACE CODE OF B/L ISSUE	X(5)		C
9	PLACE OF B/L ISSUE	X(35)		C

Record 16	SHIPPER FIELDS			C
No.	Field name	Format	Annotation	Mark
1	RECORD ID	9(2)	16	M
2	SHIPPER CODE	X(13)		C/M
3	SHIPPER (1-6)	X(35)		M/C

Record 17	CONSIGNEE FIELDS			C
No.	Field name	Format	Annotation	Mark
1	RECORD ID	9(2)	17	M
2	CONSIGNEE CODE	X(13)		C/M
3	CONSIGNEE (1-6)	X(35)		M/C

Record 18	NOTIFY-1 PARTY FIELDS			C
No.	Field name	Format	Annotation	Mark
1	RECORD ID	9(2)	18	M
2	NOTIFY-1 CODE	X(13)		C/M
3	NOTIFY-1 (1-6)	X(35)		M/C

Record 41	CARGO FIELDS			M
No.	Field name	Format	Annotation	Mark
1	RECORD ID	9(2)	41	M
2	CARGO SEQUENCE NO.	9(3)		M
3	CARGO CODE	X(3)		C
4	NUMBERS OF PACKAGES	9(6)		M

(continued)

(continued)

Record 41	CARGO FIELDS			M
No.	Field name	Format	Annotation	Mark
5	PACKAGES KIND CODE	X(2)		M/C
6	PACKAGES KIND	X(35)		C/M
7	CARGO GROSS WEIGHT	9(7).9	kg	C
8	CARGO NET WEIGHT	9(7).9	kg	C
9	CARGO MEASUREMENT	9(5).999	m^3	C

Note if the B/L is empty, the record is a virtual record, which appears only once, and the format is 41:0::0:0′

Record 43	DANGEROUS AND REEFER FIELDS RECORD			C
No.	Field name	Format	Annotation	Mark
1	RECORD ID	9(2)	43	M
2	DANGEROUS			C
2-1	CLASS	X(5)		M
2-2	PAGE	X(7)		M
2-3	UNDG NO.	9(4)		M
2-4	LABEL (1-3)	X(32)		M
2-5	FLASH POINT	X(5)	Celsius	C
2-6	EMS NO.	X(6)		C
2-7	MFAG NO.	X(4)		C
2-8	EMERGENCY CONTACT	X(35)		C
3	REEFER			C
3-1	TEMPERATURE ID	X(1)	C = Celsius F = Fahrenheit	C
3-2	TEMPERATURE SETTIN	X(5)	See Note	C
3-3	MIN. TEMPERATUR	X(5)	See Note	C
3-4	MAX. TEMPERATUR	X(5)	See Note	C

Note except for the plus (+) minus (−) sign and decimal point, only three digits are allowed in temperature

Record 44	CARGO MARKS INFO.			C
No.	Field name	Format	Annotation	Mark
1	RECORD ID	9(2)	44	M
2	MARKS(1-5)	X(35)		C

Note Empty container has no Record 44

Record 47	CARGO DESCRIPTION			C
No.	Field name	Format	Annotation	Mark
1	RECORD ID	9(2)	47	M
2	CARGO DESCRIPTION (1-5)	X(35)		C

Note Empty container has no Record 47

Record 51	CONTAINER FIELDS RECORD			M
No.	Field name	Format	Annotation	Mark
1	RECORD ID	9(2)	51	M
2	CTNNO.	X(12)		M
3	SEALNO.	X(10)		C
4	CTN. SIZE & TYP	X(4)		M
5	CTN. STATUS	X(1)	E = Empty F = Full L = LCL	M
6	CTN. NUMBERS OF PACKAGES	9(6)		
7	NET WEIGHT	9(7).9	The net weight of goods in the container	C
8	TARE WEIGHT	9(7).9		C
9	CTN. CARGO MEASUREMENT	9(5).999		C
10	OVER-LENGTH FRONT	9(4)	cm	C
11	OVER-LENGTH BACK	9(4)	cm	C
12	OVER-WIDTH LEFT	9(4)	cm	C
13	OVER-WIDTH RIGHT	9(4)	cm	C
14	OVER-HEIGHT	9(4)	cm	C
15	STOWAGE LOCATION	9(7)		C
16	CTN. OPERATOR CODE	X(13)		M/C
17	CTN. OPERATOR	X(35)		C/M

Record 99	TRAILER RECORD			M
No.	Field name	Format	Annotation	Mark
1	RECORD ID	9(2)	99	M
2	RECORD TOTAL OF FILE	9(6)	Including head record and trailer record	M

Record structure:

00 HEAD RECORD	M1
01 OTHER RECEIVERS	C1
10 VSL. & VOY. FIELDS	M1
11 VSL. & VOY. FIELDS (DETAIL RECORD)	M1
12 FIRST RECORD OF 1 B/L	M9999
13 PLACE INFO. OF 1 B/L	M1
16 SHIPPER FIELDS	C1
17 CONSIGNEE FIELDS	C1
18 NOTIFY-1 PARTY FIELDS	C1
41 CARGO FIELDS	M999
43 DANGEROUS AND REEFER FIELDS RECORD	C1
47 CARGO DESCRIPTION	C9
51 CONTAINER FIELDS RECORD	M999
99 TRAILER RECORD	M1

3.2.3 Information Verification of Import Containers

It is known that import container information is the common content of import BAPLIE and import manifest, and the information verification of import containers refers to the checking of the container data in the import BAPLIE and import manifest. Since many operations are involved from the generation of the BAPLIE and manifest to the input into the system, any negligence in the operation may cause the deviation of information, such as misstatement from freight forwarders, errors in manual input or via EDI transmission, and so on. Therefore, information verification must be carried out before the actual operation, so as to timely contact the shipping company for information correction, effectively improve the quality of the import container data, and ensure the validity.

A special module is configured in the operation system of the container terminal to realize this verification work, namely the module of "BAPLIE and manifest verification," As shown in Fig. 3.13, it is the BAPLIE and manifest verification window for the vessel ASJ5, 0927.

As can be seen from the figure above, there are two situations in the verification of BAPLIE and manifest: inconsistency and over-loading. Inconsistency between the BAPLIE and manifest means some information for the same container record is inconsistent in the BAPLIE and manifest, which may be any one or more pieces of information other than the container no. Over-loading refers to the situation that a container (or some) is included in the BAPLIE, but the corresponding record does

Fig. 3.13 Information verification of import containers

not exist in the import manifest, or there is a container (or some) in the manifest but not in the BAPLIE, or both.

If the over-loading only appears in the BAPLIE, it means there is something missing in the input of the manifest. The container no. of the missing container shall be recorded, and the record shall be added in the manifest, and then the verification shall be continued until it is passed.

If the over-loading only appears in the manifest, it means that there is a lack of containers in the BAPLIE, which has been input in the manifest. In this case, it shall be checked. If it is a problem of the BAPLIE, the corresponding record shall be added to the BAPLIE in the input interface and continue to check until it is correct.

If the two problems above both appear, it may be the superposition of the two problems but more likely to be caused by the wrong input of a container no. At this moment, the missing or over-loaded container no. should be checked and verified to correct the wrong container no., by which the over-loading problem in the BAPLIE and manifest can be solved at the same time.

After the verification, all container information in the final BAPLIE and manifest should be exactly the same, with no inconsistency or over-loading, providing effective guarantee for the subsequent unloading operations.

3.3 Planning and Scheduling of Vessel Unloading Operations

3.3.1 Storage Plan of Vessel Unloading Operations

3.3.1.1 Contents and Process of Storage Plan

The storage plan in the yard for the import containers in vessel unloading operations (hereinafter referred to as the storage plan) is to group all the import containers unloaded from the vessel and specify a target area for each group in the yard. The main purpose of the storage plan is to reasonably allocate resources in the yard, to classify and manage different types of import containers from different vessels and to lay an effective foundation for the smooth operation of container delivery in the future. After the information of import containers is checked and verified to be correct, the storage plan of all import containers must be made before the actual unloading operations.

As shown in Fig. 3.14, this is the general operation process of making the storage plan. The responsibilities of all the parties involved are described in Table 3.2.

3.3.1.2 Common Modes for the Storage Plan

The storage plan can specify the exact location for each container, or coarsely to a bay or a row, or even more coarsely to only a certain block (with several bays or rows). The practice has shown that too specific plan will reduce the unloading speed,

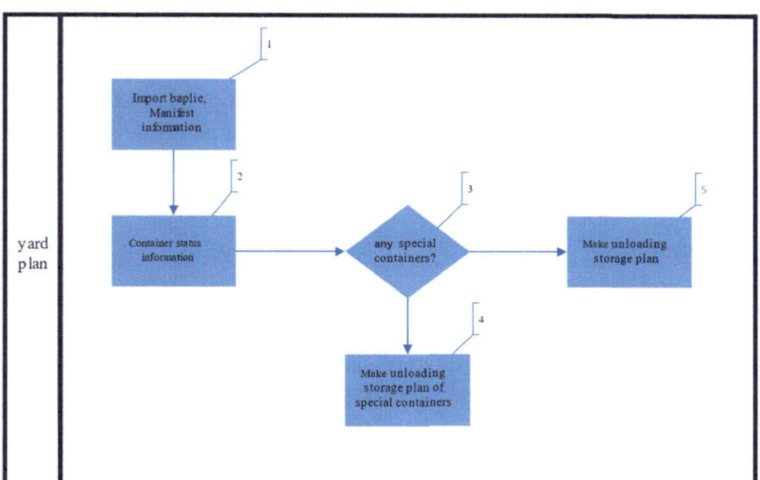

Fig. 3.14 Process of making the storage plan

Table 3.2 Import storage plan content

No.	Department involved	Description
1	Department of Information	The import information clerk receives the information of import BAPLIE and manifest transmitted via EDI platform and verify them
2	Storage Planning	Receiving and checking the state of each container and counting containers
3	Storage Planning	Checking whether there are special purpose containers involved
4	Storage Planning	If there are special purpose containers, the storage plan for special purpose containers should be arranged
5	Storage Planning	If there are no special purpose containers, the storage plan for import containers can be arranged

which is not conducive to the improvement of the unloading efficiency. Only a small number of special purpose containers, such as over-limited containers, etc., need to be designated to relatively specific storage locations.

According to the specification degree, the storage plan can be divided into one-to-one planning and fuzzy planning.

1. **One-to-one planning**

The one-to-one planning is to designate the specific location in the yard for each container to be unloaded. In this planning mode, the planner needs to arrange the loading and unloading sequence for each bay in advance, and then according to the principle of operating from landside to seaside for unloading, determine the vessel unloading sequence list, which reflects the unloading sequence for the containers on each operation line and the corresponding planed storage locations in the yard. This planning mode seems to be in good order, but since the operations in the terminal are highly random mixture of loading and unloading instead of metronomic flow operations, the one-to-one planning will often cause a lot of mechanical waiting and logistics imbalance and need to be frequently adjusted at any time so that "the plan cannot catch up with the change." Therefore, with the increasing throughput of the terminal, this planning mode has been gradually replaced by fuzzy planning mode.

1. **Fuzzy planning**

The fuzzy planning refers to grouping the containers to be unloaded and then allocating one or more areas in the block (composed of consecutive rows) for each group of containers. The feature of the fuzzy planning is to extend the minimum unit in the plan to groups with multiple containers and many locations (which is referred to as many-to-many planning), that is, a group of containers is designated to an area consisting of multiple storage locations. As a result, it has not only reduced the workload of the yard planner but also postponed the time to allocate the specific yard locations to the moment when the containers are actually unloaded from the vessel, which has greatly improved the flexibility of location allocation and real-time operation. To ensure the real-time performance of location allocation, however, it is not realistic to rely on manual work. Instead, the real-time task capture and automatic location selection should be carried out by the server, which means such planning mode should be supported by the location selection algorithm for the containers to be unloaded.

The following contents in this chapter mainly focus on fuzzy planning mode.

The planning processes for vessel unloading operation can be divided into four steps, namely: (1) sorting, (2) distributing, (3) allocating, (4) designating, the meaning and content of each step are described as follows.

1. **Sorting**

Sorting refers to the grouping of the import containers in the BAPLIE, which is the first and most important step in the planning process and will determine the grouping and storage mode of the unloaded containers in the yard. As mentioned above, in order to improve the unloading efficiency and reduce the reshuffle rate when delivering containers, as well as to facilitate terminal management, the unloading storage plan is generally made according to the types of containers. Therefore, all import containers of the selected voyage need to be sorted first. Due to the different size and management mode of the yard, there are some differences in the way of grouping storage, but the basic grouping principles are the same. The general grouping principles can be summarized as follows.

- The empty and full containers should be stored separately.
- The transit containers should be stored according to their discharging ports
- The import empty containers should be stored according to their holders.
- The reefer containers, dangerous goods containers, and over-limited containers should be stored separately.

As shown in Fig. 3.15, the sorting of the import containers of a vessel for certain voyage is to group the containers according to the properties such as size, weight,

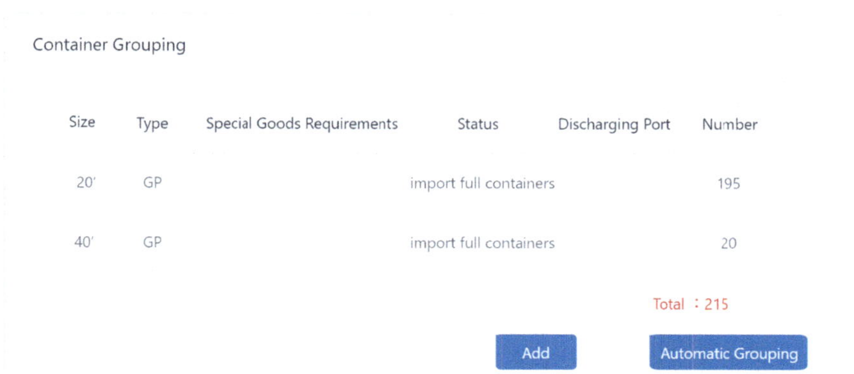

Fig. 3.15 Sorting of import containers

empty or full, special requirements of goods, the container holder, and the discharging port. The import empty containers should be sorted according to their holders, while the transit containers should be sorted according to the discharging ports. The special containers such as the reefer containers, over-limited containers, and dangerous goods containers should be sorted separately.

2. **Distributing**

Distributing is to retrieve the areas in the existing storage plan and their distribution in the yard for a certain group of containers as plan reference. The planner has to determine the number of blocks for each group of containers according to their quantity and distribution in the vessel. If the quantity of a certain group of containers is relatively small, they should be stored together. If the amount of a certain group of containers is medium with concentrated distribution in the vessel, one or two blocks can be allocated in the storage plan. If the quantity of a certain group of containers is large with wide distribution in the vessel, the storage plan should be made in as many blocks as possible, which may facilitate the subsequent facility scheduling and the concurrent operations in the yard, which is referred to as the principle of "distributing and concentrating."

The distributing result of the 40 ft, GP import full containers of a certain voyage is shown in Fig. 3.16. One hundred and thirty-six containers will be distributed in four blocks of A4, B2, B4, and B9.

3. **Allocating**

Allocating is to select the target block for a certain group of containers, which need to make storage plan or to add planned area. Besides the principle of "distributing and concentrating," it is also necessary to balance the task load in different blocks or areas. It is not only the balance in quantity but also in the occupation rate of each block and coverage of the yard cranes. As shown in Fig. 3.17, there is only one planned row in block B4, and it is relatively empty in B4, so B4 can be allocated as the target block.

Fig. 3.16 Distributing of import containers

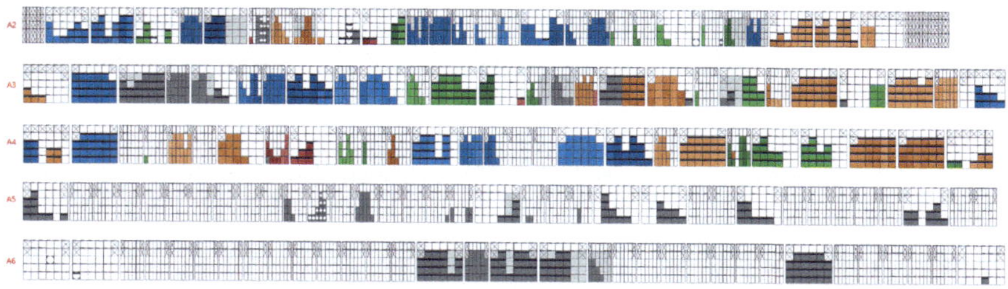

Fig. 3.17 Block allocating for unloading storage plan

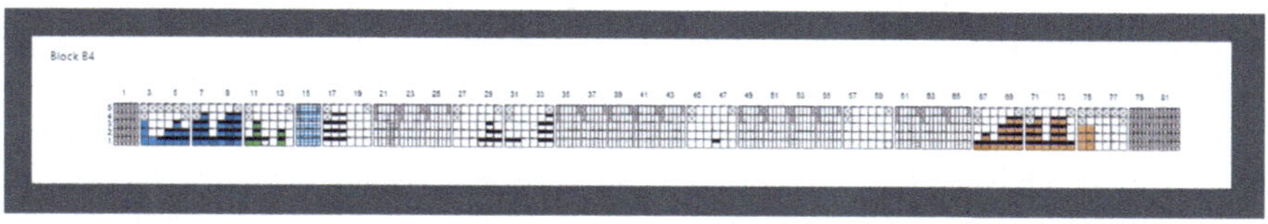

Fig. 3.18 Making unloading storage plan

4. **Designating**

Designating is to determine the rows and bays in the target block for a certain group of containers as the planned storage area. According to the detailed level, designating includes bay designating and row designating.

(1) Bay designating
The container yard is divided into blocks, bays, rows, and tiers. Bay designating is to take a bay as the plan element in unloading storage plan. A specific container can be designated to a certain bay in the block, i.e., B406. This is the common practice at most ports. The content of the "planned position in the yard" contains the block no. and the bay no., as shown in Fig. 3.18.

(2) Row designating
Row designating is to take a row as the plan element in unloading storage plan, namely designating a certain row in the block to a container, i.e., B4061. This designating mode is more detailed than bay designating and suitable for the case of small quantity of containers to plan. The content of the "planned location in the yard" contains row no. besides the block no. and the bay no.

3.3.1.3 Scheduling of Operation Lines

Operation scheduling in the container terminal includes scheduling of operation lines, scheduling of container trucks, and scheduling of yard facilities. The scheduling of operation lines is the core of the scheduling work in the terminal, and the scheduling of yard facilities and container trucks is based on the operating status of the current operation line, so as to ensure the smooth operations of vessel loading and unloading.

3.3.1.4 The Meaning and Main Task of Operation Line Scheduling

The loading and unloading operation lines in the container terminal are usually defined according to the QCs; therefore, the main task of operation line scheduling is the scheduling of QCs. The process of QC input is the process of collecting information about vessel form, current vessel stowage, stowage for vessel loading, and QC input to schedule and reschedule the idle QCs, that is, to arrange the designated QC to load/unload a certain bay of a vessel.

As shown in Fig. 3.19, it is a miniature of the QC scheduling plan. It can be seen that operation line scheduling is also a kind of resource allocation. The resources are the limited QCs, and the target of allocation is the vessel at the quay, waiting for loading or unloading, or to be more precise, it should be several bays in the vessel under operation.

The allocation of the QCs is a decision-making process, and its main decision-making contents include:

1. Determining the number of QCs designated

The number here refers to the number of QCs designated to a vessel for operation. As shown in Fig. 3.19, GUOTA and XPING are both designated with two QCs, which mainly depend on the distribution of loading and unloading tasks, the length of the vessel, the density of the vessel schedule, lead time, and other factors.

2. Estimating the operation volume

It is easy to designate the QCs to operate, but it is difficult to balance all the operation lines. The operation lines should be balanced to ensure that the operation volume for each line is about the same, which needs to estimate the approximate operating time for each bay in each cabin, and the extra time in operation due to complex cabin structures. Meanwhile, steel wire rope application during handling the large and important containers should also be considered. Then, we can roughly estimate and assign the operation area of each QC.

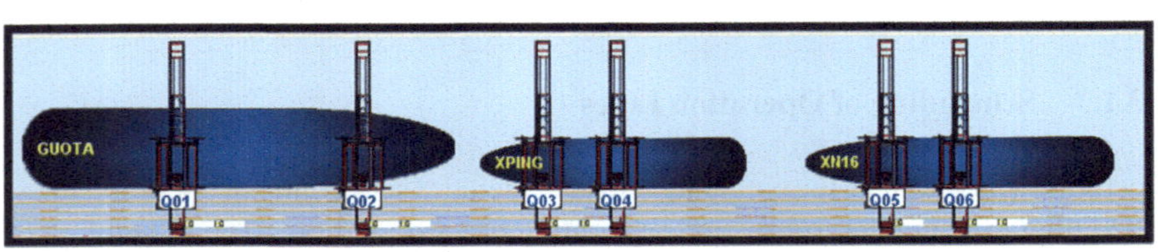

Fig. 3.19 Scheduling of operation lines

3. Arranging the operating sequence

QCs have to share tracks during operation, which means the relative positions of the QCs along the tracks cannot be changed. Therefore, the cross operation of QCs (the left QC operates at the right bay, and the right QC at the left bay) must not be allowed in the arrangement of loading and unloading tasks. In addition to the position sequence, the operating sequence should also be considered, to make sure the distance between two QCs is not shorter than the size of a large bay to avoid the "fight" of the two adjacent QCs.

3.3.1.5 The Meaning of Operation Line Scheduling

The scheduling of operation lines in the container terminal is an important part of the whole terminal operation system. The quality of the scheduling directly affects the efficiency of the loading and unloading operations, the berth date of the container liner, and the reputation of the port enterprises. Meanwhile, the scheduling of the operation lines is a highly technical job. Under the same conditions of vessel, containers, and yard, the schedulers with different technical ability and professional quality will have great difference in scheduling performance.

The process of operation line scheduling is the process of implementing and adjusting the loading and unloading operation plan. Its main work is to rationally allocate the handling facilities at the quay (QCs) so that all the QCs can work concurrently to the maximum and avoid suspension of QCs caused by human factors. Operation line scheduling is a crucial step in the process of loading and unloading operations in the container terminal, which involves all aspects of the terminal management. Figure 3.20 is a schematic diagram of information flow in the container terminal, from which it can be seen that operation line scheduling is an intersection of information flow. Therefore, operation line scheduling in the container terminal is directly related to the operating efficiency and operation level of the terminal.

It can be seen from Fig. 3.20 that in the overall environment of the container terminal operation system, the operation line scheduling module plays the role as a connection. The information required in this module is obtained based on the previous steps. The import container information is obtained by receiving the EDI platform file of the import BAPLIE and importing it into the system so that the import containers can be recorded into the operation line scheduling module. The export container information is processed via the export containers yard plan, port and tonnage grouping, electronic container load plan pre-input, then the actual storage in the yard. After the customs clearance, the containers are stowed, and the stowed containers can be recorded into the operation line scheduling module. It can be concluded that the previous work is to serve the subsequent vessel loading and unloading operations. The operation line input directly affects the actual loading and

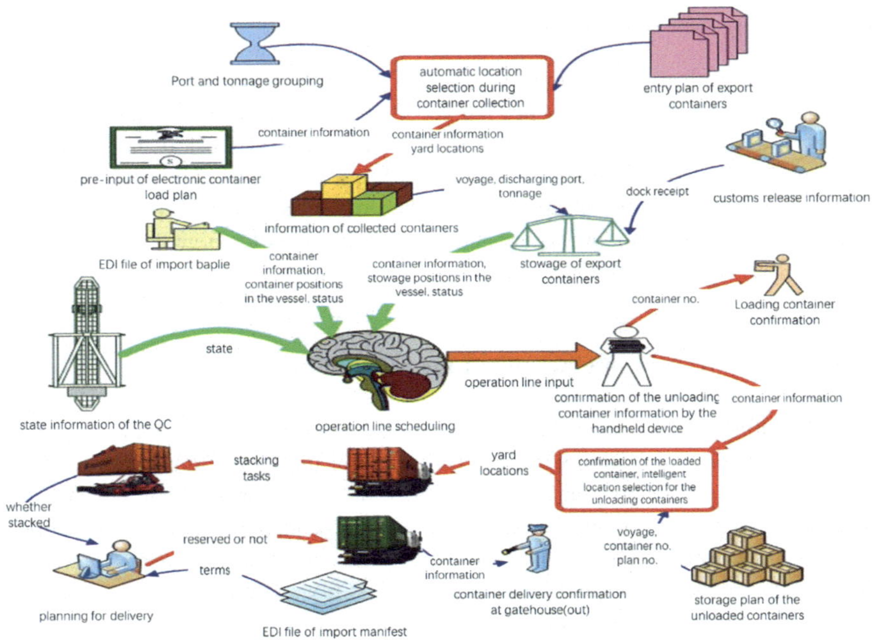

Fig. 3.20 Information flow in the container terminal

unloading operations, and the rationality (balance) of the operation line input will greatly affect the overall loading and unloading operating time.

3.3.1.6 Information Required in Operation Line Scheduling

The process of operation line input is the process of collecting information about vessel form, current vessel stowage, stowage for vessel loading, and QC input to arrange the operation line for the idle QCs. Before the QC designating, the vessel control scheduler must have the following information:

1. **The vessel information**

(1) The maximum bay number—refers to the total number of the bays in the vessel, generally proportional to the vessel length.

(2) The bay combination information—this information is the bay combination defined by the user when the vessel form is defined, such as bay 01 and bay 03 are combined on deck, bay 05 and bay 07 are combined thoroughly, and then, the corresponding bay combination information should be: 02D, 06D, 06H. The bay combination can be displayed in the vessel side profile.

(3) The engine room location—the engine room location is necessary information for the operation line scheduling. For the standard container vessel, two QCs only with the distance not shorter than a large bay can operate at the same time, but if the adjacent two large bays are sandwiched by an engine room, the two large bays can also be simultaneously operated.

(4) The berthing direction of the vessel—the berthing direction of the vessel is divided into left berthing and right berthing. When berthing on the left, the shore is taken as a reference, and the bay no. increases from left to right when facing the seaside. On the contrary, when berthing on the right, the shore is taken as a reference, and the bay no. decreases from left to right when facing the seaside. Since the position crossing cannot occur in the QCs operation, that is, the relative positions of left and right between the QCs should be consistent with the relative positions between their operating bays; therefore, the berthing direction of the vessel is key information for the operation line scheduling.

2. **Loading and unloading operations information**

As shown in Fig. 3.21, the loading and unloading operations information is mainly reflected in the diagram of containers distribution, which is mainly composed as follows:

(1) Bay no. of the operation—represents the bay no. of an operation, which accurately describes the position of the operation.
(2) Whether the operation is on deck or in the cabin.
(3) Attribute of an operation—refers to whether the operation is a vessel unloading operation or a vessel loading operation. I stands for vessel unloading operation, and O stands for vessel loading operation.
(4) The large bay no. of the operation—this information refers to the no. of the bay combination where the operation is located. If there is no bay combination, the no. of the large bay is the same as the bay no. of the operation.

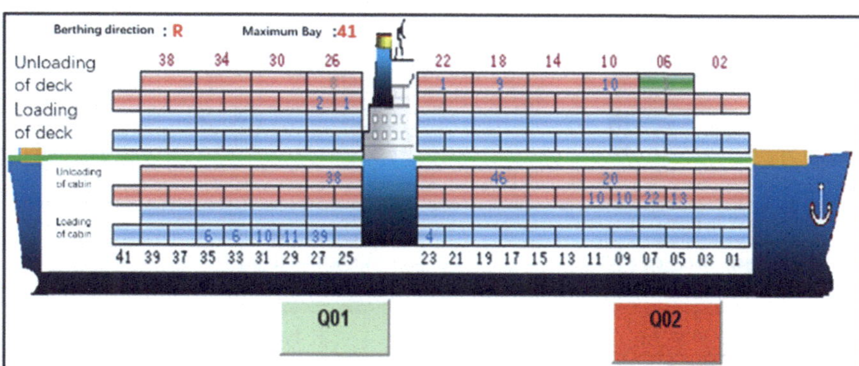

Fig. 3.21 Containers distribution in loading and unloading operations

(5) Number of containers in an operation—this information describes the number of containers to be loaded or unloaded in an operation, which reflects the operation volume, and the anticipated operating time is mainly estimated according to this.

2. QC information

As shown in Fig. 3.22, the QC information of the operation line scheduling mainly includes:

① The number of the QCs that can be designate;
② The serial no. of the QC;
③ The current statec of the QC, including loading/unloading, idle, under maintenance, failure, etc;
④ Skill level of the QC driver;
⑤ Basic attributes of the QC: single-or double-hooked, whether capable of handling double 40 ft containers, etc.

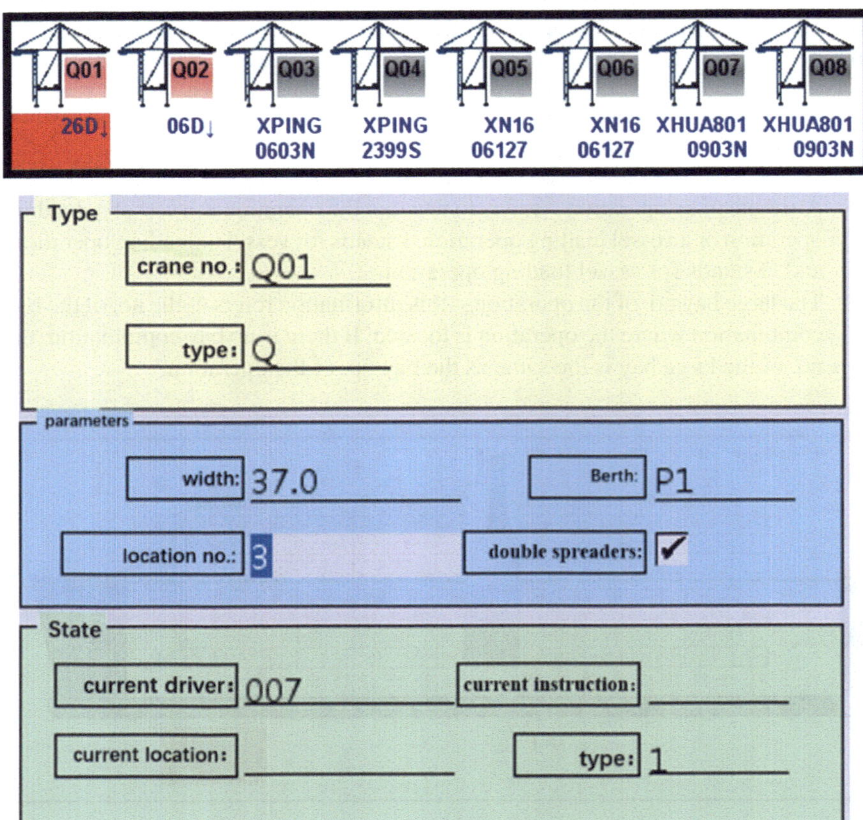

Fig. 3.22 QC information in the operation line scheduling

3.3.1.7 The Basic Principles of Operation Line Scheduling

1. The main factors affecting operating time

At present, the scheduler generally calculates the operating time based on the number of the containers, that is, assuming that the operating time for each container is the same. Therefore, the scheduler usually sums the number of the containers with the calculator, then roughly divides the operation into several pieces, each one corresponds to a QC, and then estimates the scheduling results of the relatively balanced operation lines based on experience.

Currently, most terminals seldom accurately calculate the operating time required for each container according to the container type, special requirements of goods, yard location, and other information and obtain the time required for each bay operation, so as to design the operation line scheduling plan. In fact, this method is not feasible based on the following factors.

① It is not the time required for each operation but the time ratio between each operation that really affects the scheduling results of the operation lines.
② The yard locations of the import containers may not have been planned at the time of implementing operation line scheduling. At present, more and more container terminals have transferred the unloading plan module from the planning department to the central control office, that is, to greatly improve the dynamic property of the unloading plan.
③ In the yard, the container delivery for vessel loading operation is slower than the container collection for the vessel unloading operation. The reason is that there is basically no need to reshuffle the containers in the container collection. However, the container delivery operation is restricted by many factors, from port and tonnage grouping to container collection to the stowage, each step has a direct impact on the efficiency of container delivery operation. However, at present, the reshuffle caused in vessel loading operation cannot be completely avoided, so the impact of reshuffle on the vessel loading efficiency is relatively great, and it is impossible to determine how many containers need to be reshuffled when loading a certain container into the vessel.
④ Dynamically, retrieving information of each container from the database takes up a lot of system resources and puts a lot of pressure on the server.
⑤ It is impossible to predict the effect of container delivery and collection on the vessel loading and unloading operations in the yard at present. Therefore, the operating time for each container cannot be accurately predicted.

Based on the factors mentioned above, since the main factors affecting the time ratio between various operations are the number of the containers and their loading and unloading attributes, the most vessel control schedulers in the terminals estimate the operating time required according to the number of the containers and their loading and unloading attributes.

2. Basic principles of operation line scheduling

① The unloading operation is prior to the loading operation at the same large bay. This principle is the operation sequence required by the loading and unloading techniques. If the loading operation is carried out when the unloading operation is not finished, the container to be unloaded may be buried under the loaded containers, or the position for the container to be loaded is still occupied by the container to be unloaded.
② The unloading operation on deck is prior to the unloading operation in the cabin. This principle follows the objective logic. If there are a certain number of containers on deck to be unloaded, the operation in the cabin can only begin after all the containers on deck have been unloaded.
③ The loading operation in the cabin is prior to the loading operation on deck. This principle ensures the correct logic of the vessel loading process. When there are a certain number of containers in the yard to be loaded in the cabin at certain bay in the vessel, if the loading operation on deck at this bay has already begun, the hatchway must have been covered, which makes it impossible to operate in the cabin any more.
④ The concurrent operations between the QCs should be kept with a certain distance. Generally speaking, for a normal container vessel, two QCs can operate concurrently at an interval greater than the size of a large bay. Therefore, operation line scheduling should ensure that the distance between the concurrent QCs is greater than the length of a large bay.
⑤ The operating positions of the QCs should not be crossed, namely the left QC should not be allowed to load or unload the right bay, for example, while Q01 is operating at bay 18, Q02 cannot be arranged to operate at bay 10.
⑥ The operation division with larger volume has the priority in QC allocation. This principle is used to determine the allocation of QCs when multiple operation divisions need to work concurrently, that is, to give priority to the division with large unloading volume and allocate more QCs to it.
⑦ Operation division between the concurrent operation divisions should be done first. In order to facilitate the operation, the operation with large volume will be divided into many small pieces; meanwhile, the pieces that can be operated concurrently should be done later, so as to improve the dispersion of the operations in time, which is conducive to the implementation of the subsequent operations as well as dispatching more operation lines.
⑧ The even bay at the large bay has the priority. The reason is that one of the principles of stacking containers in the cabin is that "40 ft containers can be stacked on 20 ft containers, while 20 ft containers cannot be stacked on 40 ft containers." Considering the factor of double-container QC, the advantage of double-container QC in operating efficiency can be fully explored by operating the even bay first,

and the remaining 20 ft containers that cannot be handled by the double-container QC can be left to be handled with the containers at the small bay.

3.3.2 *Scheduling of Container Trucks*

3.3.2.1 The Meaning of Container Truck Scheduling

The scheduling of container trucks refers to the scheduling of the internal trucks, which are applied for horizontal movement of the containers in the terminal. The scheduling of the container trucks is one of the important contents of the facility scheduling in the container terminal. The containers unloaded by the QC have to be moved by the container trucks to the yard and operated by the yard crane to finally store in the yard. Therefore, the QC scheduling has to be followed by the container truck scheduling, as shown in Fig. 3.23.

The scheduling of container trucks in the container terminal includes configuration of operation lines, selection of container truck pool, determination of the numbers of container collection trucks and loop trucks. The specific meanings and settings of each part are as follows:

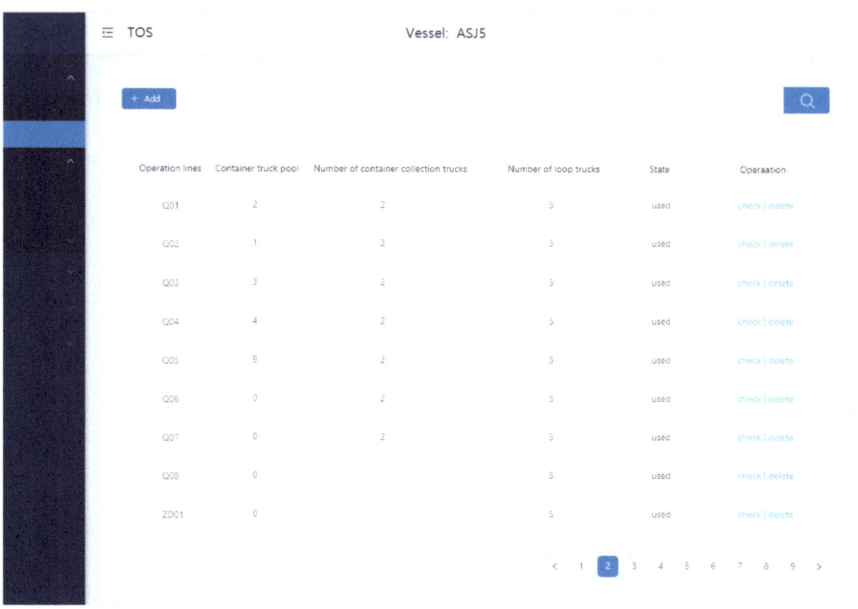

Fig. 3.23 Scheduling of container trucks

1. **Operation lines**

The operation line is the combination of QCs, container trucks, yard facilities, and control schedulers, to complete the operations of vessel loading and unloading or the combination of container trucks, yard facilities, and control schedulers, to complete the operation of container shifting. It can be divided into loading/unloading operation line and shifting operation line. To add an operation line, click the "add" button, enter the operation line name and other relevant information, check the "apply to" bay, and then click the "save" button to save the settings.

2. **Container truck pool**

The container truck pool refers to the set composed of all container trucks assigned to a certain operation line, and the no. of the set is called the no. of the container truck pool. Therefore, each operation line has its no. of the container truck pool. If two or more operation lines have the same container truck pool, it means these operation lines can share the trucks in the same pool. If the no. of the container truck pool of a certain operation line is unique, the trucks on the line cannot be used by other lines (QCs or shifting operations).

3. **Number of container collection trucks**

The number of the container collection trucks refers to the number of trucks expected to be ensured at the quay, that is, to ensure the number of trucks at the quay is not less than the number specified for the container collection trucks. When the number of trucks at the quay is less than this value, the operation line has a priority to occupy the truck resources, and when the number of trucks at the quay is equal to this value, the system will not continue to assign new empty truck to collect container.

4. **Number of loop trucks**

The number of loop trucks refers to the maximum amount of the trucks in the operating loop, that is, the number of trucks in the operating loop shall not exceed this value, otherwise the trucks shall be dispatched to satisfy other operating lines in priority.

3.3.2.2 The Common Modes of Container Truck Scheduling

1. **Fixed scheduling**

The fixed scheduling refers to the arrangement for each operation line of a specified number of trucks, which are only responsible for the operations within the line and do not intervene in the operations in other lines. In some terminals with limited resources, there is no wireless TPS receiver installed in the truck. The truck drivers are instructed by the central control office through interphones. In this case, the fixed scheduling mode is the most commonly used.

2. **Shared scheduling**

The shared scheduling refers to configuring the container truck pool to share the trucks in some operation lines and operate nearby. It is commonly used in this situation: for two adjacent QCs, one is loading, and the other is unloading. If they can share the trucks, the trucks in the loop can be loaded all the time, reducing the probability of no load. In this mode, the wireless TPS receiver must be installed in the truck to ensure the flexibility and real-time scheduling.

3. **Full cycle scheduling**

The full cycle scheduling refers to the shared scheduling of all the trucks. The trucks do not need to be arranged for a certain operation line but are assigned and directed by the server after automatic calculation. At present, this scheduling mode has been used in some large and modernized terminals.

3.3.3 Scheduling of Yard Cranes

3.3.3.1 The Purpose and Main Task of Yard Crane Scheduling

The scheduling of yard cranes is to assign a block to operate for each yard crane. There are usually limited number of yard facilities in the terminal. Ideally, it is better to ensure that at least one yard crane is designated in each block to be responsible for collecting and delivering containers. The difficulty in scheduling for the yard control scheduler depends not only on the volume of operations but also on the number of the yard cranes. The more yard cranes there are, the more options there are, and the simpler and more convenient the scheduling will be. Conversely, the fewer yard cranes there are, the more difficult it will be to guarantee the coverage of the yard cranes in the yard, and the more difficult the scheduling will be.

For the vessel unloading operations, the purpose of yard crane scheduling is to arrange yard cranes for the blocks with established unloading storage plan, so as to ensure that there is a yard crane responsible for collecting the containers in the yard after they are unloaded by the QCs from the vessel and moved horizontally to the yard via container trucks, by which the container trucks can be released in time so that they can be put into new operations.

As shown in Fig. 3.24, the blocks A3 and B3 are covered by the yard cranes L03 and L04, while the blocks A6 and B6 are covered by the yard crane L01.

3.3.3.2 The Yard Crane Scheduling in the Vessel Unloading Operations

For the vessel unloading operations, yard control schedulers generally have to carry out the following operations to ensure the reasonable allocation of the yard facilities.

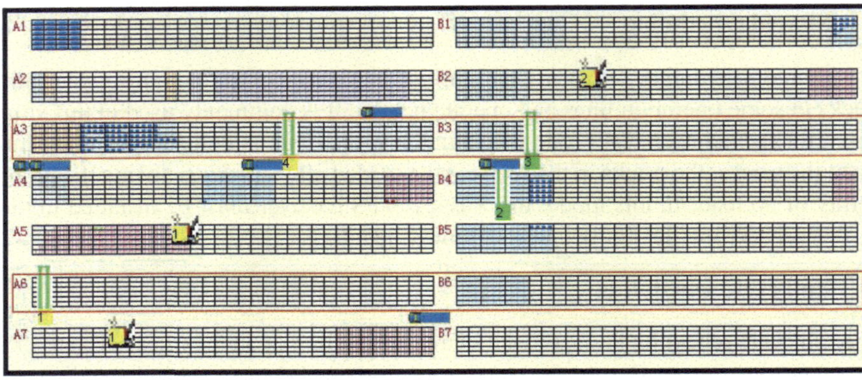

Fig. 3.24 Operation coverage of the yard cranes in the block

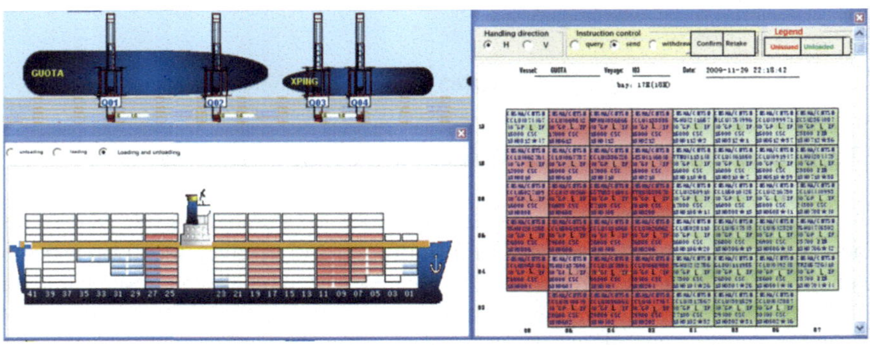

Fig. 3.25 Current information of containers for unloading checked by the yard control scheduler

① Select the vessel in operation, check the current operating bay of the corresponding QC, as shown in Fig. 3.25.
② Further check the bay view of the current bay under unloading operations to determine the group of the containers to be unloaded. The group is sorted by the planner when making the unloading storage plan, as shown in Fig. 3.25.
③ Check the yard storage plan of the containers waiting for unloading, mainly for the distribution of the blocks, as shown in Fig. 3.26.
④ Determine the blocks that need to be opened (see Fig. 3.27). If the number of such containers is small, one block will be enough. If several cranes are unloading such containers, as many blocks as possible should be opened to ensure the concurrent operation of the facilities in the yard.
⑤ Check whether each block that needs to be opened is covered by a yard crane. If there is no yard crane, select the idle yard crane nearby and assign it to operate in the block, as shown in Fig. 3.28.

Fig. 3.26 Block distribution of the yard storage plan

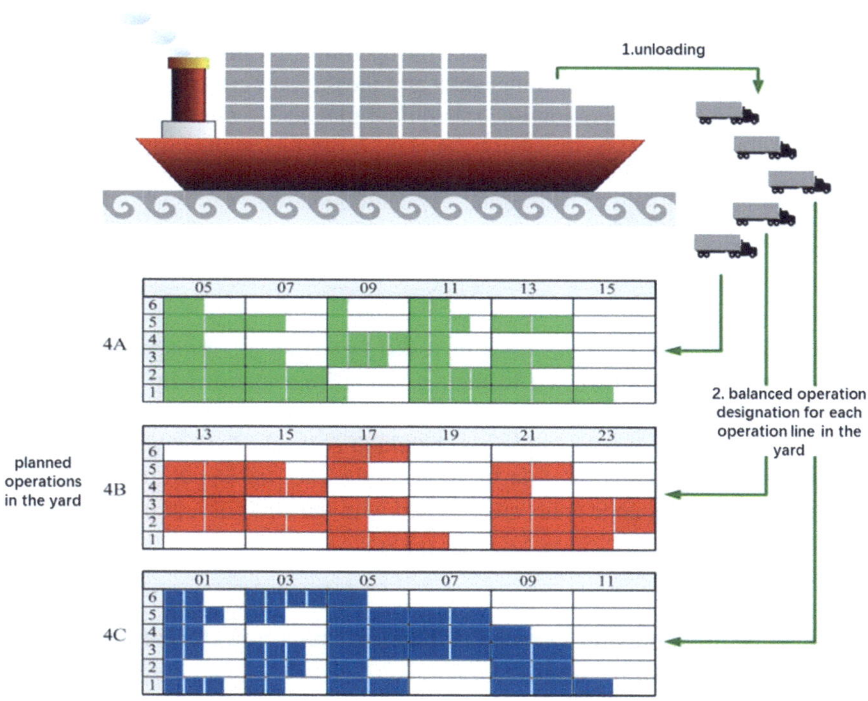

Fig. 3.27 Opened blocks for the containers to be unloaded

Fig. 3.28 Arrangement of operating block for the yard crane

3.3.4 Sending Unloading Operation Instructions

3.3.4.1 Preparations Before Sending Instructions

Sending the unloading instructions means the actual beginning of the operation at a certain bay. Before sending the instructions, it must be ensured that the facilities are ready, the vessel has berthed, and the site preparation has been completed.

1. Responsibilities of the central control office

The vessel control scheduler shall dispatch the operation line and assign the container trucks, and the yard control scheduler shall arrange the yard crane for the designated block according to the operation line scheduling plan and vessel unloading storage plan and inform the yard facilities to be in place according to the instructions.

2. The sailor room

The crew in the sailor room will berth the vessel.

3. The vessel handler

Safety measures for all the positions in vessel operation are arranged and implemented. The QCs are instructed to their positions and guarded by the operators beside the QCs to ensure their safe traveling. The lashing for the containers to be unloaded is arranged, and the implement situation is checked. The twist lock boxes are checked and placed at the safe location. In the end, the import data, awkward cargo data, dangerous goods data are checked with the chief officer (or the deputy) in the vessel.

3.3.4.2 The Meaning and Purpose of Sending Unloading Operation Instructions

Sending unloading operation instructions refers to establishing the operating sequence for the containers to be unloaded at the current operating bay, which is implemented by the vessel control scheduler. As shown in Fig. 3.29, these are the

Fig. 3.29 Sending unloading instructions

instructions sent for bay 09 in the cabin of vessel GUOTA. At this bay, the unloading instructions of 15 containers on the right top have been sent, and the sequence no. is shown at the lower right corner of each container.

The main purpose of the unloading operation instructions is to control the operation tempo and make it convenient for the tally clerk beside the QC to select the container. The vessel control scheduler can flexibly control the contents displayed in the handheld device of the tally clerk beside the QC through instruction sending and receiving. The containers with sent instructions are displayed in the information list by their sequence no., with which the vessel control scheduler can assist the tally clerk to rearrange the container sequence to easily select the container (on the top of the list).

The factors that the vessel control scheduler needs to consider when sending the instruction sequence are

1. Hatchway cover

As shown in Fig. 3.29, if it is a two-piece translation-type hatchway cover, all the containers in the cabin can only be divided into two groups for unloading, so instructions cannot be sent only by tier. The key is to consider the operation sequence in the actual unloading.

2. Yard storage plan

If there are special purpose containers in the cabin, or the containers in the cabin do not fall into the same group, the unloading instructions for one group should be sent first, and the instructions for another group should be sent later, to avoid the frequent cross unloading of the two groups of containers.

3.4 Actual Operations of Vessel Unloading and Kanban Confirmation

3.4.1 Vessel Unloading Confirmation by the QC

1. Main contents of vessel unloading confirmation

The main contents of vessel unloading confirmation include two parts: confirmation of the container truck information and confirmation of the unloading container information, by which a one-to-one relationship is established between the unloading container and the container truck.

(1) Confirmation of the container truck information
Confirmation of the container truck information mainly refers to the confirmation of the container truck no., as shown in Fig. 3.30.
(2) Confirmation of the container information
Confirmation of the container information mainly refers to the confirmation of the container no. to be unloaded, as shown in Fig. 3.31.

2. Operation processes of vessel unloading on the scene

The on-site operations of vessel unloading for import containers involve multiple jobs of multiple positions, and the general operation processes are shown in Fig. 3.32.

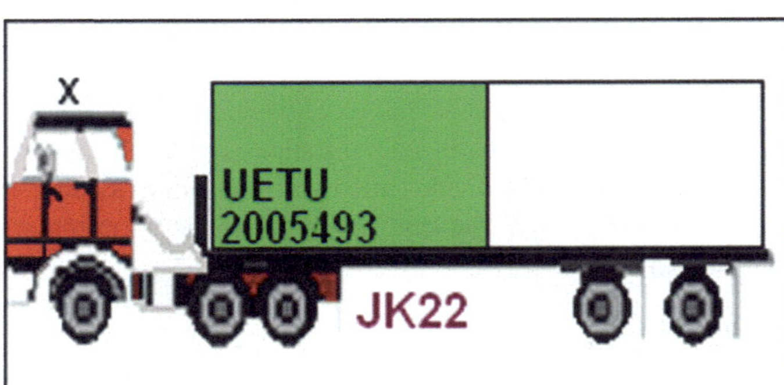

Fig. 3.30 Confirmation of the container truck information

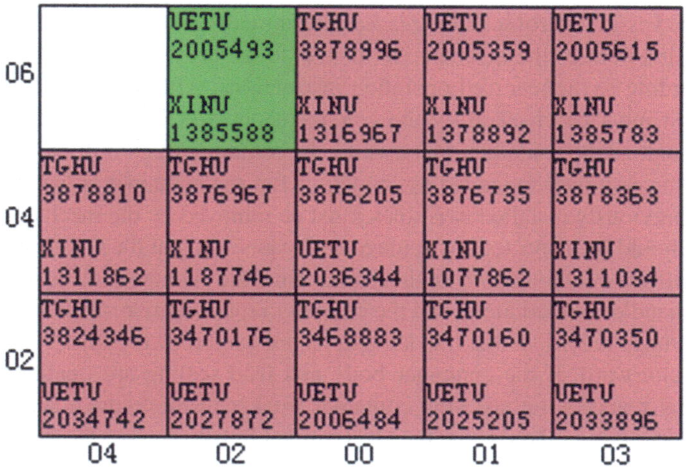

Fig. 3.31 Confirmation of the container information

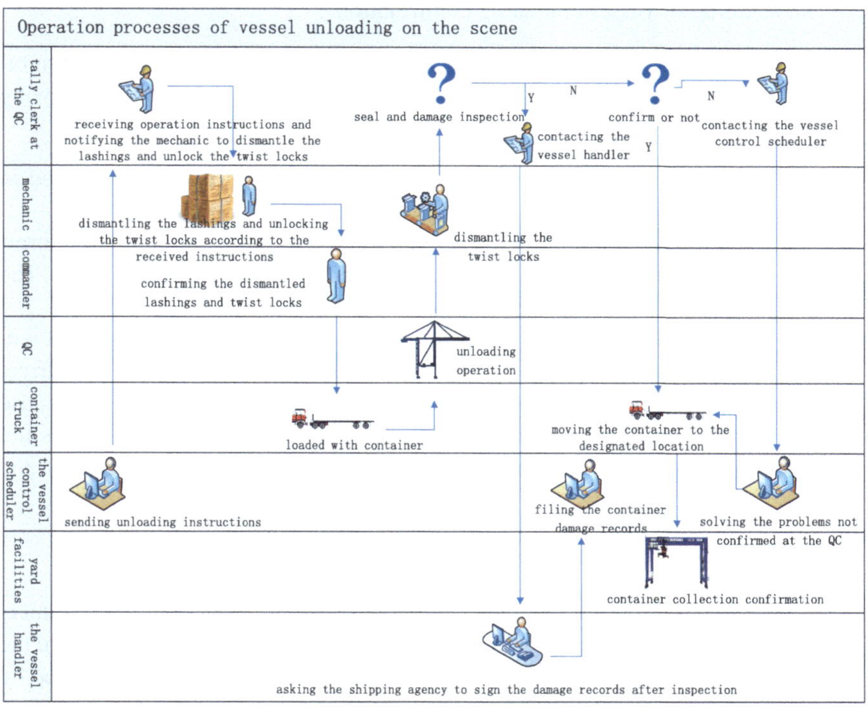

Fig. 3.32 On-site operation processes of vessel unloading

After the vessel control scheduler issues the unloading instructions, the tally clerk at the QC first notifies the mechanic to dismantle the lashings and unlock the twist locks according to the received operation instructions. After the completion of the operation, it must be checked by the command operator before the QC driver is notified to unload the container. Meanwhile, the container truck is driven to the quay to receive the container unloaded by the QC. After the container is loaded onto the truck, the previously installed twist lock must be removed by the mechanic, and the container should be inspected for damage and lead sealing at the QC. If there is any problem with the container body or the lead sealing, the damage shall be inspected by the vessel handler immediately, and the shipping representative shall be required to sign the damage record so that the damage record can be filed by the vessel control scheduler afterward. If the container body and lead sealing are qualified and the QC terminal receives the task, the container truck will move the container to the designated location in the yard. If the QC terminal fails to receive the task normally, the vessel control scheduler shall be informed, and the unconfirmed problem shall be solved before the container is moved to the yard.

3. The meaning of vessel unloading confirmation

The confirmation of vessel unloading plays a very important role in the import container unloading operations in the terminal, which mainly includes the following three aspects:

① In the terminal operation system, the vessel unloading tasks are recorded in real time, and the progress of unloading can be tracked.
② The unloading stacking task is generated. Each unloading stacking task corresponds to a container truck no. and the corresponding unloaded container no. The server automatically selects the location by capturing the task.
③ It has avoided the unloading of transit containers into the terminal, which will cause unnecessary loss to the terminal. Due to the limited vision, the QC driver sometimes performs the unloading operation on the transit containers because he cannot see the specific information on the containers clearly. The unloading confirmation can detect the error in time and put the containers back into the vessel.

4. Common situations in vessel unloading confirmation

The unloading operations are flexible, and the unloading sequence of the containers at the same bay can be changed for the purpose of convenience and rationality in the actual operations, rather than strictly following the unloading sequence specified by the vessel control scheduler. To change the unloading task, the unloading container no. should be entered in the input box of the container no. correction. For example, if the task "UETU2005493" is to be changed to "TGHU3878810", input "018" in the reverse order from the end of the container no. After the container is found, the prompt box will pop up automatically, and the unloading task can be changed after confirmation, as shown in Fig. 3.33.

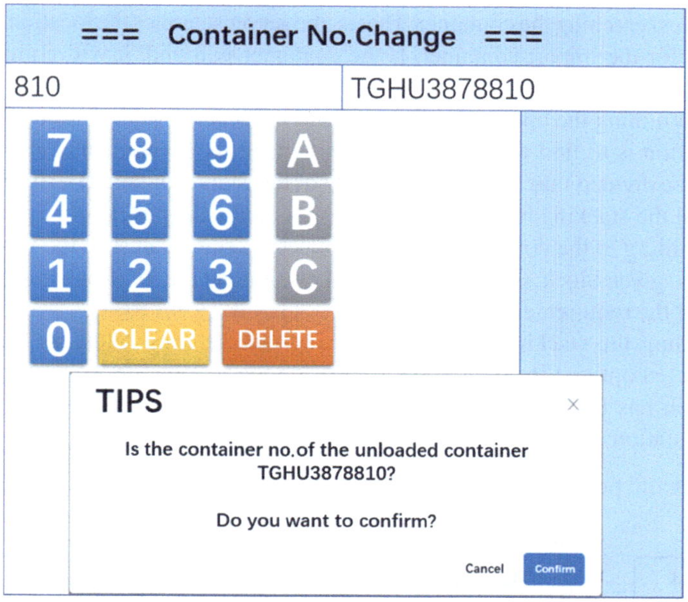

Fig. 3.33 Changing the unloading task

Another situation that may appear in the vessel unloading confirmation is that the no. of the unloading container cannot be found in the system. There are two possible reasons for this. ① It is caused by omissive or wrong input of the BAPLIE information, so it is necessary to contact the central control office immediately to add or modify the BAPLIE records. ② The unloading container is a transit container, which should not be unloaded at this port. Then, the container should be promptly loaded back to the vessel, to avoid the discharging at the wrong port. This is one of the main reasons for the confirmation of the vessel unloading.

3.4.2 Automatic Location Selection in the Yard for the Unloaded Container

1. The steps of automatic location selection in the yard for the unloaded container

The automatic location selection in the yard for the unloaded container refers to the process in which the management system automatically selects a specific location in the yard for the container that has been unloaded by the QC and generates a stacking task which can be seen at the yard crane terminal.

Automatic location selection for the unloaded container can be divided into the following three main steps:

(1) Server searching the container. That is, the server searches the location selection task for the import container. If the container is found, it will go to the next step;
(2) Determining the optimal location. The determination of the optimal container location is to find a best stacking location for the found container. This step can be divided into: ① check the group of the found container; ② further determine the stacking block in the yard (vessel unloading plan) of such container according to the container group; ③ select the row nearest to the yard facility in the given block; ④ select an optimal one among the locations in the row that meet the requirements of unloading.
(3) Sending the stacking task. The stacking task sending refers to sending the selected optimal location to the TPS receiving terminal in the container truck which has undertaking the task, so as to guide the truck driver the moving orientation.

The specific process of the automatic location selection is shown in Fig. 3.34.

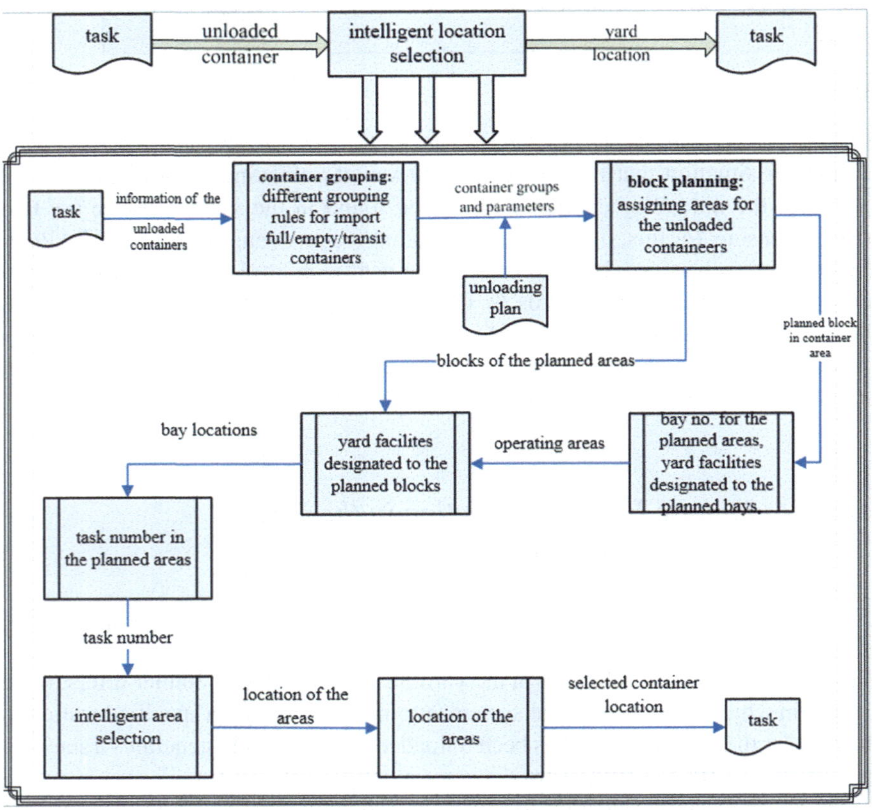

Fig. 3.34 Process of automatic location selection

2. Storage modes and stacking principles of the unloaded containers in the yard

(1) Storage modes of the unloaded containers in the yard
Generally speaking, there are two main storage modes of the unloaded containers in the yard.

① The import containers are mixed in the blocks. In this storage mode, the unloaded containers are stacked randomly in the import container blocks regardless of the vessel's name and voyage. In this mode, the efficiency of vessel unloading can be improved under the condition of sufficient yard facilities, but the reshuffle rate during delivery is relatively high.
② The import containers are stored by vessel's name and voyage. In this mode, the import containers of the same vessel's name and voyage are stored in multiple blocks according to the groups of the unit containers. In this mode, the utilization rate of the facilities can be improved, and the waiting time of the container trucks and the reshuffle rate be reduced, but the utilization rate of the blocks is relatively low.

(2) Stacking principles of the unloaded containers in the yard
At present, there are six rows and four tiers in a bay in the block, and the total storage capacity is 24 unit containers. Considering the need of reshuffle (containers in the same bay do not be reshuffled to other bays), the actual maximum storage capacity is 21 unit containers.

The common stacking principles of the unloaded containers in the yard include

① Tier priority principle, namely the arrangement in tier is in priority. The containers are stacked tier by tier from bottom to top, as shown in Fig. 3.35. The operating trajectory of the RTG's spreader is an arc, as shown in Fig. 3.36. This method used to be often applied for the purpose of safe operation, but it is seldom used now.

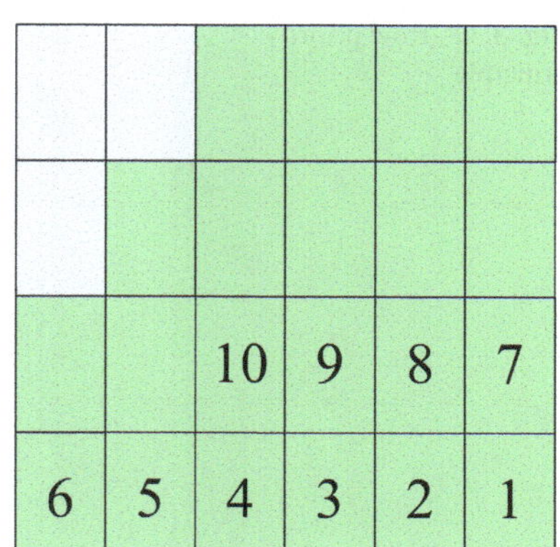

Fig. 3.35 Tier priority principle

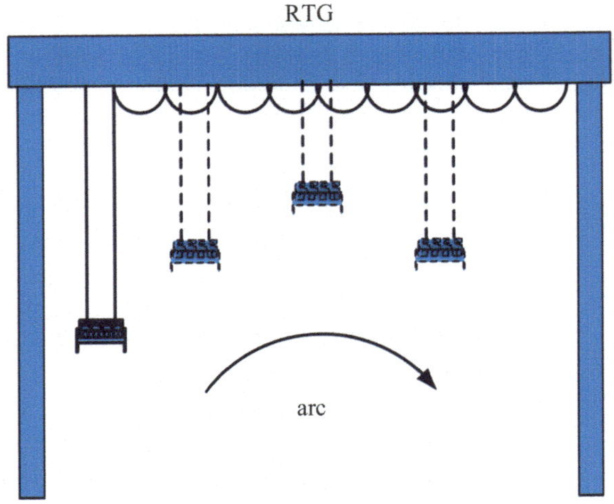

Fig. 3.36 Operating trajectory of the RTG's spreader

② Row priority principle (then tier priority), that is, container arrangement in row is in prior. The containers are stacked row by row, and in one row from bottom to top. In principle, the row far away from the lane should be first filled with containers, and then, the other rows are filled with containers successively in the direction relatively close to the lane, as shown in Fig. 3.37. Since the operating trajectory of the RTG's spreader consists of straight lines with 90 degree turns, which is safe to operate, well planned, and easy to manage with the same type of containers. Therefore, it is a widely used stacking principle at present, as shown in Fig. 3.38.

③ The random stacking principle, with which the containers are stacked randomly and the yard crane driver can change the stacking location of the container at any time according to the busyness condition of the operation line, resulting in high speed in container collection operation. However,

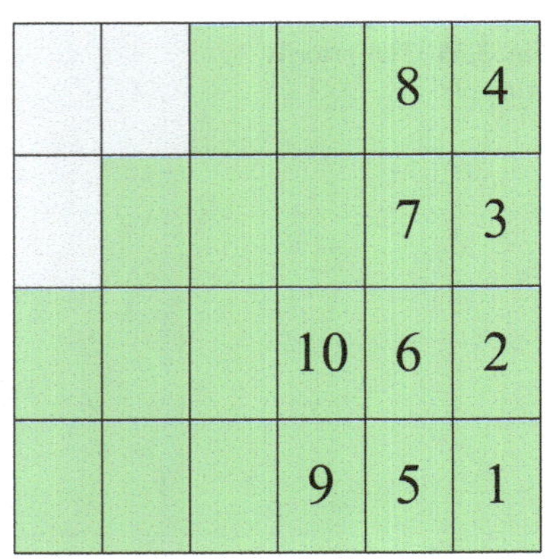

Fig. 3.37 Row priority principle

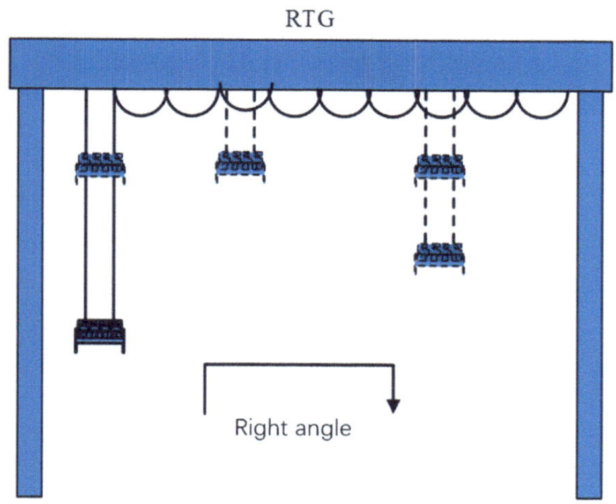

Fig. 3.38 Operating trajectory of the RTG's spreader

the operation is poorly planned, so this method is seldom used in actual operation.

3.4.3 Container Stacking Confirmation by the Yard Crane

1. The main content of the container stacking confirmation

The container stacking confirmation by the yard crane refers to the process in which the yard crane driver stacks the container into the designated location according to the instructions issued by the yard control scheduler. The main content of the container stacking confirmation is to determine the final storage location of the unloaded container so that the yard control scheduler can timely query the storage status of the yard.

2. Steps in stacking confirmation operation

(1) Selecting a task
After the container truck enters the yard, the yard control scheduler selects the stacking task according to the truck no. The stacking task list is shown in Fig. 3.39.
In order to avoid the mistakes caused by the yard crane driver's operation on the tasks that do not belong to the block, the visibility of the tasks is set by

Fig. 3.39 Stacking task list

Task List

Operation type	Plate No.	Container No.	Size	Stack
Loading sequence	JK12	CCLU3190462	20	A34334
	JK15	CCLU2611314	20	A34314
		GESU2097553	20	A34313

the system, that is, only the stacking tasks in the block that the current yard crane is responsible for are displayed in the task list.

(2) Checking the container location
Container location checking means that the operator checks whether the designated location of the corresponding container on the selected container truck can be used for the actual stacking operation. In general, the stacking can always be smooth, but sometimes, due to the influence of the entry order of the container trucks, the stacking location may have been occupied or the container below has not been stacked. If the container location is occupied, the operator shall change the location of the container in the block plan according to the actual situation and directly change it to the location where it can be stacked. If the container below has not been stacked, it can be stacked directly to the location below, or when the stacking task of the container below is right behind, the container below will be stacked first.

(3) Stacking the container
A one-to-one relationship is established between the container and the actual storage location in the yard, and the storage status of the yard is monitored accordingly. On the completion of the stacking task, the container truck corresponding to the task is released at the same time. The server can capture the container truck and put it into the new task.

The stacking confirmation of the import containers can avoid the phenomenon of "losing container" in the terminal. The so-called losing container refers to the situation that the actual storage location of the container is not the same as the designated location, and the container cannot be searched by yard location if it is not confirmed.

3.5 A Comprehensive Case of the Vessel Unloading Operations

This chapter mainly introduces the whole processes of the vessel unloading operations of the import containers, from information collection and processing to the statistics of the handling operation data. In order to help the readers better understand and master the operation processes of the terminal operation system as well as the operating steps of the system, the voyage 0927 of vessel ASJ5 is selected as an example to explain in details the operating steps of the vessel unloading operations in the container terminal, and all the details and possible misoperations are clearly presented to the readers.

1. **Input of the import BAPLIE**

At present, the document import with EDI has been adopted to input BAPLIE in many container terminals, but the addition, deletion, and modification of individual record still need to be done manually. Login to the main interface of the container terminal operation management system, select the submenu of "voyage management" under the menu of "vessel and voyage" in the left navigation bar and set the vessel ASJ5, 0927 as the current voyage in the pop-up interface of "voyage management", as shown in Fig. 3.40.

After selecting the voyage, click the "add" button in the interface to add a record, input, and save all the information one by one according to the BAPLIE data at hand. The BAPLIE for the vessel unloading of this voyage after input is shown in Fig. 3.41.

Fig. 3.40 Login interface of the manual input and modification of the BAPLIE

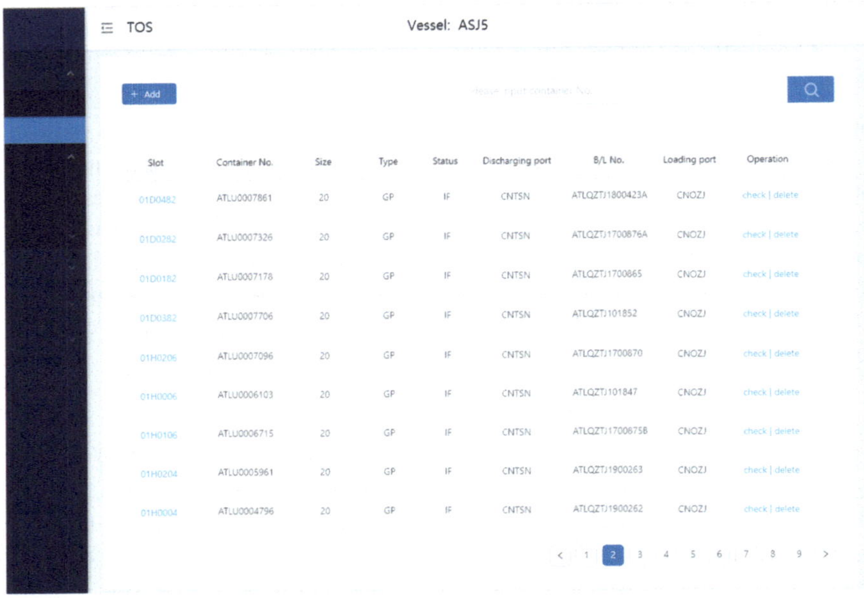

Fig. 3.41 Interface of the manual input and modification of the BAPLIE

In order to accurately and efficiently complete the input of the BAPLIE, there are several announcements to mention.

① When a container is a general container, many fields can be left as default.
② When a container is a transit container, select "status" and input the destination port of the container.
③ For the container weight, it is assumed as empty when the weight of a 20 ft container is less than 2.3 ton and when the weight of a 40 ft container is less than 3.8 ton. If the "container weight" field is skipped during input, the system will take "empty" as default. If it is a direct delivering container, information should be input in the "direct delivering unit" field.

Click the "container position in the vessel" in an existing record to view the profile of this bay, and the position of the container in the BAPLIE will be displayed specially. Take the record of the container no. "ATLU0007178" as an example, click the container position "01D0182," the position of the container is highlighted in blue, and its bay profile is shown in Fig. 3.42.

Click the "view" link in the action bar, the comprehensive information of the container can be queried. Similarly, take the record of "ATLU0007178" as an example, the query interface of the comprehensive container information is shown in Fig. 3.43.

As can be seen from the figure above, the voyage code of the container with the no. of "ATLU0007178" is "I000000013." It is a 20 ft general flat container, the

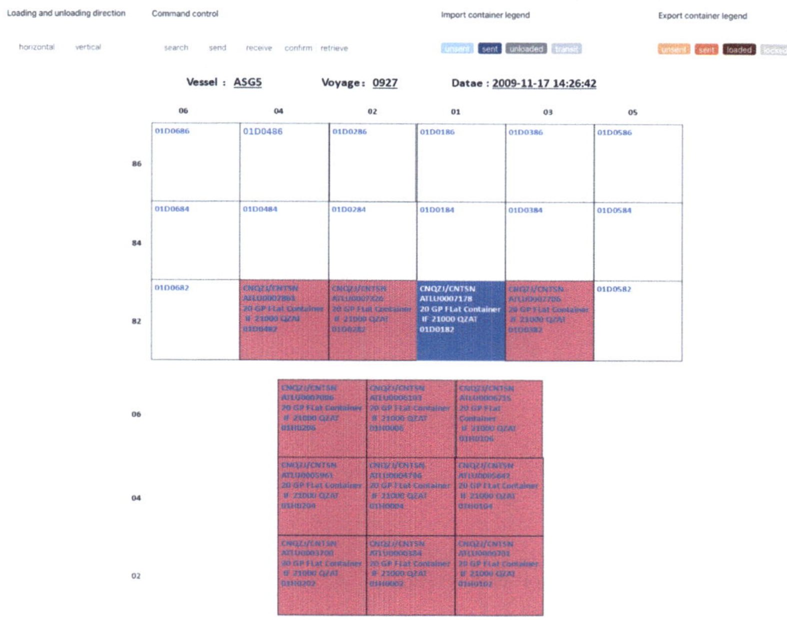

Fig. 3.42 Profile view of a bay

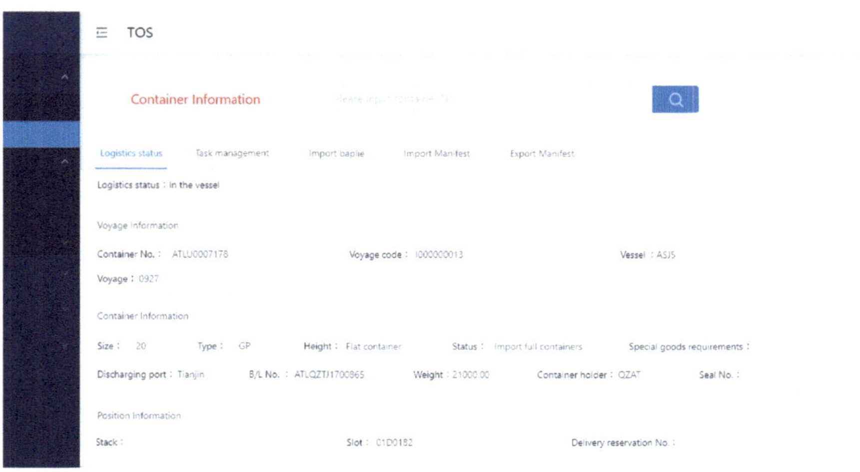

Fig. 3.43 Query interface of the comprehensive container information

discharging port is "Tianjin," and the container is still waiting for unloading in the vessel.

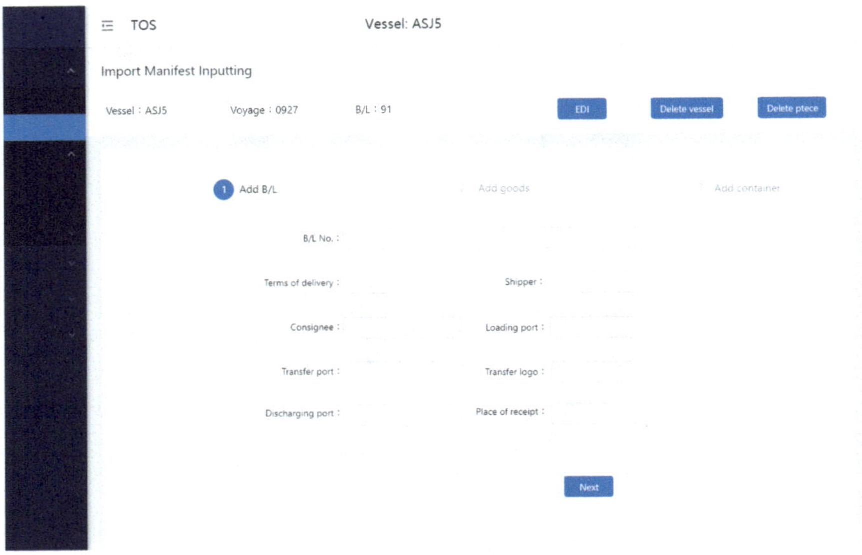

Fig. 3.44 Login interface of the import manifest input

2. **Input of the import manifest**

The import manifest input as well as the import BAPLIE input and modification are in the same module of import vessel unloading, and the sub-module of import manifest input is shown in Fig. 3.44.

After logging in, there are three steps to input the new bill of lading:

First, in "add a bill of lading," add one record of bill of lading, input the bill of lading no. and other relevant information according to the paper document, and click "the next."

Second, in "add new cargo," successively input the cargo serial no., cargo code, and other information and save. If one bill of lading contains multiple types of cargos, multiple cargo records shall be added and saved successively. Then, click "the next".

Finally, in "add container" for a certain type of cargo, enter the corresponding information of all containers packed with this type of cargo and submit it.

For now, the information input of a bill of lading is completed. When there is more information in the bill of lading, the EDI file import method is often adopted in the terminal, and the input of manifest information can be completed quickly by clicking the EDI button for selection. The import manifest of this voyage after the input is shown in Fig. 3.45.

3. **Verification of BAPLIE and manifest**

The verification of BAPLIE and manifest is an indispensable step in vessel unloading operations. The module of verification of BAPLIE and manifest after logging in is shown in Fig. 3.46.

Fig. 3.45 Interface of import manifest input

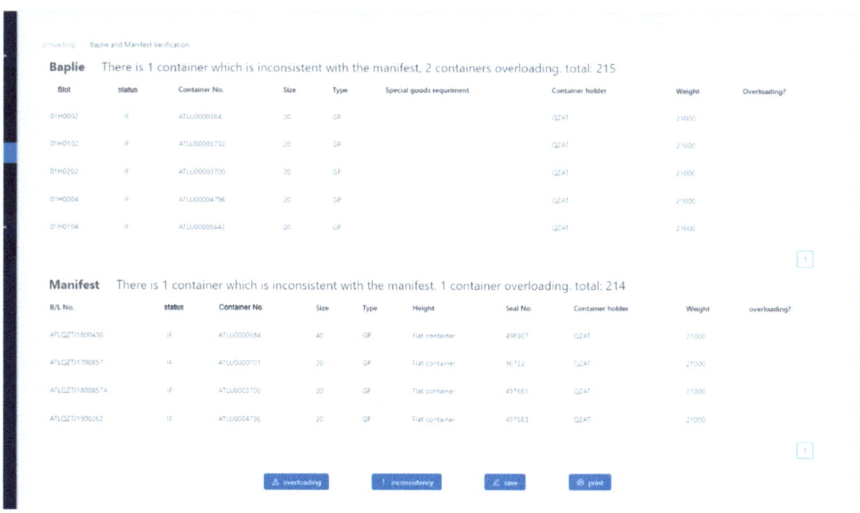

Fig. 3.46 Verification of BAPLIE and manifest

After logging in, the system will automatically display the verification results, and if the BAPLIE is completely consistent with the manifest, the verification will be approved. For the purpose of illustration, both over-loading and inconsistency exist in this example.

As shown in Fig. 3.46, there is one piece of container information in the BAPLIE that is inconsistent with the manifest, as well as two over-loaded containers. In the manifest, there is one container inconsistent with the BAPLIE, as well as one over-loaded container. All mismatches are highlighted in yellow. To display the inconsistency and over-loading more clearly, click the inconsistency and over-loading buttons on the right of the interface.

(1) Inconsistency of the BAPLIE and the manifest
Click the inconsistency button, and the interface will be transformed into the one shown in Fig. 3.47.

As can be seen from the figure, the size of the container with the no. "ATLU0000384" in the BAPLIE is inconsistent with that in the manifest. At this time, the operator needs to find the original file or contact the shipping company for checking, to determine whether the BAPLIE is wrong or the manifest and correct the error information after confirmation and save it.

(2) Over-loading in the BAPLIE and the manifest
Click the over-loading button, and the interface will be transformed as shown in Fig. 3.48.

As can be seen from the figure, the containers with the no. "ATLU0000702" and "GESU3398124" only exist in the BAPLIE, while the container with the

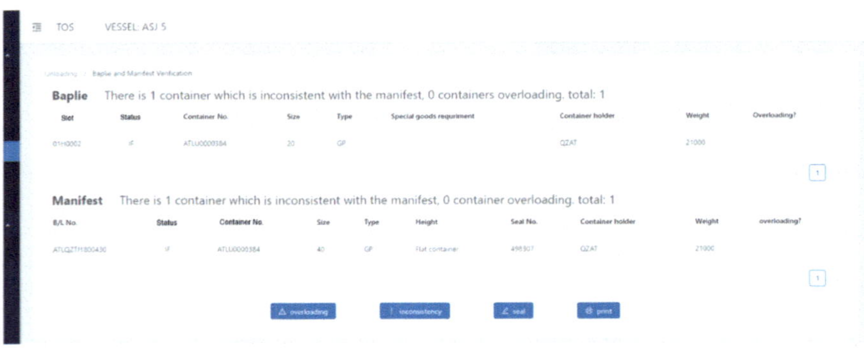

Fig. 3.47 Inconsistency of the BAPLIE and the manifest

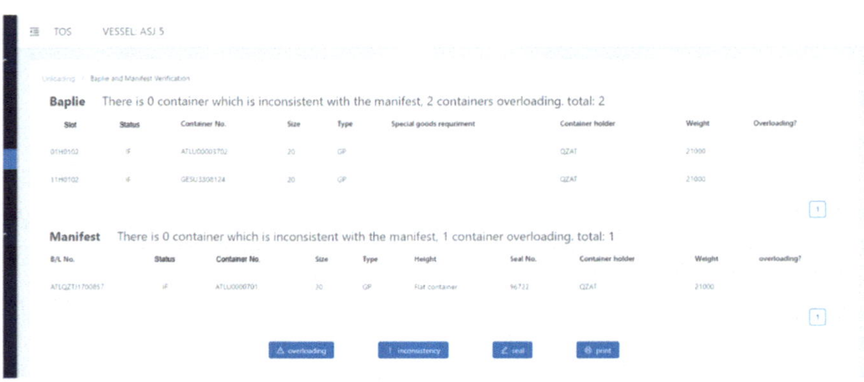

Fig. 3.48 Over-loading in the BAPLIE and the manifest

no. "ATLU0000701" only exists in the manifest. Through comparison, it is not difficult to see that there is only one difference between "ATLU0000702" and "ATLU0000701," which is likely to be caused by a wrong input of a certain container no. After checking the information and correcting the wrong container no., the corresponding over-loading situation in the BAPLIE and manifest can be eliminated at the same time. The problem of the container with the no. "GESU3398124" is due to the manifest omission, which needs to be solved by adding a record in the manifest input interface after confirming with the shipping company.

4. **The storage plan of vessel unloading**

The storage plan of vessel unloading is an effective ensurance for the smooth implementation of actual operations, and its operating interface is shown in Fig. 3.49.

After logging in, there is no information in the "container grouping" sub-window on the upper left of the interface, indicating that no containers have been made with the yard storage plan. Click the "automatic grouping" button to sort the containers of this voyage. Containers in this example are sorted as 20 ft general import full containers and 40 ft general import full containers. The steps of making storage plan for vessel unloading are as follows:

① Click the record of the selected 20 ft containers;
② Select "plan to bay" as the control mode in the "plan control";

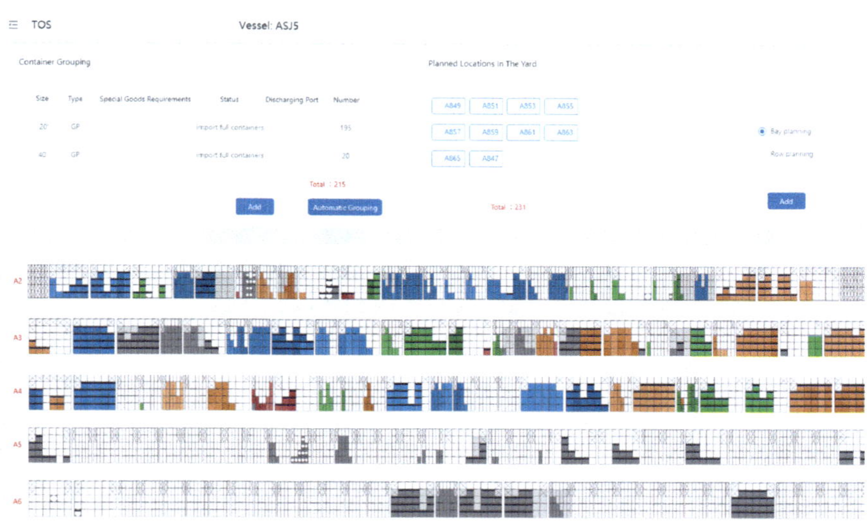

Fig. 3.49 Making the storage plan for vessel unloading

③ Right click at any position of block A8 to enlarge the display of the block and drag the mouse to specify the planned locations for the selected containers in the yard, as shown in Fig. 3.49;

④ Click the "save" button to save the plan.

In this way, the 20 ft containers' yard storage plan for this voyage has been designated to bays 49–65 in block A8, and the 40 ft containers' yard storage plan can be completed in the same way. To modify the plan, select the container location in the "yard plan location" that needs to be canceled first, then click the "delete" button, and save the modification. The steps for respecifying the container locations are the same as before.

5. **Scheduling of QCs for vessel unloading**

After the storage plan for vessel unloading is made, the central control operators will start the QC scheduling for vessel unloading and log in the QC scheduling module.

Select the vessel ASJ5, 0927, and the vessel profile for this voyage is displayed in the interface. According to the berth position of the vessel and the status of the idle QCs, Q08 is arranged for the vessel unloading operation of this voyage. Take bay 22 on deck as an example, first click and select Q08, then click the deck position of bay 22, and Q08 is displayed in different forms at the same time under the vessel's side view and on the vessel's top view, as shown in Fig. 3.50.

When the QC is designated, the sub-window of the container trucks assigned to the QC is displayed in the interface at the same time. Since the container trucks have not been arranged, the sub-window is empty. When the container truck scheduling is

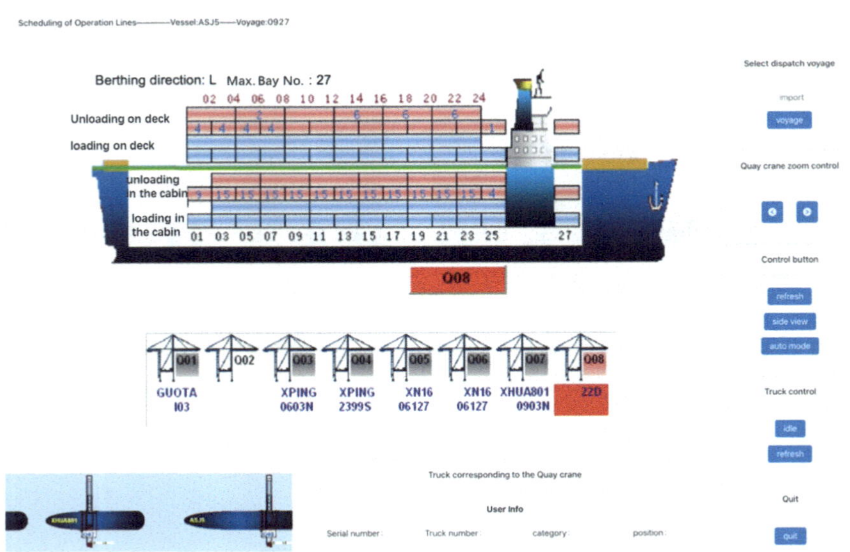

Fig. 3.50 QC scheduling of vessel unloading

completed, the information of container trucks assigned to the QC will be displayed. The currently available container trucks can also be queried by clicking the "idle" button before scheduling.

6. **Scheduling of container trucks**

The container truck scheduling is an operation following the QC scheduling. The containers unloaded by the QC must be moved to the yard via the container trucks. The container truck scheduling module after logging in is shown in Fig. 3.51.

After logging in, an operation line can be added in "add," and the operation line is named as "Q08" according to the name of the designated QC. Since this operation line will not share container trucks with other lines, the value of the container truck pool will remain "0" as the default value, the container collection truck number and loop truck number are set to 2 and 5, respectively, and the operation line status is set to "in use." After the operation line information is entered, click "save" button to save the record. Meanwhile, select "arrange vehicles" from the operation options to make route arrangement, as shown in Fig. 3.52. Select the "designated" check box of JK16, JK22, and JK24, respectively, and dispatch the three container trucks to the operation line Q08. Click "OK" button to save the operation and complete the container truck scheduling process at the same time. Now, check the QC scheduling interface, JK16, JK22, and JK24 are displayed in the small window of the container trucks designated to the QC.

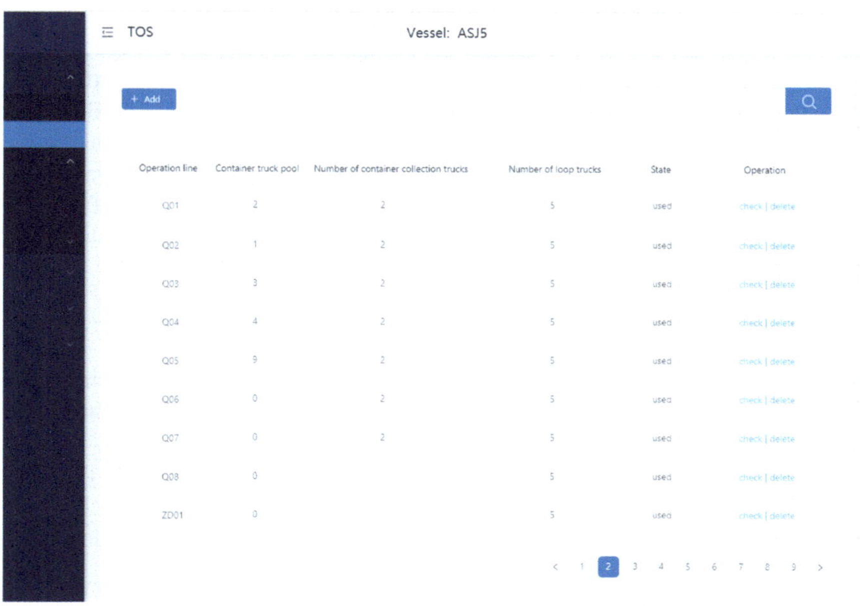

Fig. 3.51 Operating interface of container truck scheduling

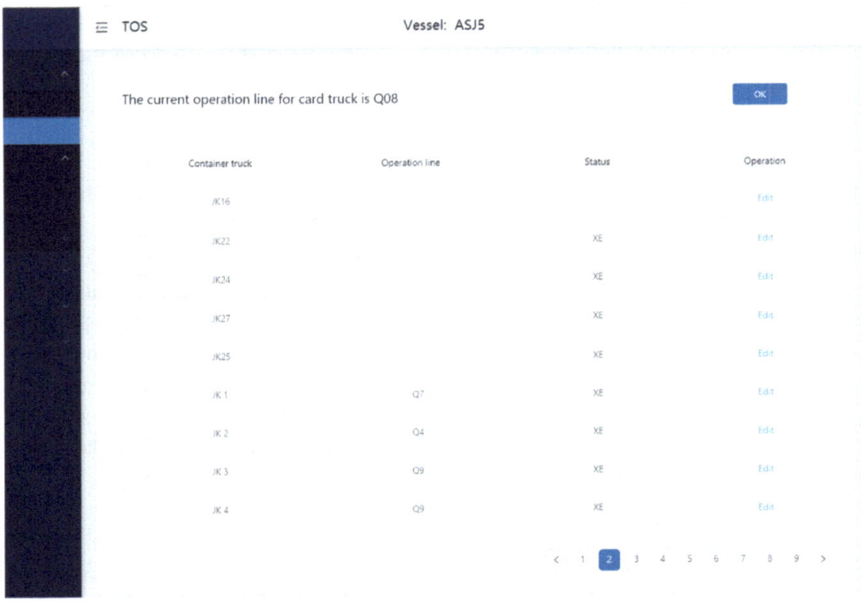

Fig. 3.52 Operation interface of container truck scheduling with arrangement

7. **Scheduling of yard cranes**

After the completion of the container truck scheduling, log in the yard crane scheduling module, as shown in Fig. 3.53, to continue to designate the yard cranes for the unloading yard. First of all, select the idle yard cranes in the drop-down list box of "crane no." on the left top of the interface and pick L03. When making the storage plan for vessel unloading, block A8 has been designated to this voyage, thus click block A08 in the operation mode of "by block," and click "OK." Then, the yard crane scheduling for the vessel unloading operation of this voyage has been completed.

8. **The vessel unloading operation of QC**

After all the mechanical scheduling is completed, the actual vessel unloading operations will start, and the system operation scheduling program will be initiated first, as shown in Fig. 3.54.

The first is the vessel unloading operation of the QC, which is shown in Fig. 3.55 after logging in the module. The unloading operation will start after checking the container position information in the vessel. The operating steps are as follows:

① Select the container with the no. "GESU5816861" in the profile view of the vessel.

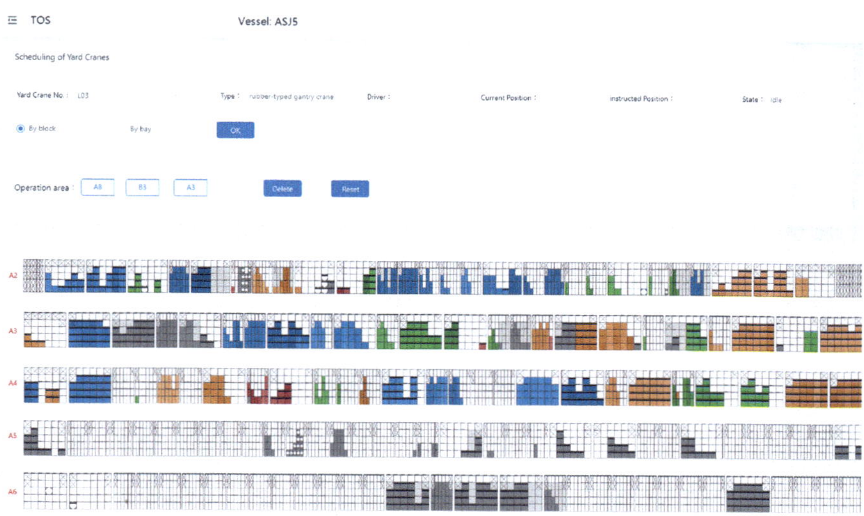

Fig. 3.53 Scheduling of yard cranes

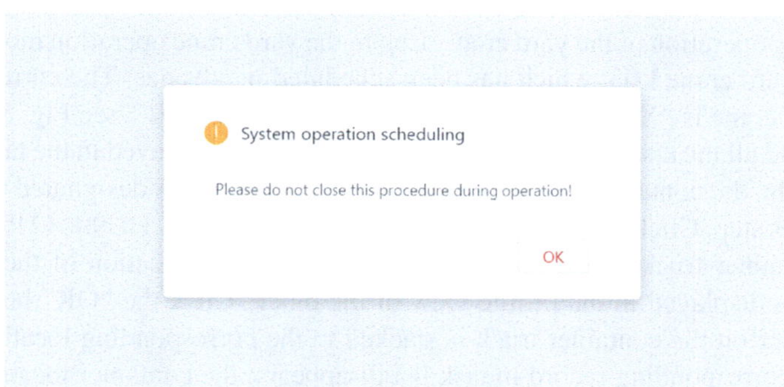

Fig. 3.54 System operation scheduling

② Select the container truck JK22, and click to confirm the selection of the container truck.

③ Click and then the container in the profile view will disappear. Then, if you need to cancel the operation, click the "cancel" button.

④ Click to complete the QC's unloading operation, and the container truck will move the container to the planned location in the yard. Repeat the operation until all the containers at this bay are unloaded.

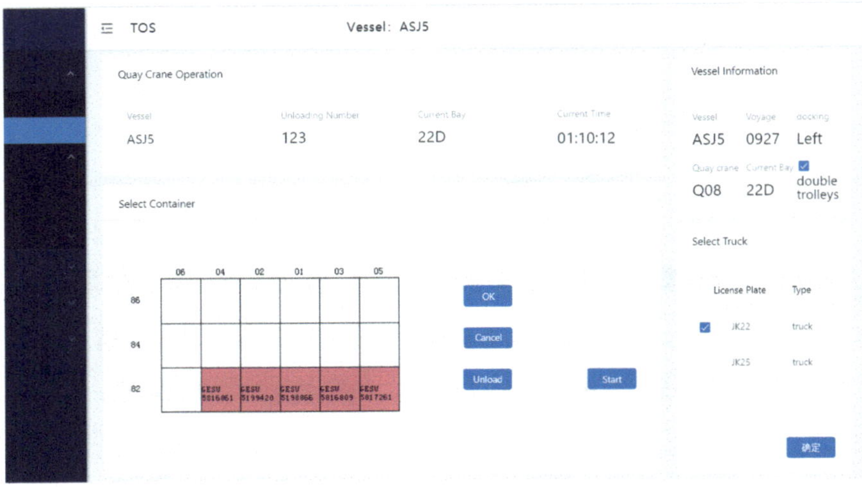

Fig. 3.55 Interface of the QC operation

9. The stacking operation of the yard crane

The containers unloaded by the QC are moved to the yard by the container trucks for the stacking operation of the yard crane. Log in the yard crane operation module, and select the yard crane L03, which has been scheduled in advance. The current driver of the yard crane is "Xu Jianhua," whose employee no. is "007", see Fig. 3.56.

Click and all the tasks waiting for stacking by L03 are displayed in the task list. In this example, the container trucks involved are the three trucks designated to Q08 in the previous step. Click the first task, namely the container "CLHU9064219" carried by the container truck JK24, the corresponding container location of the selected container is displayed in the profile view of the block. Click the "OK" button, and the container on the container truck is stacked to the corresponding location in the yard. The corresponding record in task list disappears; the container location in the yard is displayed in dark red, meaning the completion of the stacking operation of the yard crane, as shown in Fig. 3.57. Repeat the above operation until all the stacking tasks in the list are completed.

After the stacking operation of the yard crane in the interface is completed, continue to return to the interface of the vessel unloading of the QC. At this time, the container trucks designated to the QC have completed the yard operation and returned to the quay, waiting for the unloading operation of other containers in the vessel. Repeat the operations of the QC and the yard crane until all the containers at this bay are unloaded.

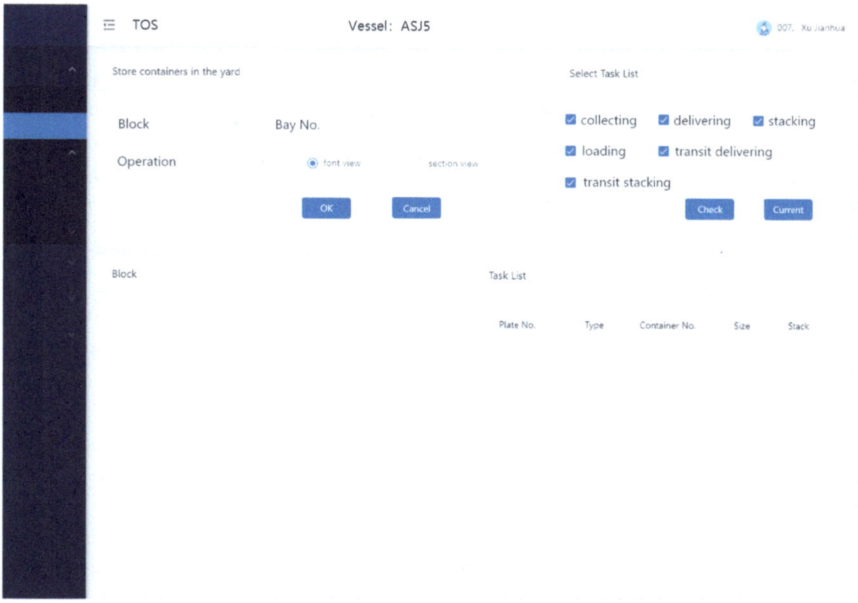

Fig. 3.56 Interface of stacking operation of the yard crane after logging in

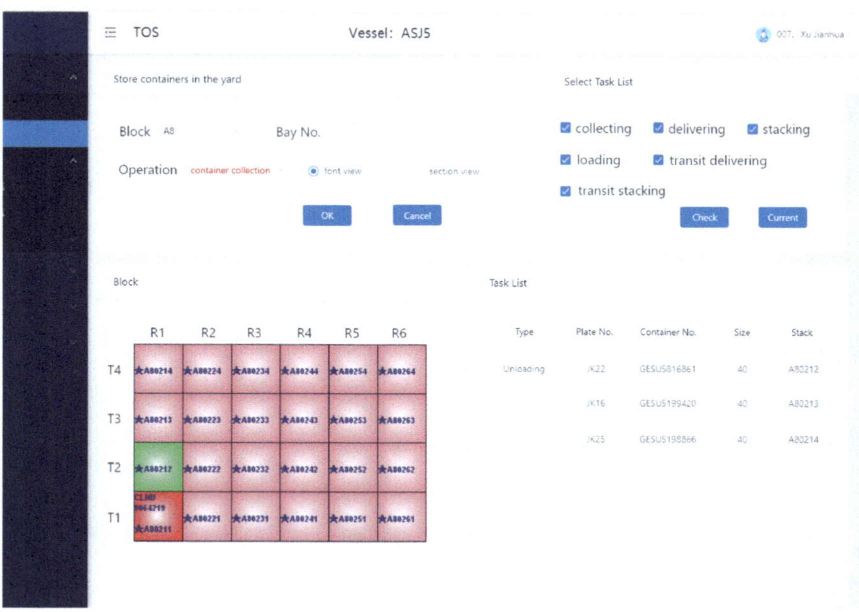

Fig. 3.57 Stacking operation of the yard crane

10. **Statistical report of the vessel unloading**

During or after the vessel unloading operations, relevant statistical reports can be viewed and printed, including the report of actual over-unloading/short-unloading, summary of vessel unloading, separate bay view BAPLIE, general overview BAPLIE, summary of import single vessel, etc. This example gives a brief introduction to the report of the actual unloading and the summary of vessel unloading.

(1) Report of actual over-unloading/short-unloading
Since we have only completed the unloading operations of twelve 40 ft full containers for this voyage, in the report of actual over-unloading/short-unloading, all the containers are shown as "short-unloaded" except the unloaded containers whose status of over-unloading/short-unloading is shown as "unloaded," as shown in Fig. 3.58. When all the containers of this voyage are unloaded, the amount of over-unloading/short-unloading shall be zero.

(2) Summary of vessel unloading
In the process of vessel unloading, the status of the unloading containers can be viewed. In this example, twelve 40 ft containers are unloaded for this voyage. Therefore, the actual number of the unloaded containers in the vessel unloading summary is twelve 40 ft full containers, while 195 20 ft containers and eight 40 ft containers to be unloaded, as shown in Fig. 3.59. Upon completion of the unloading operations of all containers of this voyage, the unloading plan

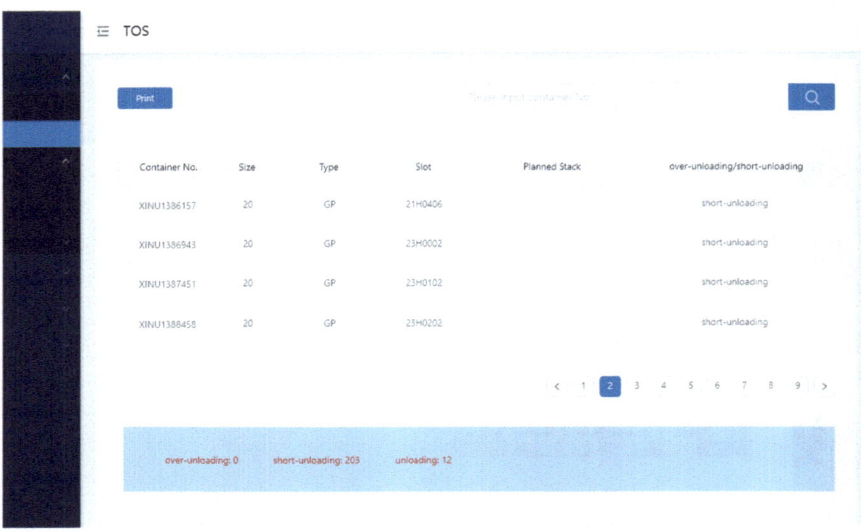

Fig. 3.58 Report of actual over-unloading/short-unloading

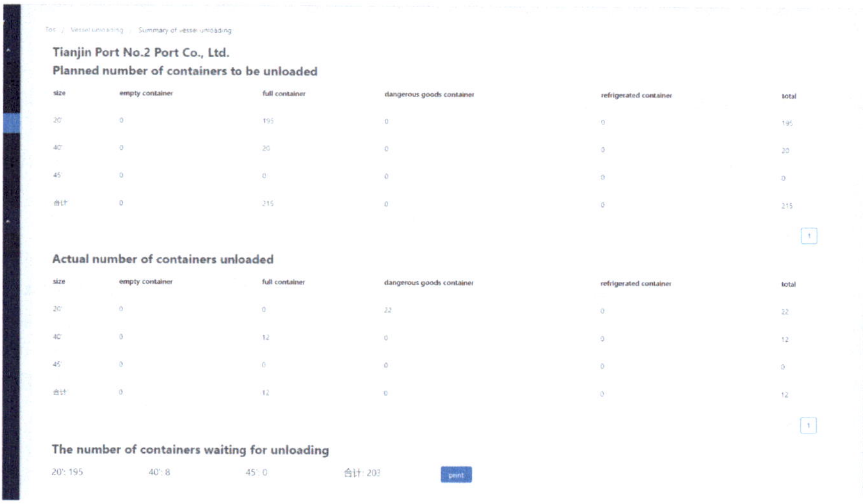

Fig. 3.59 Summary of vessel unloading

and the actual unloading can be checked, and the summary can be printed after verification by clicking the "print" button.

11. **Closing the import voyage**

When all the vessel unloading operations of the voyage are completed, the terminal should close the voyage in time to prohibit the modification of the import data, release the occupied resources in time, and record the results of the vessel unloading operations.

3.6 Intelligent Vessel Unloading System

3.6.1 Definition of Intelligent Vessel Unloading System

Intelligent vessel unloading, whose full name is intelligent location selection system of vessel unloading in container terminal, is one of the necessary and key steps in the intelligent transformation of the container terminal. Intelligent vessel unloading system is designed in view of the operating mode of simultaneous loading and unloading in the conventional container terminal, to automatically and dynamically allocate yard location for the unloading container, under the premise that the computer system considers the facility state, task distribution, storage condition, etc., according to the location allocation principles of "distributing and centralizing" in the yard without pre-made artificial yard plan.

3.6.2 Development Situation

In recent years, with the new trend of the industry 4.0, the traditional production mode has been evolving to the direction of automation and intelligence. The continuous improvement of new processes and technologies in container terminals and the increasing throughput of the terminals are driving the port and terminal industry to join the industry 4.0.

In the future, with the gradual formation of the global liner transportation network, more and more container handling will be concentrated to the hub ports, thus exacerbating the problem of insufficient storage yard resources at the hub ports. Therefore, it is particularly important to improve the management level of the container terminals, that is, to maximize the utilization of limited yard resources by optimizing the allocation of container locations in the yard. As one important type of containers in the container terminal, the rationality of stacking locations of the import containers in the yard directly affects the efficiency of vessel unloading and the number of reshuffle operations during delivery. Therefore, it is the most critical technical step in the process of intelligent transformation of the traditional artificial terminal to realize the automatic location selection of the vessel unloading.

For now, the intelligent location selection for the import containers in the vessel unloading operations in container terminal is still at its initial stage. The vast majority of such system has a certain degree of dynamic properties and relatively reduced the intensity of the operators, but the effect on the port operating efficiency improvement is not very obvious, and it is limited by many factors.

In the process of intelligent transformation of the traditional artificial terminals to realize the automatic location selection for the unloading containers, the manually made yard plan has been abandoned, which has greatly liberated the labor force. According to the attributes of the import containers, it conducts real-time full-yard search with the terminal resources (such as container trucks and yard cranes) to obtain the optimal yard location, so as to realize dynamic and balanced location selection and improve the unloading efficiency. CMICT has improved its operating efficiency and greatly reduced its operation cost after applying the intelligent location selection system during vessel unloading. It can be summarized as follows:

- The vessel unloading operating time has been shortened.
- The labor cost of vessel unloading has been greatly reduced.
- The utilization rate of the yard has been increased.
- The yard cranes shifting distance has been reduced.
- The reshuffle rate during delivery has been reduced.

In the case of relative shortage of yard resources, yard becomes the bottleneck of operations in the container terminal, and the quality of the yard resource allocation and utilization will directly affect the overall operating efficiency of the terminal. Through the intelligent yard arrangement of vessel unloading operations, the limited yard resources, including storage space resources and handling facility resources, can be rationally utilized to improve the efficiency of the terminal operations.

3.6.3 The Significance of the Intelligent Vessel Unloading

The intelligent location selection of vessel unloading is mainly to further improve the degree of refined management within the enterprise and provide intelligent decision support for the managers, keep the terminal in a good stacking state, and improve the efficiency of loading and unloading. The specific research objectives are as follows:

(1) To dispense the manually made yard plan and save the labor

The intelligent vessel unloading is no longer with the concept of making plans. The yard plan for unloading containers is changed to: the unloading of each import container shall be calculated by the computer in the background once for location allocation, which should be completed within 1–3 s.

(2) To realize dynamic and balanced location selection and improve efficiency

Aiming at the special operation mode of loading and unloading at the ends of the automatic terminal, the intelligent vessel unloading location selection system automatically and dynamically allocates the location of the unloading container by taking into account the facilities situation, task distribution, storage status, and other factors.

(3) To realize the automatic stacking by B/L and reduce the reshuffle rate during delivery

Under the condition of intelligent location selection by the computer, it has the ability to control the stacking of the containers with the same bill of lading no. by rows or by bays as far as possible with the dynamic location selection algorithm. For the B/L with large container amount, the reshuffle rate in the later delivery can be greatly reduced.

3.6.3.1 Main Features

- No manual grouping of containers is required
- No manual storage plan for unloading is required
- Dynamically selecting location and dynamically adjusting storage plan.

3.6.3.2 Comparison with the Traditional Mode

Mode	Traditional mode	Intelligent mode
Illustration	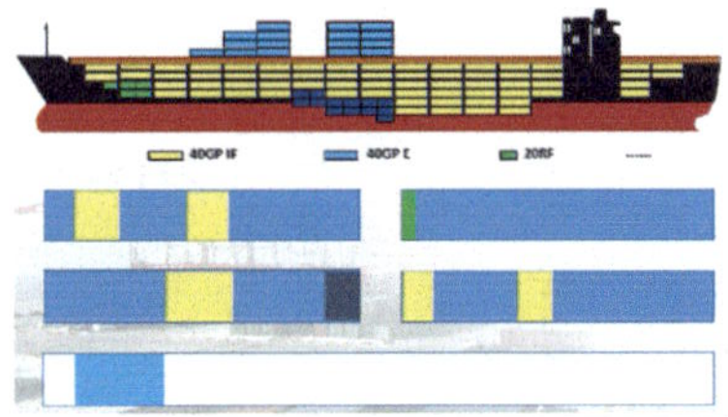	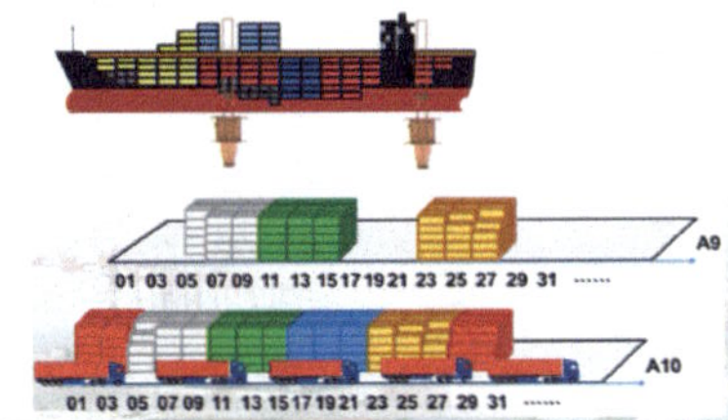
Process	Manual grouping → manual plan → location allocation by the computer	Location allocation directly by the computer
Implementation	Manual plan	Dynamically, calculating and adjusting the unloading plan by the computer
Operation mode	Making yard plan of vessel unloading for a single vessel	Taking the whole terminal as one object, the unloading containers are automatically sorted and grouped by the computer, and the yard location is dynamically calculated according to the amount of the unloading containers and the amount of unloaded containers as well as the storage state, the facility condition in the yard, and other factors

(continued)

(continued)

Mode	Traditional mode	Intelligent mode
Comparison	In the traditional mode, the location selection of unloading containers requires manual storage plan in the yard for the unloading import containers in advance. The storage yard plan refers to the yard plan for grouping all the import containers of vessel unloading and designating one or more blocks as target storage areas for each group in the yard. The purpose is to reasonably allocate the yard resources according to the attributes of the import containers and meanwhile to facilitate the smooth and efficient completion of the container delivery operations	The intelligent vessel unloading no longer requires manual yard plan. Instead, the existing resources of the storage yard are uniformly allocated. At the moment when the container is loaded on the container truck, the system dynamically allocates a location in the yard to the container according to the resources of the storage yard and the attributes of the container
Management solution	Following the plan	Mainly based on dynamic allocation, breaking through the "limitation" of yard plan and greatly improving the dynamic performance of location selection decision
Planning scope	Whole vessel containers 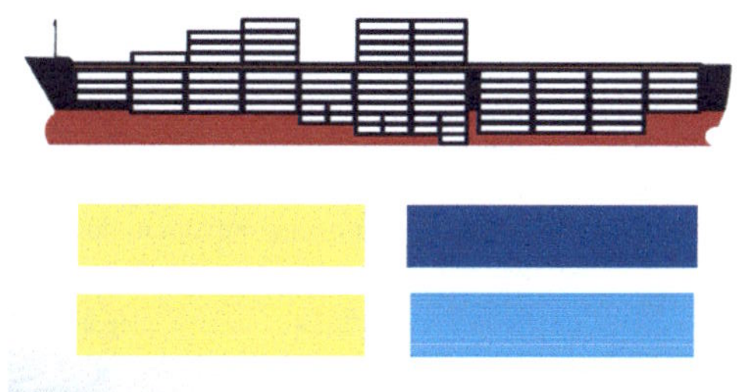 The planner manually makes the storage plan for the whole vessel unloading containers	Certain groups of containers 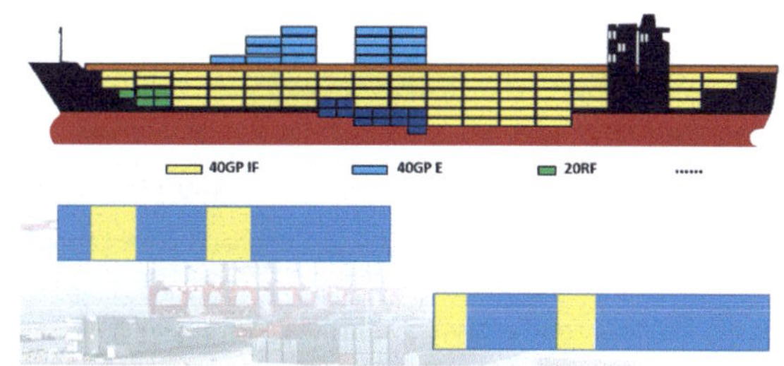 In the intelligent vessel unloading mode, the computer arranges/allocates the yard plan of certain container groups (such as one B/L) according to the attributes of the unloading containers

(continued)

(continued)

Mode	Traditional mode	Intelligent mode
Planning frequency	Once: After the planner makes the whole vessel unloading plan, if it needs to be adjusted in the process of loading and unloading, the central control scheduler usually cancels part of the plan and manually modifies the yard arrangement	Dynamic, multiple times: In the intelligent vessel unloading mode, the computer arranges and adjusts the yard plan of vessel unloading in real time according to the operation situation of vessel unloading, instead of designating the blocks for unloading containers at one time
Considering the facility location	Facility location cannot be considered when planning: In the traditional mode, the location selection of unloading containers is all according to the attributes of unloading containers to find the yard locations of the corresponding container group, which is difficult to be dynamically adjusted and will cause more facility shifting and waste of resources when the facility is busy 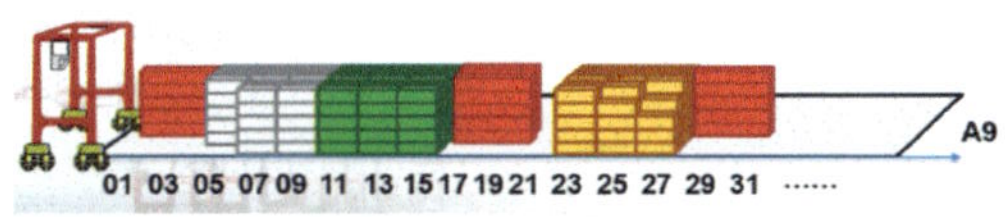 	Container location selection follows the facility location to reduce facility shifting: Since the intelligent vessel unloading is not limited by the yard plan, the container location can be dynamically selected according to the current operation location of the yard crane, so as to reduce unnecessary shifting of the yard crane and improve the facility utilization rate 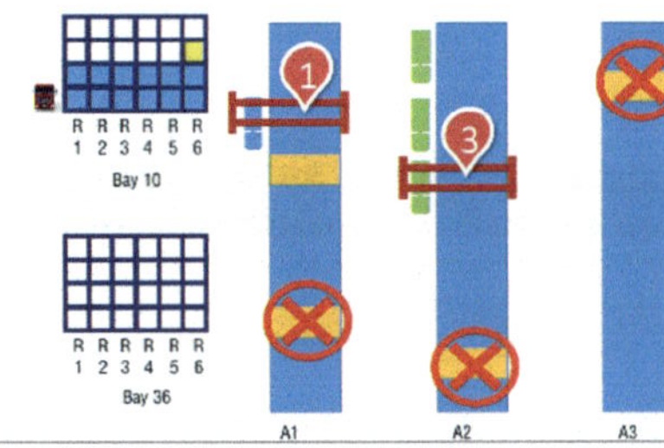

(continued)

(continued)

Mode	Traditional mode	Intelligent mode
Mixing stacking of containers of multiple vessels	Under the traditional planning mode, a single vessel is generally taken as the object, and multi-vessel mixing is not considered, which will lead to waste of space resources of the storage yard and occupy a lot of yard facilities	In the intelligent vessel unloading mode, the current vessel unloading operations of the whole terminal can be considered altogether to realize multi-vessel mixing. For containers of miscellaneous groups, they can also be mixed in the same row. With the facilities as the guidance, intensive control of the intensity of the operations of the yard cranes can be realized to improve the facility operating efficiency 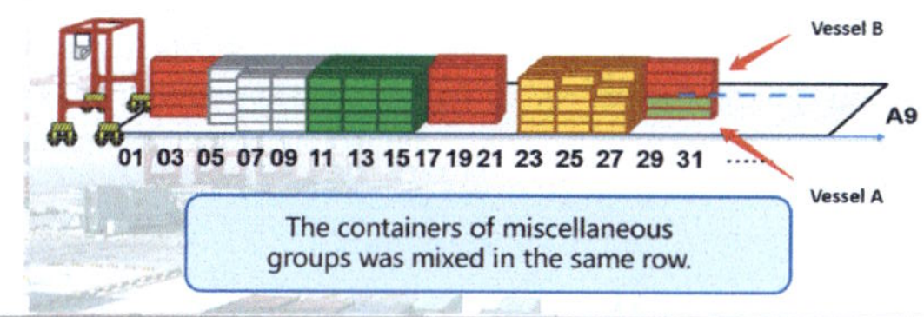

(continued)

(continued)

Mode	Traditional mode	Intelligent mode
Stacking by B/L	If several container groups of the same B/L are stored separately when manual yard plan is made, the workload is relatively high, and automatic stacking by B/L can only be realized when the container amount of the B/L reaches a certain level 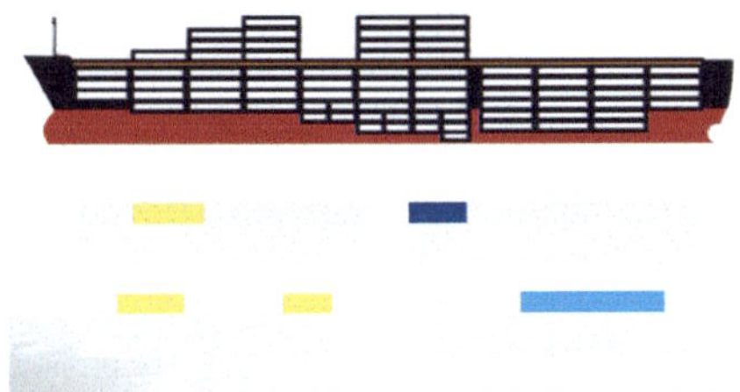	In the intelligent vessel unloading mode, the yard location can be dynamically calculated by the computer, with stacking by B/L applied. Up to four containers can be stored in one row 
Mode of location selection	In the traditional mode, the yard location of the unloaded container is also dynamically selected after the confirmation of the tally clerk but only among the locations within the planned blocks	In the intelligent vessel unloading mode, theoretically, any block that can be stacked for import containers may be the target block of vessel unloading containers, and the system will dynamically decide the block and yard location that the container will be stacked according to the principles of location selection

Chapter 4
Management of Container Delivery in the Container Terminal

4.1 Introduction of Process of Container Delivery Operations

Container delivery operations in the terminal generally refer to the import full containers, which are the main targets container delivery operations in the container terminal. For the import full containers, the container delivery operations are the second stage of the whole import operation processes, and the first stage is the vessel unloading operations. The terminal is a pivot point of "collection and distribution," and the import operations are processes from "collection" to "distribution," while "distribution" mainly refers to the process in which the containers in the storage yard are transported outside of the terminal at the stage of container delivery, so the logistics characteristics of the container delivery process are random and discrete.

1. **Delivery process for the import full containers**

The delivery process for the import full containers generally can be divided into three stages, namely (1) planning stage, (2) charging stage, and (3) operation stage. During the three stages, parties involved include the customers, business service center, banks, the carriers, the gatehouses, the yard cranes, etc., which are closely related and dependent on each other as shown in Fig. 4.1.

The delivery operation process for the import full containers is shown in Table 4.1.

2. **Delivery process for empty containers**

The empty containers are commonly delivered in batches. When the shipping company needs to use or dispatch empty containers, the empty container application form will be issued to the terminal first. The customer applies for empty container delivery with the effectively stamped shipping company contact form. The yard planner in the terminal checks the B/L with releasing stamp, consignee stamp, and customs stamp offered by the customer, accepts the application after verification, and makes an empty container plan according to the container holder and size. Then it is the

N. Zhao et al., *Digital Management of Container Terminal Operations*,
https://doi.org/10.1007/978-981-15-2937-5_4

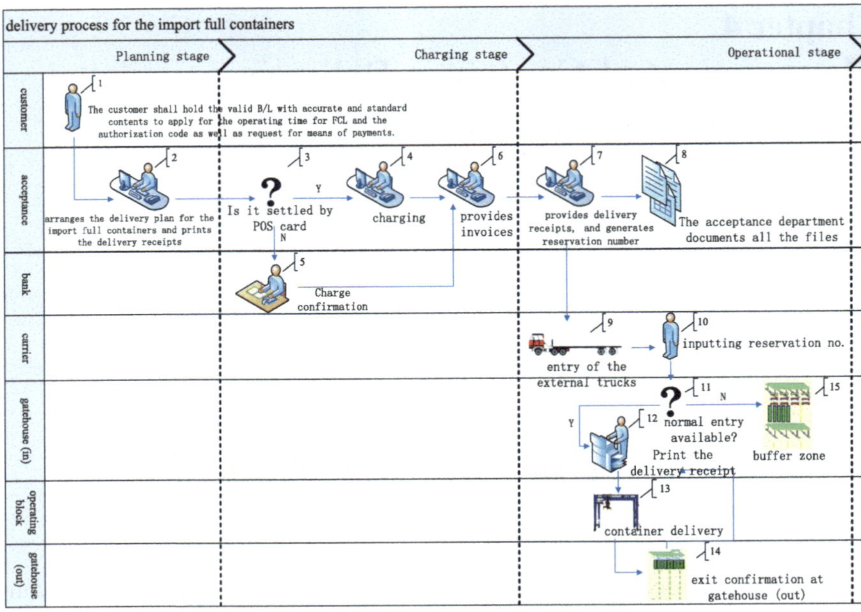

Fig. 4.1 Delivery process for the import full containers

settlement process. There are agreements for monthly settlement between some terminals and shipping companies. For the monthly settled customer, the reservation no. will be directly provided. Otherwise, the customer needs to pay on the spot, and the reservation no. can only be provided after payment confirmation. Then the actual operations of the delivery are similar to those of the import full containers.

3. **Delivery process for shutout containers**

The shutout containers refer to the export containers which fail to be released for loading by the customs for some reasons. The shutout containers can be delivered to the customer after the customer completes the shutout procedures such as "application form for taking unloaded goods out of the customs surveillance zone." The process is shown in Table 4.2.

4. **Delivery process for the transferring containers**

Transferring refers to the process of transferring the import containers from the internal storage yard to the professional storage yard outside the terminal when the storing days of the containers after unloading reach the days specified. The transferring plan in the terminal is usually made by the vessel plan department. Firstly, according to the vessel information, the planner enquires whether the storing days of the containers in the yard exceed the prescribed time and determines the containers to be transferred. Then, the planner arranges the transferring plan for

Table 4.1 Descriptions of the delivery operation process for the import full containers

No.	Parties involved	Descriptions
1	Customer	The customer shall hold the valid B/L with accurate and standard contents to apply for the operating time for FCL and the authorization code as well as request for means of payments
2	Business service center	The clerk of business service center arranges the delivery plan for the import full containers and prints the delivery receipts
3	Business service center/bank	Customer pays directly by credit card in the business service center or at the bank counter
4	Business service center	The clerk of business service center settles all the fees and provides invoices
5	Business service center	The clerk of business service center checks the related receipts, provides delivery receipts and generates reservation no.
6	Business service center	The clerk of business service center documents all the files
7	Carrier	The external container truck arrives at the gate of the terminal
8	Carrier	The driver enters the reservation no. at the gatehouse
9	Gatehouse (in)	The gatehouse clerk checks whether the truck can enter the terminal normally
10	Gatehouse (in)	If the truck can normally enter the terminal, the gatehouse clerk prints the loading receipts according to the reservation no.
11	Operating block	The yard crane driver operates on the containers according to the operation sequence
12	Gatehouse (out)	The gatehouse clerk checks the information in the system, equipment interchange receipt and the actual situation of the full container; collects the equipment interchange receipt and loading receipt after verification; confirms the exit of the container in the system, and issues the exit certificate
13	Gatehouse (in)	If the truck cannot enter the terminal normally, it will go to the buffer zone for the container delivery procedures

the containers, provides the container reservation no. to the logistics company, and informs the logistics company to arrange the transportation. The following delivery operations are similar to those of import full containers.

Table 4.2 Descriptions of the delivery process for shutout containers

No.	Parties involved	Descriptions
1	Customer	The customer provides the valid shipping company contact form and the application form for taking the unloaded goods out of the customs surveillance zone issued by the customs
2	Business service center	The clerk of business service center makes the delivery plan for the shutout containers
3	Business service center	The clerk of business service center checks whether the payment can be made with POS machine
4	Business service center	If the customer can pay with the POS machine, a special fast channel t will be opened to charge with bank card
5	Bank	If the customer cannot pay with the POS machine, the customer shall pay at the bank counter
6	Business service center	The clerk of business service center settles all the fees and provides invoices
7	Business service center	The clerk of business service center checks the related receipts, provides delivery receipts and generates reservation no.
8	Business service center	The clerk of business service center documents all the files
9	Carrier	The external container truck arrives at the gate of the terminal
10	Carrier	The driver enters the reservation no. at the gatehouse
11	Gatehouse (in)	The gatehouse clerk checks whether truck can enter the terminal normally
12	Gatehouse (in)	If the truck can normally enter the terminal, the gatehouse clerk prints the loading receipt according to the reservation no.
13	Operating block	The yard crane driver operates on the containers according to the operation sequence
14	Gatehouse (out)	The gatehouse clerk checks the information in the system, equipment interchange receipt and the actual situation of the container; collects the equipment interchange receipt and loading receipt after verification, confirms the exit of the container in the system, and issues the exit certificate
15	Gatehouse (in)	If the truck cannot enter the terminal normally, it will go to the buffer zone for the container delivery procedures

4.2 Delivery Reservation Plan

Delivery reservation refers to the process in which customers go to the terminal to make a delivery reservation plan with valid delivery certificates. Taking import full containers as an example, the reservation service is provided by the business service center in the hall of the terminal, which is mainly responsible for the window service of the container terminal to the external.

The premise for the customer to precede the reservation procedure is that the payment has been completed. After the reservation, the terminal system will automatically generate a reservation no. When the driver enters the terminal to take delivery of the container, he needs to input the reservation no. at the gatehouse so that the system can select the corresponding container and the central control office arranges the facility.

There are two ways to make a delivery reservation according to the amount of the containers: single container reservation and batch reservation. Single container reservation generally refers to making a reservation according to the container no., while batch reservation includes reservation by the B/L no., reservation by the container holder and reservation in combination.

4.2.1 Single Container Reservation

Single container reservation is generally for the goods in one container; that is, a shipment of goods are all packed in one container; in this case, the cargo owner only needs to take delivery of one container. For the clerks of the business service center in the terminal, in order to facilitate the operation, single container reservation is generally made according to the container no., as shown in Fig. 4.2.

The basic operation of the reservation by container no. includes (1) to find the container, (2) to check the information, (3) to guide the payment, (4) to make the reservation plan. The clerk should first search the container information in the system according to the container no. specified on the B/L provided by the customer. As shown in Fig. 4.2, it is the process that the clerk inquires the container by entering the container no. If the target container has been unloaded and not yet reserved, then the clerk can see the container information in the operation system, as shown in Fig. 4.3.

After finding the container, the clerk also needs to check the container information in the system and in the B/L provided by the customer. After verification, the clerk can guide the customer to pay the fees. After payment, the reservation plan can be submitted, and delivery receipt be printed, in which the reservation no. automatically generated by the system is also displayed for the truck driver to input when taking the delivery of the container.

Fig. 4.2 Reservation by container no.

Fig. 4.3 Container query in single container reservation

4.2.2 Batch Reservation

Batch reservation refers to making one reservation plan for multiple containers, which can be delivered in no particular order. Therefore, the reasonable arrangement of batch delivery plan and automatic container selection at the gatehouse can greatly reduce the probability of container reshuffle when operating in the yard. For example, if a customer wants to take delivery of 30 containers, some of which are stacked together in the yard, there must be containers stacking under others. If single container reservation is made, there will be a high possibility that the container stacked below would be delivered firstly, causing unnecessary container reshuffle. However, if batch reservation is made, the terminal could deliver the container stacked on the top to the customer first, to avoid container reshuffle.

For different containers and different customer needs, there are several modes of batch delivery reservation, namely reservation by B/L no., reservation by container holder, and reservation in combination.

1. **Reservation by B/L no.**

For the import full containers, if the customer has one B/L with multiple containers (one B/L no., multiple container no.), the clerk in the business service center can generally make a reservation by the B/L no., as shown in Fig. 4.4.

The general process of reservation by the B/L no. includes

Fig. 4.4 Reservation by B/L no.

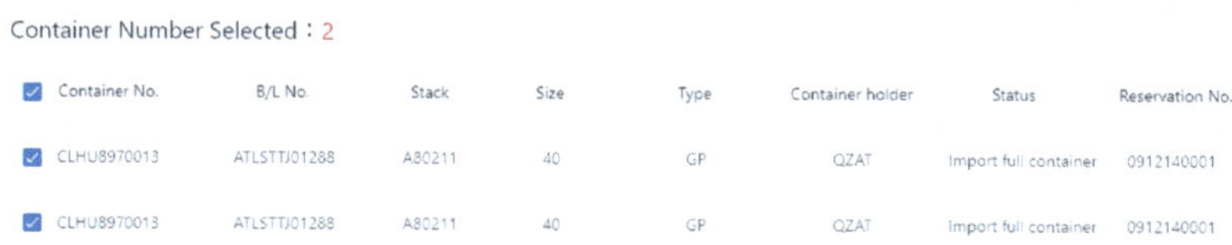

Container Number Selected : 2

Container No.	B/L No.	Stack	Size	Type	Container holder	Status	Reservation No.
CLHU8970013	ATLSTTJ01288	A80211	40	GP	QZAT	Import full container	0912140001
CLHU8970013	ATLSTTJ01288	A80211	40	GP	QZAT	Import full container	0912140001

Fig. 4.5 Container query by the B/L no.

① Container information query. According to the customer's B/L, select the vessel's name and voyage and input the B/L no., so as to find the unloaded yet unreserved import containers in the yard, as shown in Fig. 4.5.
② Container selection. According to the container no. in the B/L required by the customer, select multiple containers from the query list of containers shown in Fig. 4.5. When the amount of containers is large, the container amount should be checked.
③ Container information verification. Check whether the container information displayed in the system is the same as the container information in the paper B/L.
④ Payment guidance. The clerk guides the customer to pay the fees.
⑤ Making reservation plan. After payment, the reservation plan of this B/L can be submitted, and the delivery receipt of the containers can be printed for the driver to use when taking the delivery of the containers.

2. **Reservation by container holder**

For the empty containers, there are no goods inside, so the customers applying to take delivery of them are usually the container holding companies or the shipping companies. They tend to take delivery of empty containers of specified quantity without specifying the container no.; so for the terminal operator, it is to search for empty containers in the yard of a certain holder and of a certain size and type and make the empty container plan. Some container holding companies will require following the principle of "first-in-first-out" when they are taking the delivery of containers. Therefore, when making the empty container plan, the empty containers with a relatively long storage period in the yard should be arranged in priority.

The operating process of this reservation mode is similar to that of reservation by B/L no., except that when searching the containers, the holder, size and type of the containers should be input. If the size or type of the container is not input, it means to take delivery of containers of any size or any type. For the other steps, please refer to "reservation by B/L no.".

3. **Reservation in combination/fuzzy reservation**

To improve the flexibility of service, most terminals also support the reservation in combination (fuzzy reservation) mode, since it is possible for a customer to take multiple B/Ls to take delivery of the containers at the same time. If the terminal can sort the containers of different B/Ls into a group and make a unified plan, it will definitely facilitate the operations of the container delivery in the terminal.

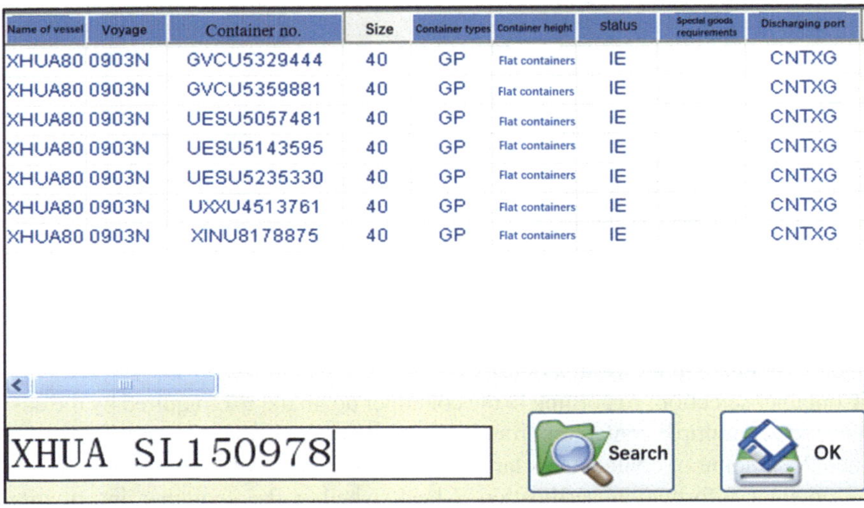

Fig. 4.6 Selecting containers in a fuzzy method

In such reservation mode, more flexible retrieval method is needed when selecting the containers instead of searching containers by the B/L no. or the container holder or the container type, etc. Shown in Fig. 4.6 is the query with the keywords "XHUA" and space and "SL150978". It is to find containers unloaded from the vessel XHUA with the B/L no. beginning with SL150978 for delivery, where the space in the keywords means the logical condition of "and."

The verification, payment, reservation, and other operations after container selection in the fuzzy mode can refer to "reservation by B/L no.".

4.3 Entry Control at Gatehouse in Delivery Operations

After the customer makes a reservation, he can get the delivery receipt printed by the terminal with the delivery reservation no. and then entrust the carrier to take delivery of the container in the terminal at an appropriate time. Before entering the terminal yard to take delivery of the container, the vehicles (container trucks) first need to go through the entrance gatehouse to get registered.

1. **The main work in the vehicle registration**

The entry formalities of vehicles shall be handled by the tally clerk at the entrance gatehouse of the terminal. For the entry of delivery vehicles (empty vehicles), the tally clerk at the gatehouse shall assist the truck driver to complete the following operations.

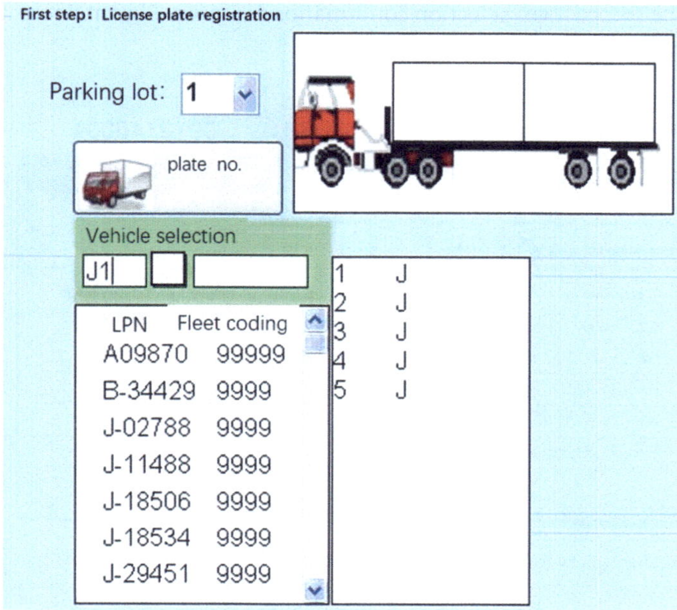

Fig. 4.7 Registration of the delivery vehicle

(1) Registering the vehicle license plate number
The terminal needs to record the main information of the delivery operations, including the container truck information. At present, in most terminals of seaports in China, the automatic registration method is adopted, such as IC card and RFID technology. The identified truck license plate number will be automatically recorded in the gatehouse operation module of the terminal operation system, so as to reduce the input work of the tally clerk at the gatehouse. Of course, in a small number of terminals, the truck license plate number still needs to be manually entered by the tally clerk at the gatehouse, as shown in Fig. 4.7.

(2) Finding the reservation record
The customer must make a reservation in the terminal before taking the delivery of the container, and the tally clerk at the gatehouse or the truck driver shall search the reservation record according to the container no. or the reservation no. on the delivery receipt. At present, most of the terminals in China need the tally clerk at the gatehouse to complete this work. If the customer has made a single container reservation, the tally clerk at the gatehouse can input either the container no. or the reservation no. to query the reservation record. If the customer has made a batch reservation, the tally clerk can only input the reservation no. to query, as shown in Fig. 4.8.
Some terminals have implemented intelligent gatehouse operation systems, and the automated processing equipments similar to ATM machines are installed at

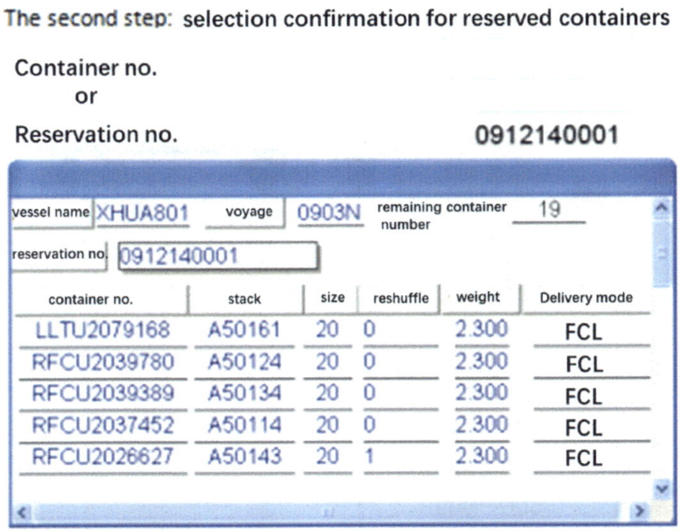

Fig. 4.8 Container query by reservation no.

the entrance gatehouse, where the container truck driver can enter reservation no. or directly scan the bar codes on the receipt.

(3) Calculating the container reshuffle
The container reshuffle is calculated automatically by the system. This step can be skipped if it is a single container reservation. In the case of batch reservation, as shown in Fig. 4.9, each delivery container queried from the reservation record has its own reshuffle. Container reshuffle refers to the number of containers to be turned over to pick up a specific container in the current stacking state, which is the number of containers that are currently stacked above the container. If the container reshuffle is 0, it means the container is stacked on the top and can be picked up directly. Therefore, to reduce the container reshuffle rate in the operation, the tally clerk at the gatehouse needs to select the container with 0 or the smallest container reshuffle to dispatch to the container truck entering the terminal, which also embodies the advantages of batch reservation.

(4) Generating delivery task and printing receipt
The tally clerk at the gatehouse selects the container with the smallest container reshuffle as shown in Fig. 4.9, and the system will automatically query the information of the container and generate a delivery task as shown in Fig. 4.10. Meanwhile, the printed receipt with the delivery task is delivered to the truck driver, so that the driver can drive the truck to the specific bay in the block according to the receipt to wait for the yard facility for delivery operation.

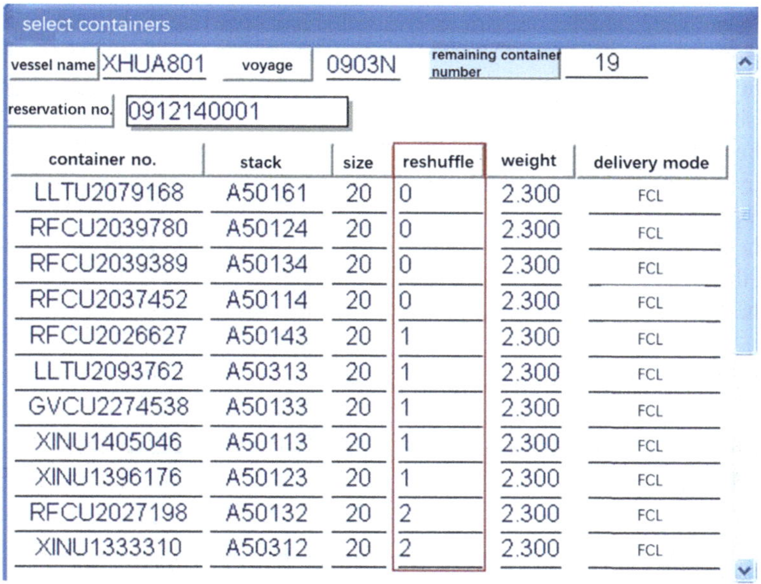

select containers

vessel name XHUA801 voyage 0903N remaining container number 19

reservation no. 0912140001

container no.	stack	size	reshuffle	weight	delivery mode
LLTU2079168	A50161	20	0	2.300	FCL
RFCU2039780	A50124	20	0	2.300	FCL
RFCU2039389	A50134	20	0	2.300	FCL
RFCU2037452	A50114	20	0	2.300	FCL
RFCU2026627	A50143	20	1	2.300	FCL
LLTU2093762	A50313	20	1	2.300	FCL
GVCU2274538	A50133	20	1	2.300	FCL
XINU1405046	A50113	20	1	2.300	FCL
XINU1396176	A50123	20	1	2.300	FCL
RFCU2027198	A50132	20	2	2.300	FCL
XINU1333310	A50312	20	2	2.300	FCL

Fig. 4.9 Calculating the container reshuffle

Fig. 4.10 Generating delivery task

2. **The significance of delivery vehicle registration at the gatehouse**

The information flow of container delivery operations in the container terminal is a closed loop, in which the registration of delivery vehicles at the gatehouse is an indispensable part. The significance includes:

(1) The delivery vehicles are strictly controlled and only released with reservation records.
(2) The container reshuffle is pre-calculated to reduce unnecessary container reshuffle.
(3) The delivery task is generated so that the central control office and the yard crane driver in the designated block can get the information in advance and make the preparation.
(4) The receipt is printed for the container truck driver to inform him the destination in the yard.
(5) The delivery operation is recorded for the future query and statistics.

4.4 Delivery Operations in the Yard Based on Kanban Management

4.4.1 Facilities Arrangement for the Delivery Operations

According to the delivery reservation plan, the operator will verify the information of the vehicle to take delivery of the container at the gatehouse and automatically search the location of the container in the yard. Meanwhile, a delivery task will be generated in the system, which is the main reference for the yard control scheduler to arrange facility for the delivery operations.

In order to avoid the long waiting time and improve the service, the yard control scheduler must arrange the yard crane in time for the current delivery task. If a yard crane is idle in the block where the container is to be delivered, it shall be arranged to operate. If there is no idle yard crane available, idle yard crane from other block nearby should be dispatched. As shown in Fig. 4.11, the container delivery task is in block B9, and there is no yard crane available, so the yard control scheduler needs to dispatch the no. 1 yard crane from block A6 to operate.

4.4.2 Delivery Confirmation by the Yard Crane

After receiving the instructions from the yard control scheduler, the driver of the yard crane operates the machine to the designated position and selects the corresponding task in the task list according to the license plate number of the delivery vehicle, as shown in Fig. 4.12.

After a task is selected, the yard location of the corresponding container displayed in the terminal will be flashed in green, and the yard crane driver will operate the yard crane to transfer the corresponding container from the actual container location

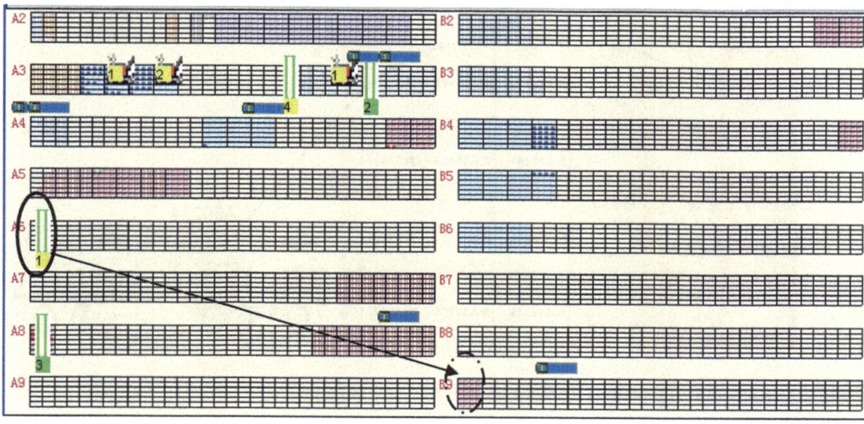

Fig. 4.11 Facility arrangement for the delivery operation

Fig. 4.12 List of delivery tasks

task list

operation	plate no.	container no.	size	stack
container delivery	J-02788	CLHU 8970013	40	A80224
	J-11488	GESU 5198866	40	A80214
	J-18506	CLHU 9064219	40	A80211
	J-18534	GESU 5199420	40	A80213
	J-29451	GESU 5816861	40	A80212
	J-34925	CLHU 8980639	40	A80223
	J-42929	GESU 5817261	40	A80222

in the yard to the delivery vehicle. The confirmation process of container delivery is shown in Fig. 4.13.

In the process of container delivery, if the container is stacked under other containers, the priority should be given to consider whether it is a batch reserved container. If it is a batch reserved container, other container in the same batch that causes no reshuffle should be delivered in priority. If container reshuffle cannot be avoided, the final yard location of each reshuffled container shall be confirmed in accordance

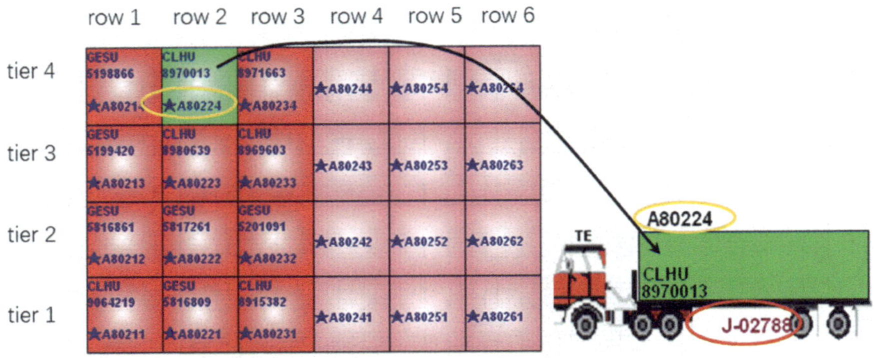

Fig. 4.13 Delivery confirmation by the yard crane

with the actual location after reshuffle. The priority of reshuffle goes to the container reshuffle within the same bay, and then the container reshuffle in the bay nearby.

Delivery confirmation by the yard crane is an indispensable step in the operation of container delivery. Only confirmed by the yard crane can the clerk at the gatehouse see the relevant information of the delivery vehicle in the system, so as to check the information of the delivery vehicle when it leaves the terminal.

4.5 Exit Verification of Delivery Vehicles

After the container delivery is confirmed by the yard crane, the delivery vehicle leaves the yard and prepares to leave the terminal. The delivery vehicle must be checked by the clerk at the gatehouse before exit. The verification of the delivery information includes the license plate number and the corresponding container no.

According to the actual license plate number of the vehicle for exit, the clerk at the gatehouse selects the vehicle in the system, as shown in Fig. 4.14a. The container no.

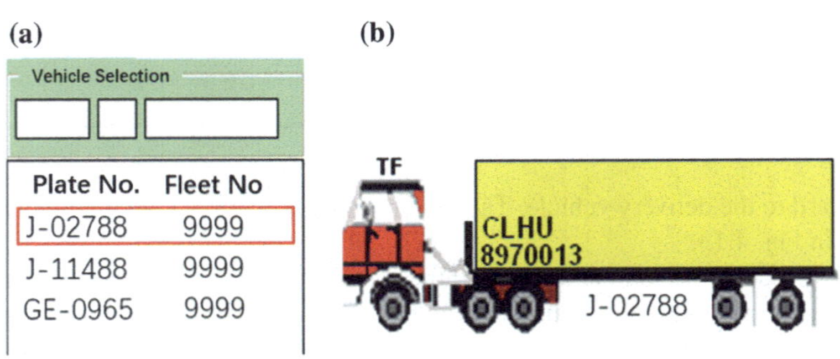

Fig. 4.14 Exit verification of the delivery vehicle

Table 4.3 Information of containers to be picked up

Container no.	B/L no.	Location in the yard
GESU5198866	ATLSTTJ01288	A8020104
CLHU9064219		A8020101
GESU5199420	ATLSTTJ01282	A8020103
GESU5817261		A8020102
GESU5816861	ATLSTTJ01298	A8020202
CLHU8970013	ATLQZTJ101856	A8020204
CLHU8980639	ATLQZTJ101844A	A8020203

will be displayed on the container carried by the corresponding vehicle, as shown in Fig. 4.14b. The clerk needs to check the container no. in the system with the container no. actually carried by the delivery vehicle, and the vehicle can only be released after verification. If the actual container carried by the delivery vehicle is inconsistent with the designated container in the system, the vehicle shall return to the yard and take delivery of the correct container.

As the last step of the container delivery operations in the terminal, the information verification of the container delivery vehicle is of great significance to the terminal, which can avoid serious loss caused by the wrong container delivery.

4.6 Comprehensive Case of Container Delivery Operations

In this section, the vessel ASJ5, 0927 is taken as an example to illustrate the container delivery operation process of the import full containers. There are 8 containers with 5 B/Ls, and the container no., the corresponding B/L no., and the storage locations of the containers in the yard are shown in Table 4.3.

1. **Container delivery reservation**

According to Table 4.3, a reservation is made according to the B/L no. ATL-STTJ01288 for two containers, as shown in Fig. 4.15. In the delivery reservation module, ① click "add" to add a reservation record, and the system automatically generates a reservation no. with the current date plus a four-digit serial number (0001); ② select the reservation mode by "bill of lading"; ③ select the vessel name and voyage, namely ASJ5-0927; ④ input the B/L no.; ⑤ input container amount and other information; ⑥ click "save" and the system prompts that *you have made a reservation.*

When one B/L corresponds to two or more containers, it is convenient to make a reservation by the B/L. When making a reservation by the container no., the reservation mode should be set as by the container no. and the container no. is input. Other operation is the same as that of making a reservation by the B/L no. For reservation by the container no., there is no need to select the vessel name and voyage. Similarly, the delivery reservation plan is made for the other containers in Table 4.3.

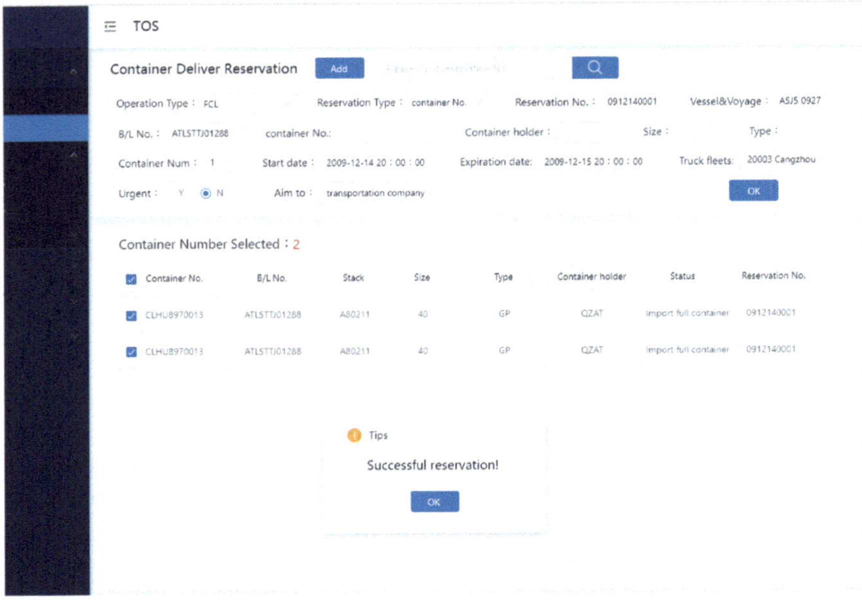

Fig. 4.15 Container delivery reservation plan

2. **Vehicle entry for delivery operation**

After the completion of the container delivery reservation, external vehicles can come to the terminal to take delivery of the containers. The first step is to precede the relevant procedures for the entry at the gatehouse. In this case, the vehicle J-02788 comes to take delivery of the container with the reservation no. 0912140004.

The clerk at the gatehouse clicks "next vehicle" in the system to select the vehicle J-02788 and enters the reservation no. of the delivery vehicle 091214004 in the "reservation no.". The system will automatically search the container corresponding to the reservation no. and display relevant information, as shown in Fig. 4.16.

Click the record in the red box shown above, and the system requires confirming the vehicle information. After verification, click "yes" to assign the delivery task to the vehicle. Meanwhile, the container information confirmation box shown in Fig. 4.17 appears, the vehicle can enter after all information is verified.

3. **Delivery operation by the yard crane**

After the delivery vehicle completes the procedures at the gatehouse, it is driven to the actual location of the container in the yard. Meanwhile, the clerk at the gatehouse will send the delivery task (including the delivery vehicle license plate number, the container no. for delivery, etc.) to the yard crane driver. When the vehicle reaches the target position, the yard crane driver starts to deliver the container, as shown in Fig. 4.18. The actual operation is similar to container operation in vessel loading, as long as "container delivery" in the operation mode is selected.

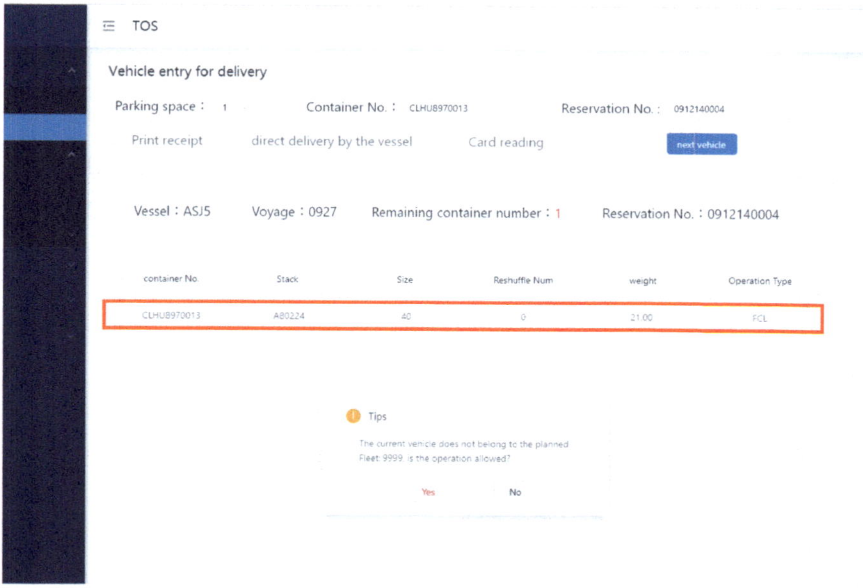

Fig. 4.16 Vehicle entry for container delivery

Fig. 4.17 Information verification for container delivery

4. **Exit verification of delivery vehicle**

After the yard crane's operation, the delivery vehicle with the container goes again through the gatehouse for exit. The clerk at the gatehouse checks the actual information of the vehicle license plate number and the container no. with the record information in the system, as shown in Fig. 4.19. The clerk clicks the

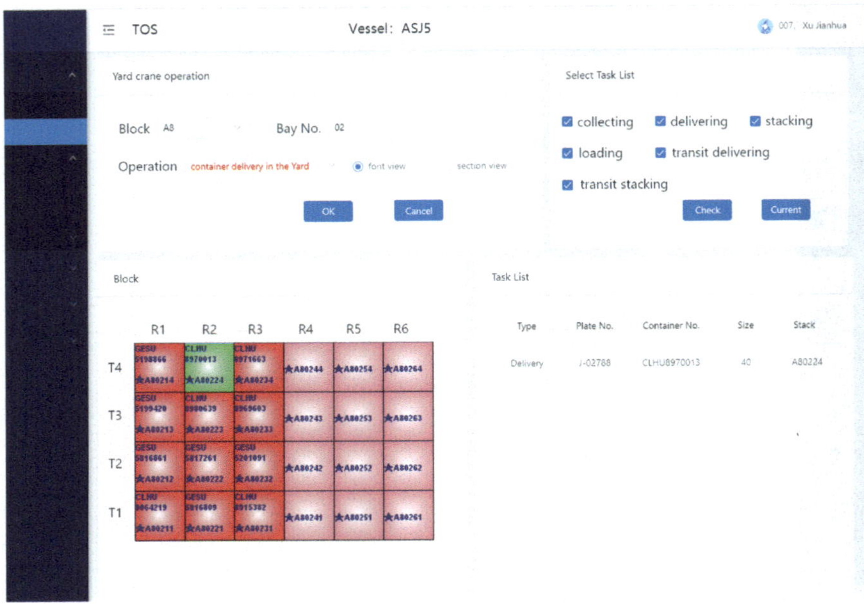

Fig. 4.18 Delivery operation by the yard crane

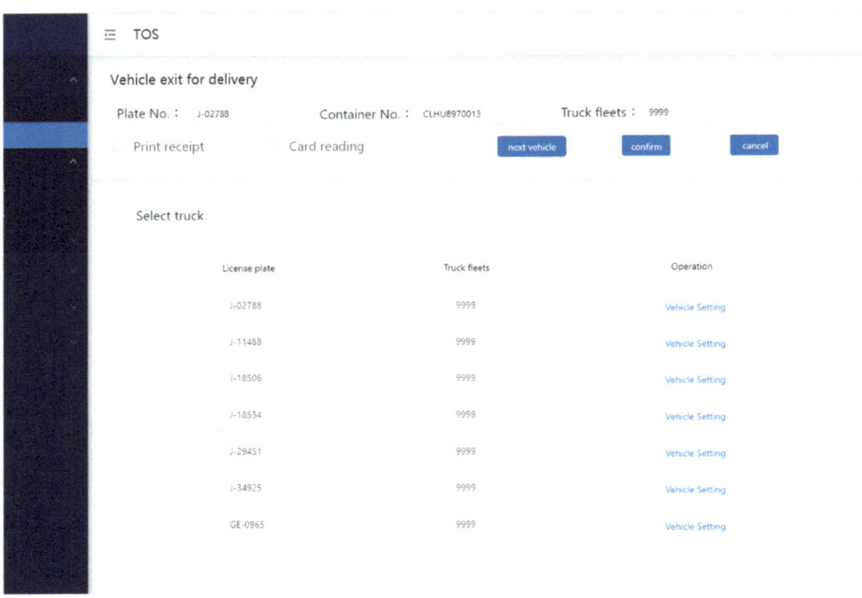

Fig. 4.19 Exit of delivery vehicle

"next vehicle" in the system to select the vehicle J-02788, and its designated container is CLHU8970013. Then he checks it against the actual container no. After verification, he clicks "confirm" to complete the container delivery; meanwhile, the vehicle is released.

Chapter 5
Management of Container Collection Operations in the Container Terminal

5.1 Introduction of Container Collection Operations

Before the vessel arrives at the port, the export containers will be successively transported to the storage yard of the port by the truck fleet, waiting for loading. Such process is usually referred to as container collection. Generally, the collection of export containers will start in 3–5 days before the arrival of the corresponding vessel, and the containers will be transported by the external container trucks to the yard successively. The container collection will be closed in 6 h before the arrival of the vessel (the container will be stopped from entering the terminal), so that the final stowage of the vessel and the loading operation plan can be made according to the information of the collected containers.

The collection operations of the export containers include pre-input of export container information, port and tonnage grouping of export containers, yard plan of export containers, location selection of the export container,container collection in the yard, etc. There are two ways to pre-input the export container information. The first one is to pre-input the export container information by the carrier outside the port, which is adopted by many large terminals at sea ports. The second one is to directly use the export manifest as the pre-input information. Since the terminal obtains the complete export manifest relatively late, the container collection can only begin in 2–3 days in advance. When the loaded truck enters the terminal at the gatehouse, there is a high possibility that the manifest of the corresponding container has not arrived yet. In this case, the clerk at the gatehouse has to input the container information manually according to the container load plan. So for the second pre-input mode, the information of the collected export containers is either from the manifest or input manually by the clerk at the gatehouse. The port and tonnage grouping are to sort all the export containers of the same voyage according to their sizes, discharging ports, and tonnage. The yard plan of export containers is to sort the export containers by the information of vessel name, voyage, size, discharging port, container type, special requirements of goods, etc., and arrange the block and locations for each group of the export containers of the same voyage. Since the export containers arrive at the

N. Zhao et al., *Digital Management of Container Terminal Operations*,
https://doi.org/10.1007/978-981-15-2937-5_5

terminal separately and are loaded together, and the corresponding vessel has not arrived when the container collection begins, the relevant information is not accurate enough. Besides, there are strict time and space restrictions for loading, so the entry plan of the export containers is relatively difficult. The location selection of the export container is to designate a specific storage location for the export container according to the specific data of the container and the entry plan of the voyage to which the container belongs. Container collection in the yard is to lift and stack the container in the yard by the yard crane diver and save the final stacking location of the container to the database of the terminal operation system (TOS) through the interface installed in the yard crane.

In the process of export container collection, the operations in sequence are dependent on each other. Only after the previous operation is finished, can the subsequent operation begin. For example, only after the location selection of the export container is finished, can the container collection in the yard begin. The basic process of container collection is shown in Fig. 5.1.

The positions involved in the process of export container collection and the corresponding work are shown in Table 5.1.

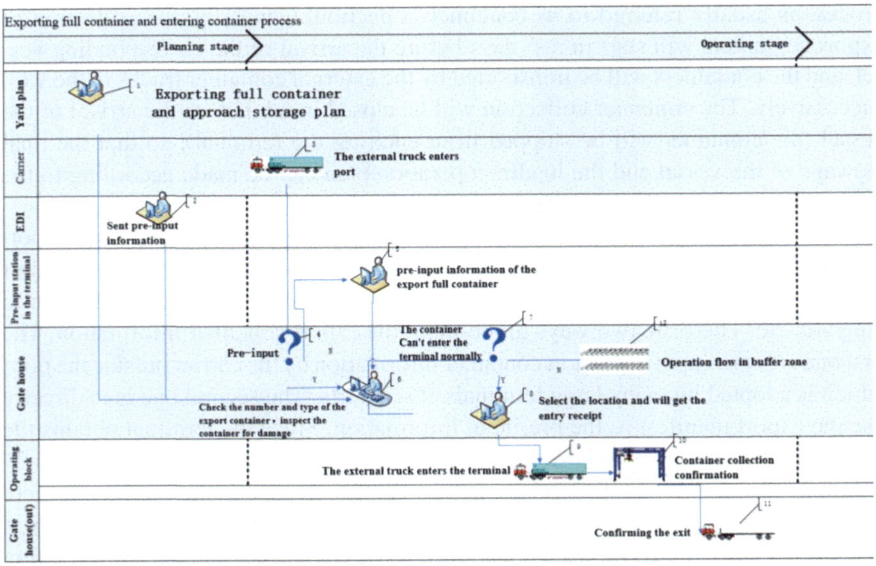

Fig. 5.1 Basic process of container collection

Table 5.1 Positions involved in the process of export container collection and the corresponding work

No.	Position	Description
1	Yard plan	Port and tonnage grouping of the export containers
2	Yard plan	Making yard plan according to the ports of call of the export vessel as well as the port and tonnage grouping of the export containers
3	EDI	Sending electronic container load plan information of the export containers
4	carrier	Transporting the export full container to the terminal
5	Gatehouse (in)	Checking the pre-input information of the export full container
6	Pre-input station in the terminal	If there is no pre-input information, the information of the export full container should be pre-input at the pre-input station in the terminal
7	Gatehouse (in)	If the container information has been pre-input, the clerk at the gatehouse can directly check the no. and type of the export full container according to the electronic container load plan information, and inspect the container for damage
8	Gatehouse (in)	Checking whether the container can enter the terminal normally
9	Gatehouse (in)	If normal entry is permitted, the system will automatically select the location and the driver will get the entry certificate
10	Gatehouse (in)	If normal entry is not permitted, the truck driver should enter the buffer zone
11	Operating block	The external truck enters the corresponding block
12	Operating block	Container collection by the yard crane and confirmation with wireless interface
13	Gatehouse (out)	Checking the container truck, collecting the equipment interchange receipt and the entry certificate, and confirming the exit

5.2 Information Pre-input of the Export Container

The main purpose of the information pre-input of the export container is to obtain various attributes of the export container according to the container no. during container collection operations, so as to facilitate the entry and location selection in the yard.

5.2.1 *Main Contents of Information Pre-input of the Export Container*

The main purpose of information pre-input is to lay a foundation for the location selection of the export container, since the location should be selected according to the specific data when the export container enters the terminal. The main contents of information pre-input include:

① Container no., which is the only identification of the container.
② Size, which is the length of the container with ft as the unit.
③ Container type, which is the type description of the container.
④ Container height, which is the height of the container.
⑤ Container holder, which is also referred to as container operator.
The detailed explanation of the items above is shown in the main composition of the container information in BAPLIE in Sect. 3.2.1.1.
⑥ Seal no., usually marked behind the container no. after the customs inspection as the no. of the lead seal. A lead seal is a device similar to a latch applied by a particular clerk after the goods are loaded into a container and the door is properly closed. Once the seal is properly locked, it cannot be opened unless it is violently broken (cut open), and the damaged seal cannot be reused. Each seal has a unique identification no. As long as the container has a complete appearance, the container door is properly closed and the lead seal is normally locked, it can be proved that the container has not been opened without permission during the transportation, and the condition of the container is supervised during the loading.
⑦ Discharging port, is the port at which a container is discharged from a vessel. The discharging port is a very important attribute of the export container. In general, there may be more than one port of call at a voyage of a vessel to discharge containers. In order to avoid container reshuffle at the intermediate port, the export containers with different discharging ports should be separated as much as possible in vessel stowage, which requires that the export containers with different discharging ports should be stacked separately in the yard before loading. Therefore, the container operations such as location selection during entry, vessel stowage plan, loading scheduling are closely related to this attribute. Once the information of the discharging port is input incorrectly, the export container will be unloaded at the wrong port, which will bring a serious loss to the carrier.
⑧ Destination port, is the port at which the export container finally arrives. Under normal circumstances, the shipping company has its fixed routes, and when the required destination port of the goods is not the same as the port at which the vessel calls, the goods will be unloaded at the nearest port to the destination port, and then transported to the destination port required by the customer by other means. So when booking space with the shipping company, it is required to clearly specify the destination port, in case that the goods might be shipped to the wrong place. In particular, when trading with inland areas, the port near to the inland city where water transportation can be arranged shall be selected as the discharging port instead of the inland city, which shall only be selected as the destination port.

⑨ Net weight of the container, is the total weight in tons of the goods in the export container, which does not take into account the weight of the container itself.
⑩ Gross weight of the container, is the total weight in tons of the goods in the export container plus the container itself. This is another very important attribute of the export container. Since heavy container stacking on light one should be avoided when loading the export containers, the export containers with different weight tonnage levels should be separated or light container be stacked on heavy one in stowage, which requires the export containers with different weight tonnage levels should be stacked separately or heavy container be stacked on light one in the yard. Therefore, the gross weight is closely related to operations such as location selection during entry, vessel stowage plan, container delivery in vessel loading, and so on.
⑪ Container state, refers to the transition state and packing state of the container. For export containers, there are generally export full (EF) containers, export empty (EE) containers, export transshipment containers, etc.
⑫ Description of goods, is the description of the types of goods in the container. This attribute is not very important for the container terminal since the terminal operation is subjected to the container, instead of the goods in the container.

5.2.2 *Receiving and Converting of Export Manifest*

5.2.2.1 Significance of the Export Manifest

At some ports of China (such as Tianjin Port), it is not necessary to arrange special module to pre-input export container information, so the export manifest is needed to replace the role of pre-input export container information module. Meanwhile, after the completion of the export container collection of a certain voyage, it is necessary to check the export manifest information with information of all the collected containers in the yard of the same voyage. If there is any inconsistency, the information of the containers in the yard will be modified after confirming with the shipping company.

5.2.2.2 Format and Conversion of the EDI File of the Export Manifest

In order to improve the operating efficiency of the terminal and reduce the misoperation rate, EDI technology is widely used at major ports to obtain export manifest information.

The container terminal operation system must have a designated module to complete the decoding of EDI file, as shown in Fig. 5.2.

The specific conversion standard of the EDI platform file of the export manifest is listed below:

sender and receiver;
shipping company, shipping agency → terminal, tally, port supervision, etc.

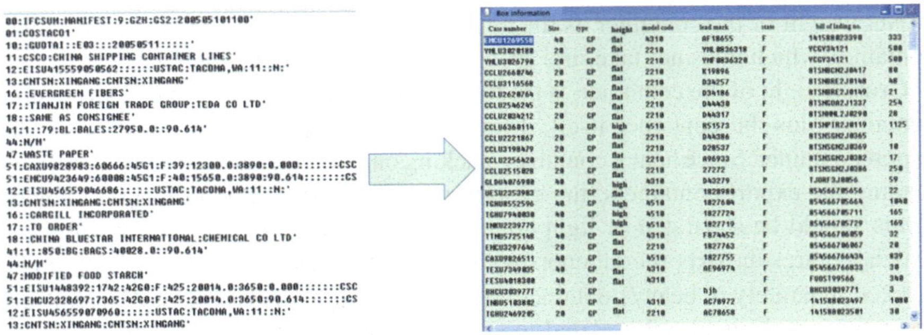

Fig. 5.2 Export manifest imported into the system by EDI

Function: The platform file corresponding to the manifest subset of IFCSUM messages provides information about the cargos for a particular voyage and details of the carrier, transportation mode, transportation facilities, equipment and intermodal cargos. The manifest is the proof that the vessel carries the container goods, it is also the necessary document to go through the import and export customs formalities, as well as the operation document for the terminal to prepare vessel loading and unloading.

Corresponding documents: export manifest, namely the export BAPLIE of the export containers.

Format of export manifest: the export BAPLIE of the export containers (PENAVICO TIANJIN EDI N.C.L. FORMAT)

Each record is of the fixed length: X(128).

Record 00	HEADER RECORD			
No.	Field name	Format	Position	Annotation
1	RECORD-ID	9(2)	1–2	00
2	DATA TYPE ID	X(3)	3–5	"NCL"
3	FILE DESCRIPTION	X(20)	6–25	"CONTAINER LIST"
4	PAYER-ID	X(2)	26–27	CODE
5	SENDER CODE	X(8)	28–35	CODE
6	RECEIVER CODE (1-4)	X(8)	36–67	CODE UNLESS ONE
7	FILE SEND TIME	9(10)	68–77	YYMMDDHHMM
8	SEND TIME CODE	X(1)	78	"L": LOCAL TIME "G": GREENWICH TIME
9	SEND PORT CODE	X(5)	79–83	CODE
10	RECEIVER PORT CODE (1-4)	X(5)	84–103	CODE UNLESS ONE
11	FILE CREATOR	X(8)	104–111	CODE

(continued)

(continued)

Record 00	HEADER RECORD			
No.	Field name	Format	Position	Annotation
12	CREATOR PORT CODE	X(5)	112–116	CODE
13	VERSION ID	X(3)	117–119	1.1
14	FILLER	X(9)	120–128	SPACES

Record 11	VESSEL & VOYAGE RECORD			
No.	Field name	Format	Position	Annotation
1	RECORD-ID	9(2)	1–2	11
2	FILLER	X(3)	3–5	SPACES
3	LINE MARK CODE	X(2)	6–7	SAME AS DOCUMENT
4	FILLER	X(3)	8–10	
5	VSL. CODE	X(6)	11–16	COSCO VESSEL CODE
6	VSL. NAME	X(20)	17–36	VSL. FULL NAME
7	VOYAGE	9(5)	37–41	VOY. NO.
8	ARRIVAL DATE	X(6)	42–47	YYMMDD
9	SAILING DATE	X(6)	48–53	YYMMDD
10	SHIP'S REGISTRY NO.	X(15)	54–68	FOR DG. CARGO(US)
11	CALL SIGN	X(6)	69–74	FOR DG. CARGO(US)
12	FILLER	X(54)	75–128	SPACES

Record 51	DETAIL RECORD-1			
No.	Field name	Format	Position	Annotation
1	RECORD-ID	9(2)	1–2	51
2	CONTAINER NO.	X(11)	3–13	SAME AS DOCUMENT
3	CONTAINER SIZE & TYPE	X(4)	14–17	SAME AS V3.1
4	SEAL NO.	X(10)	18–27	
5	CONTAINER STATUS	X(1)	28	E: EMPTY, F: FULL, P: LCL
6	DOCUMENT NO.	X(16)	29–44	SAME AS DOCUMENT
7	DISH PORT OPTION-1	X(5)	45–49	
8	PORT OF DISH	X(5)	50–54	CODE SAME AS DOCUMENT

(continued)

(continued)

Record 51	DETAIL RECORD-1				
No.	Field name	Format	Position	Annotation	
9	PORT OF DELIVERY	X(5)	55–59	CODE SAME AS DOCUMENT	
10	FISH PORT OPTION-2	X(5)	60–64		
11	PACKAGE NO.	9(6)	65–70	"999999"	
12	KIND OF PACKAGE	X(8)	71–78	"CARTONS"	
13	TEMPERATURE ID	X(1)	79	"C/F"	TEMPERATURE RANGE
14	TEMPERATURE MARK FROM	X(1)	80	"±"	
15	TEMPERATURE FROM	9(2)	81–82	"99"	
16	TEMPERATURE MARK TO	X(1)	83	"±"	
17	TEMPERATURE TO	9(2)	84–85	"99"	
18	DG. CARGO-ID	X(1)	86	"Y" OR "N"	
19	FRONT	9(3)	87–89	IN CM	OVER SIZE & OVER LENGTH
20	BACK	9(3)	90–92	IN CM	
21	LEFT	9(3)	93–95	IN CM	OVER EIDTH
22	RIGHT	9(3)	96–98	IN CM	
23	OVER HIGH	9(3)	99–101	IN CM	
24	LENGTH	9(4)	102–105	IN CM	UNCONTAINERIZED CARGO DIMENSIONS
25	WIDTH	9(4)	106–109	IN CM	
26	HIGH	9(4)	110–113	IN CM	
27	NET WEIGHT	9(6)	114–119	"999999" (KG)	
28	GROSS WEIGHT	9(6)	120–125	"999999" (KG)	
29	CNT NO. CHECK FLAG	X(1)	126	"Y" OR "N"	
30	FILLER	X(2)	127–128	SPACES	

Record 52	DETAIL RECORD-2 (FOR DANGEROUS CARGO)			
No.	Field name	Format	Position	Annotation
1	RECORD-ID	9(2)	1–2	52
2	CONTAINER NO.	X(11)	3–13	SAME AS RECORD 51
3	SHIPMENT FLASH POINT	X(3)	14–16	
	DANGEROUS ITEMS			(OCCURS 2 TIME?)
4	HAZARD CODE (IMCO CLASS)	X(7)	17–23	SAME AS DOCUMENT
5	PAGE NUMBER	X(7)	24–30	
6	UNDG NUMBER	X(4)	31–34	
7	UNDG GOODS FLASH POINT	X(8)	35–42	
8	EMSNO.	X(6)	43–48	
9	MFAG	X(4)	49–52	
10	PACKING GROUP	X(3)	53–55	CODE
11	DG GOODS LABEL MARKING	X(4)	56–67	
12	FILLER	X(61)		SPACES

Record 53	DETAIL RECORD-3 (FREE TEXT FOR REMARKS)			
No.	Field name	Format	Position	Annotation
1	RECORD-ID	9(2)	1–2	53
2	CONTAINER NO.	X(11)	3–13	SAME AS RECORD 51/52
3	GOODS DESCRIPTION	X(30)	14–43	
4	STOWAGE LOCATION	X(6)	44–49	
5	CANNCELL ID	X(1)	50	"C" OR "U"
6	REMARKS	X(75)	51–125	TEXT
7	FILLER	X(3)	126–128	

Record 99	TRAILER RECORD			
No.	Field name	Format	Position	Annotation
1	RECORD-ID	9(2)	1–2	99
2	DATA TYPE ID	X(3)	3–5	"NCL"
3	RECORD NUMBER	9(6)	6–11	INCLUDING HEADER AND TRAILER
4	FILLER	X(117)	12–128	SPACES

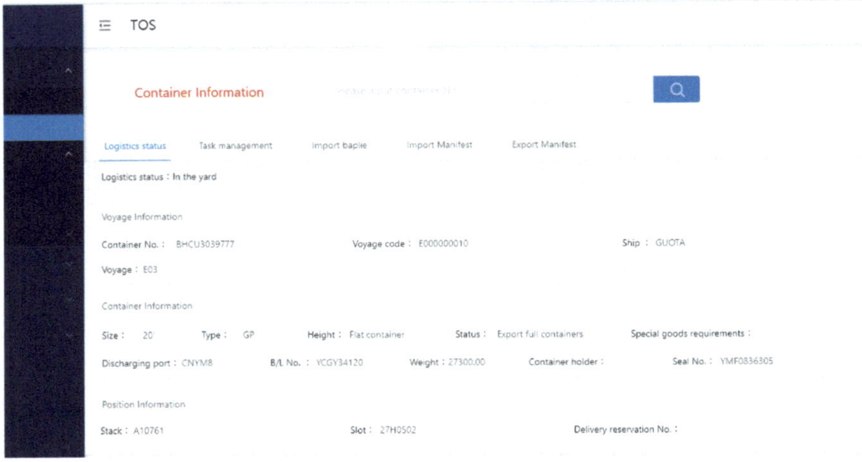

Fig. 5.3 Manual input of the export manifest

Theoretically, the export manifest is assumed to be received and saved in the database through EDI, but manual input is still necessary in actual operations, since through manual input of export manifest, the export container data can be checked, made statistics, and modified appropriately, as shown in Fig. 5.3.

5.3 Port and Tonnage Grouping of Export Containers

5.3.1 Function of Port and Tonnage Grouping

Before making the yard plan of the export containers for each export voyage, according to the discharging ports along the route of the voyage, the export containers that are expected to enter the port are sorted into groups (port grouping), and then the export containers of each discharging port are sorted into groups according to their weights (tonnage grouping). This operation is referred to as port and tonnage grouping. During stowage of the vessel, it is necessary to ensure good stability and proper trim and to avoid container reshuffle at intermediate discharging ports; therefore, the export containers with different discharging ports and tonnage levels should be separately stacked as much as possible in the vessel, and try to keep heavy containers under the light ones, which requires the export containers to be stacked separately in the yard according to their discharging ports and tonnage levels. Therefore, the port and tonnage grouping of the export containers is the foundation of making yard plan and vessel stowage. If the port and tonnage grouping is not done well, the export containers of each voyage will be stacked randomly in the yard, which will bring

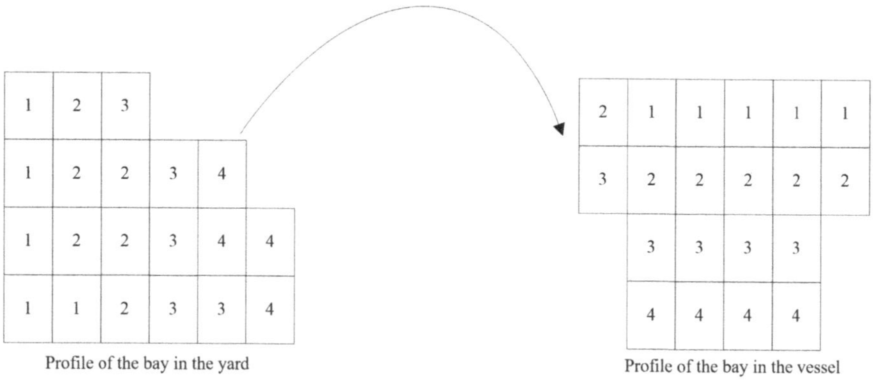

Fig. 5.4 Container stacking illustration in the yard and in the vessel with different tonnage levels

great trouble to the stability of the vessel and the unloading of containers at the intermediate ports of the vessel. As shown in Fig. 5.4, the containers stored in the yard are grouped into four tonnage levels, namely 1, 2, 3, and 4. In the yard shown in the figure, containers of different tonnage levels are stacked at the same bay but different rows, or the container with high tonnage level is stacked above of the one with low tonnage level. It should be noted that the tonnage level distribution in the figure is only for illustration of the tonnage grouping. The stacking strategy of each terminal may be different, but the principle of container stacking in the yard is generally that heavy container is stacked above light container, and containers of the same tonnage level are stacked in the same row.

5.3.2 *Main Work in Port and Tonnage Grouping*

The main work in port and tonnage grouping of export containers is to sort all the export containers of each voyage into groups according to container sizes, discharging ports, and tonnage. The main steps are as follows.

Step 1: in the container terminal operation system, query the sizes and discharging ports of all the export containers for this voyage through the export manifest, and select a specific combination of size and discharging port, as shown in Fig. 5.5.
Step 2: add tonnage levels for the records of selected size and discharging port. Each level represents a certain range of tonnage, such as level 1 for 5–10 t and level 9 for 40–50 t. The range of tonnage of each level is not fixed. The range of tonnage of each level may be different for different voyages, and the range of tonnage of the same level may be different for the same voyage with the same size but different discharging ports. In the tonnage grouping, the lower the level is, the lighter the export containers, for example, the containers of level 2 are lighter than those of level 3. The number of tonnage levels of each size and discharging port is generally related

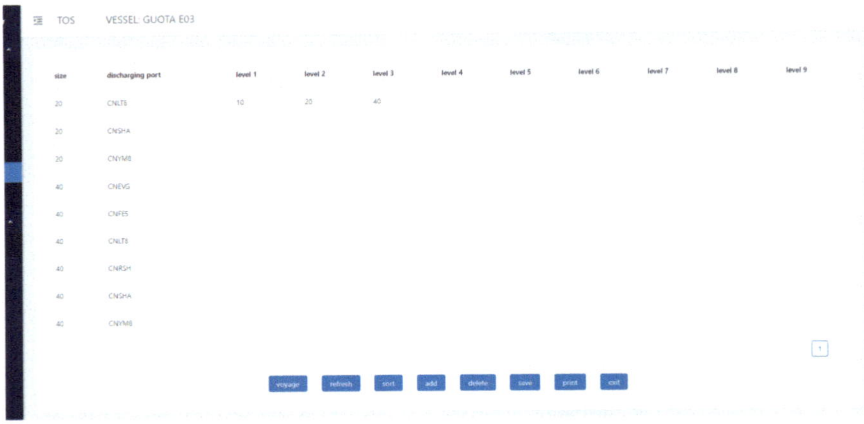

Fig. 5.5 Selecting one combination of container size and discharging port

to the container quantity and the vessel's stability requirements for the voyage. If the container quantity is small and the vessel's stability requirements are not very high, the tonnage grouping can be made roughly with fewer levels. If the container quantity is large and the vessel's stability requirements are high, the tonnage grouping should be made with more levels. The container tonnage grouping of a certain size and a certain discharging port is shown in Fig. 5.6.

Step 3: due to time constraints, the input of export manifest is not timely, or some temporary operations do not use export manifest, so the information imported from the export manifest cannot meet the actual operation requirements. In this case, the new container size and discharging port should be added in the system, and the new record should be designated with corresponding tonnage levels definition.

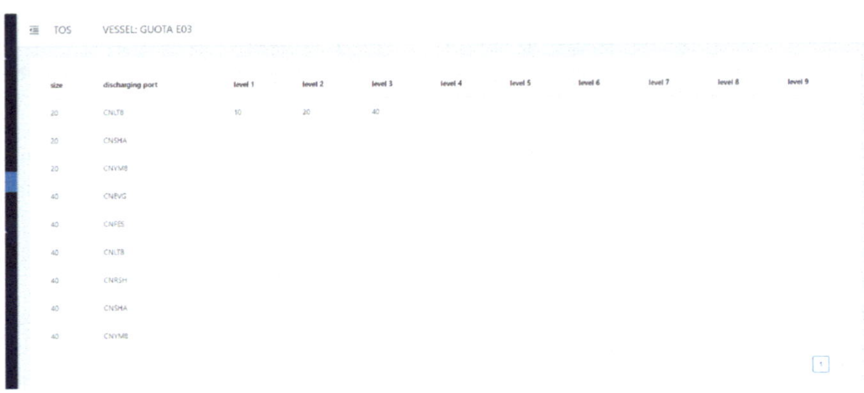

Fig. 5.6 Tonnage grouping

5.4 Yard Plan for Export Containers

5.4.1 Definition of the Yard Plan for Export Containers

The yard plan for export containers, also known as the container entry plan, refers to the decision-making process of grouping the export containers according to discharging ports and tonnage, estimating the container amount of each group, and arranging target storage area for the export containers of certain tonnage level and discharging port.

The export containers are needed to be stacked in accordance with certain rules and be ready for loading, so the planning department of the container terminal has to make a yard plan for export containers in advance. The yard plan for export containers is based on the vessel name, voyage, pre-input data of export containers, in combination with the actual stacking situation of the container terminal yard. In the process of making the yard plan for export containers, the preparation of the future vessel stowage plan and the actual occupation of the storage yard should be taken into comprehensive consideration, so as to ensure the smooth loading of the export containers into the vessel.

5.4.2 Features and Functions of the Yard Plan for Export Containers

Container liner transportation is a logistics component in trading activities, and container collection in the terminal is also a step in the process of container liner transportation. Therefore, the uncertainty of trade activities leads to the uncertainty of container liner's loading amount at each port of call, as well as the information uncertainty of container collection operations. In the process of container collection for a vessel, containers from different places will be transported to the terminal by the customer at different time points, it is impossible to forecast the final amount of the export containers of each discharging port in the yard, which is similar to a bus that the number of passengers on at each stop cannot be accurately estimated. So for the yard, the export containers arrive separately and randomly but are stacked concentratedly with rules. Therefore, the difficulty of yard plan for export containers lies in the uncertainty of information, namely the uncertainty of container amount and arrival sequence, as shown in Fig. 5.7.

The yard plan for export containers is the most basic and core part in the export container operations. The quality of the yard plan for export containers directly affects the rationality of the export container stacking, and is the basic guarantee for the smooth operations of subsequent stowage operation and vessel loading. Meanwhile, since the yard plan for export containers is made before the arrival of the export containers, it is conducive to the overall understanding of the rationality of

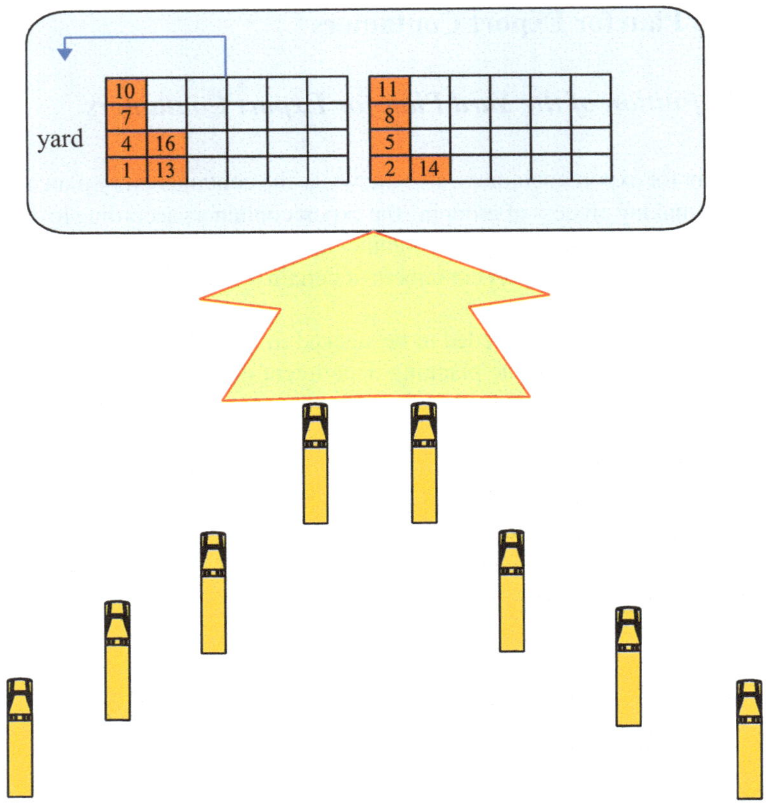

Fig. 5.7 Export containers enter the terminal separately

the container stacking, and plays a macro guiding role in the storage of export containers. Therefore, the yard plan for export containers is the premise and foundation for the effective organization of the subsequent operations of export containers. If the yard plan is made reasonably, it will bring the following benefits to the terminal, the shipping company, and the carrier.

(1) Reducing the operation cost of the terminal
A reasonable yard plan for export containers can improve the utilization rate of the yard, decrease the shifting distance and transfer times of the yard facilities, and lower the reshuffle rate, thus helping to reduce the operation cost of the terminal.
(2) Decreasing the energy consumption of facilities in the terminal
Due to the reduction of shifting distance and transfer times of the yard facilities as well as the reshuffle rate, the relevant energy consumption and pollution emission will also be decreased.
(3) Shortening the vessel's duration at port

A reasonable yard plan for export containers will be helpful to the concurrent operation of multiple operation lines during vessel loading, which will play a positive role in improving the vessel loading efficiency and shortening the vessel's duration at port.

(4) Shortening the external container truck's duration at port
A reasonable yard plan will help to reduce the queuing time of the external container truck for the yard facility, thus shortening its duration at port.

(5) Improving service in the terminal
Two important criteria for evaluating the service of the terminal are the vessel's duration at port and the external container truck's duration at port. The decreasing of these two fully indicates the improvement of the service in the terminal.

5.4.3 Constraints of Container Stacking in the Yard

Export container stacking is closely related to container type, goods type, empty/full, and yard facility.

1. **Requirements of yard facility for export container stacking**

(1) The stacking height in the block shall be lower than the working height of the yard facility.
For the operating techniques of RTG, if the maximum working height of the RTG is five container tiers high, there can only be four tiers of containers stacked in the block, as shown in Fig. 5.8.

(2) Container stacks should be reserved for container reshuffle.
When stacking export containers, in order to facilitate the container reshuffle for the yard facility inside the bay, n container stacks are generally reserved at the same bay, where n is generally assumed as the number of stacking tiers minus 1 (as shown in Fig. 5.9, where three container stacks are reserved for container reshuffle).

2. **Requirements of container type and goods type for export container stacking**

(1) Export containers of different sizes should be stacked separately.
Export containers of 20 ft, 40 ft, and 45 ft shall not be stacked at the same bay in the block, and 45 ft export containers shall generally be stacked at both ends of the block.

(2) Dangerous goods containers should be stacked in dangerous goods container block.
Each container terminal has a special dangerous goods container block, and all dangerous goods containers can only be stacked in the special block. Dangerous goods containers should be stacked separately according to isolation requirements.

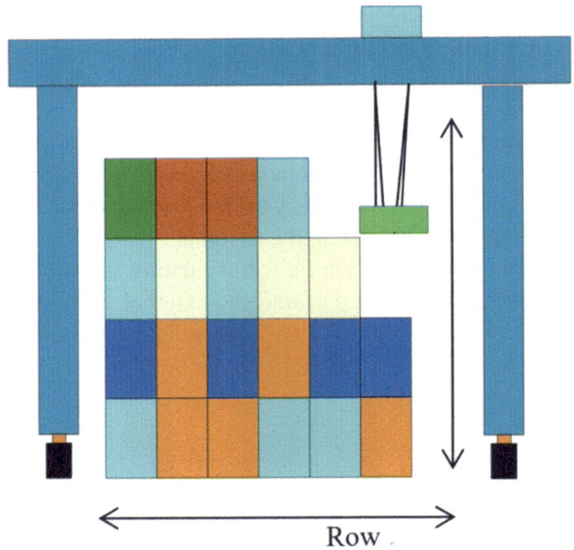

Fig. 5.8 Operating techniques of RTG

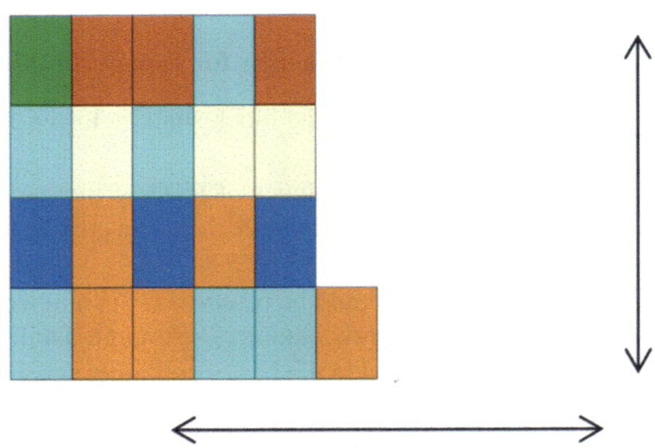

Fig. 5.9 Container stacks reserved for container reshuffle

(3) Reefer containers should be stacked in reefer container block.
Since the reefer containers need to be plugged in, each container terminal has a special reefer container block, and all the reefer containers can only be stacked in the special block. The reefer containers loaded with dangerous goods should be placed in the reefer container block. In the reefer container block, containers of different sizes can be stacked together.

(4) Over-limited containers should be stacked in over-limited container block.
The over-limited containers and rack containers should be stacked in the over-limited container block. The over-limited containers can only be stacked one tier high and can be mixed with containers of different sizes. If the oversize in width exceeds 30 cm, adjacent rows cannot be stacked with such containers,

and if the oversize in length exceeds 50 cm, adjacent bays cannot be stacked with such containers.

(5) Empty containers and full containers should be stacked separately.
The empty containers are generally stacked according to the container holder grouping. The empty containers can be generally stacked in six tiers in the block and the seventh is saved for container passing. The full containers must be stacked according to the specific rules defined by their operation attributes. The full containers are generally stacked in four tiers in the block and the fifth is saved.

3. **Stacking sequence in the block**

(1) Generally, the stacking sequence in the block is from the bay with smaller bay no. to the bay with larger no.
(2) If both ends of the block are fed into containers at the same time, one end is fed from the smaller to the larger, and the other end is fed from the larger to the smaller.
(3) In case of stacking at the same bay, the stacking sequence is from the row farthest away from the lane to the nearest row. If the lane is row 7, the containers shall be stacked from row 1 to row 7.

5.4.4 *Making the Yard Plan for Export Containers*

5.4.4.1 Lead Time and Content of the Yard Plan for Export Containers

At present, most container terminals in China have 4 days for container collection, and the progress proportion of container collection for each voyage is different every day. Generally speaking, the first day is with less container arrival, the second day with more, the third day with the most, and the fourth day with less. Most ports usually begin container collection at 9 am on the fifth day before the vessel arrives. Generally, the entry plan for the next day is made at 12:00 am. Meanwhile, the situation of containers entering and leaving the yard in the next few days should be considered comprehensively.

The yard plan should be made before the container collection, which divides the storage areas for the coming containers, namely to designate blocks to the coming containers of the specific vessel. The plan includes:

- Export container grouping
  The container grouping is based on the vessel name and voyage, container size, discharging port, tonnage level, container type, special requirements of goods, etc.
- Master block plan
  The master block plan is to select the primary block(s) designated for stacking all export containers for a particular voyage.

- Location plan: bay plan or row plan
  The bay plan is to designate certain one or several bays in the block to a certain group of export containers, while the row plan is to designate a specific row (1 stack serial) in the block to a certain group of export containers. Bay plan and row plan are two modes of yard planning in the block with different detail levels.

5.4.4.2 Target of the Yard Plan for Export Containers

Export containers can be sorted according to different properties, including empty containers and full containers, general containers and reefer/over-limited/dangerous goods containers, standard containers and special containers (such as foldable containers). Generally speaking, the proportion of empty containers, reefer/over-limited/dangerous goods containers, foldable containers, and so on is very small. There is no need to make yard plan for these special containers. The terminal will sort and stack these special containers uniformly, and the overall setting can be achieved in the TOS system. However, the general full containers take the largest proportion, including reefer containers loaded with goods that do not require temperature control seen also as general full containers. Therefore, for most terminals, the target of the yard plan for export containers actually refers to the export general full containers.

5.4.4.3 Container Grouping

The container grouping refers to that, on the basis of port and tonnage grouping, the export containers that are expected to enter the terminal are sorted into several groups according to the container size, container type, vessel name and voyage, and discharging port and tonnage level for the discharging port, so as to make the subsequent storage arrangement for different groups, as shown in Fig. 5.10.

Since the yard plan for export containers specifies the stacking area for the same group of export containers, the grouping of export containers is the basis and premise of making the yard plan. The detail level of the container grouping mainly depends on the detail level of tonnage grouping, which is related to the container amount of a certain size and discharging port expected to enter the terminal. For the manual plan, it is generally the greater the container amount, the more detailed the tonnage level definition. The more detailed the tonnage level definition is, the more complicated the process of yard planning is, and the more yard resources may be occupied. The rougher the tonnage level definition is, the less complicated the yard planning will be, but it will increase the difficulty of vessel stowage and the rate of container reshuffle during vessel loading.

The purpose of container grouping is to ensure that each container group can be stacked independently to a certain extent. If the storage space in the yard is adequate, the container groups can be stacked at different bays, respectively, and the container

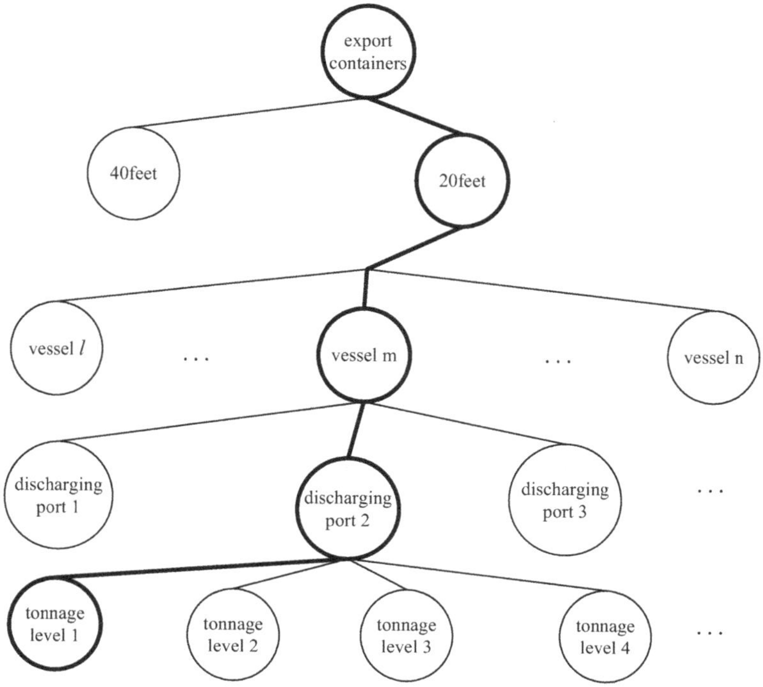

Fig. 5.10 Container grouping

groups with small amount can be stacked in rows, respectively. If the storage space in the yard is not adequate, mixed stacking to some extent is also frequently adopted, such as:

- High cubics and general containers are mixed.
- Containers of different tonnage levels are mixed.
- Containers of different discharging ports are mixed.

5.4.4.4 General Principles of Master Block Plan

(1) **Principle of proximity berthing**
The horizontal movement distance from the vessel to each container block is the total distance from the container location to the designated quay crane. The horizontal movement distance is related to both the vessel loading and unloading efficiency and the handling cost of the terminal. The shorter the distance is, the shorter the movement time of the container truck will be, the higher the vessel loading and unloading efficiency will be, and the lower the handling cost of the terminal will be, and vice versa. Therefore, the distance between the vessel's docking berth and the block should be fully considered when planning the block

designated to the export containers of a certain voyage, and the planned block(s) should be as near as possible to the berth where the vessel is docking. As shown in Fig. 5.11, if the docking position of the vessel is the berth for planning, the blocks checked are more suitable as the master blocks for this voyage.

(2) **Principle of separation of import and export blocks**

The vessel operations can be divided into vessel loading operations and vessel unloading operations, and for the container blocks, there are also export container blocks and import container blocks. When making the yard plan, the separation of import and export operations should be ensured as much as possible under the condition of sufficient container blocks, so that the export container operations should be arranged in the export container blocks as much as possible, and the unloading operations should be arranged in the import container blocks as much as possible. If the above separation is not made, it may cause inconvenience in the yard crane operation, disorder in the yard, and increased probability of conflict. Separation of import and export blocks can also avoid the interference of future customer's taking delivery of containers to the vessel loading operations.

However, if the import container blocks are all occupied, the vessel unloading operations may also be arranged in the export blocks. Similarly, if the export container blocks are full, the vessel loading operations can be arranged in the import container blocks.

(3) **Principle of avoiding berth date conflict**

The principle is that for the vessels expected to operate at the same time, the export containers of these vessels should not be planned to stack in the same

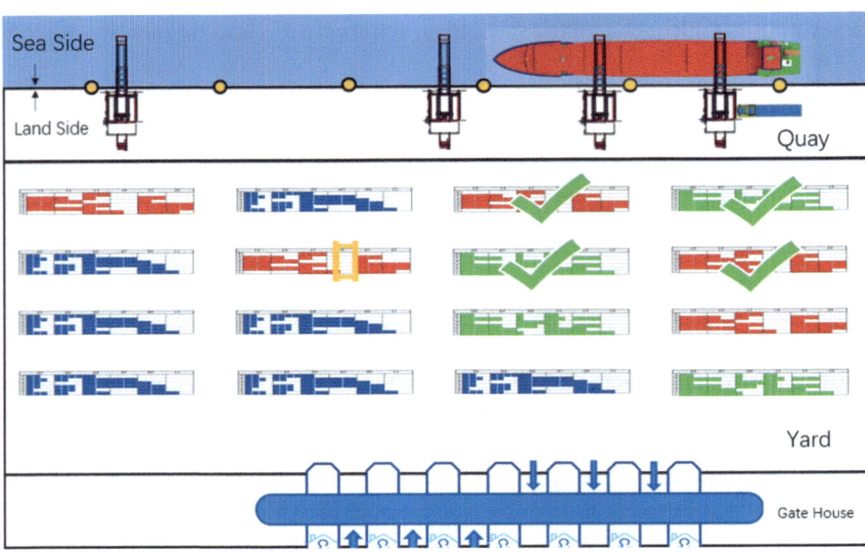

Fig. 5.11 Illustration of principle of proximity berthing

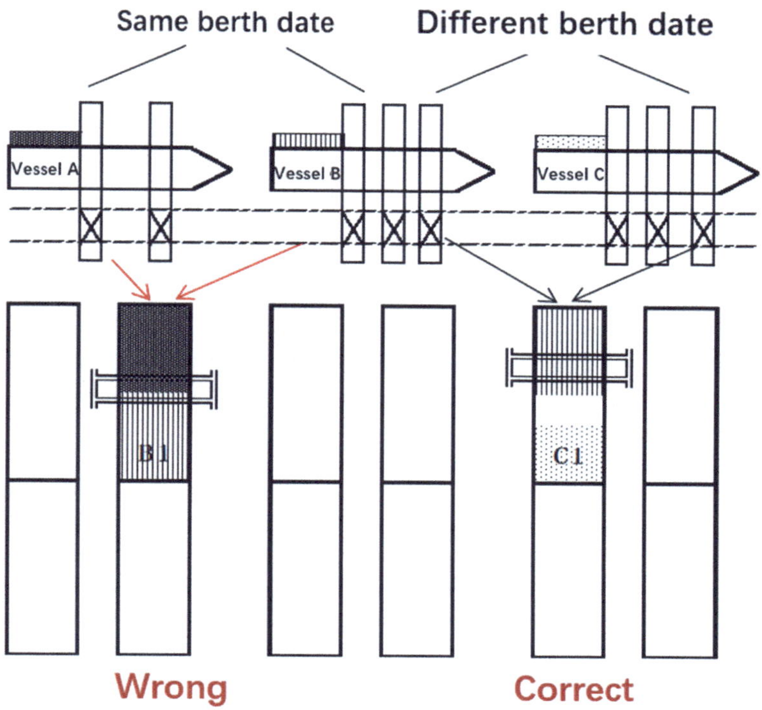

Fig. 5.12 Illustration of avoiding berth date conflict when making yard plan

container block. As shown in Fig. 5.12, when selecting the master block for vessel B, block B1 that has been planned for the export containers of vessel A should be avoided as much as possible; instead, block C1 that has been planned for the export containers of a different berth date should be selected. In addition, one block should not be planned for too many vessels; generally, two vessels are advisable.

(4) **Principle of concurrent operation of yard cranes during vessel loading**
The concurrent operation of multiple yard cranes has an important impact on improving the vessel loading efficiency. In general, the loading and unloading efficiencies of the quay crane are slightly higher than that of the yard crane. If one quay crane corresponds to only one yard crane during vessel loading, the quay crane will frequently be in a waiting state, namely, the bottleneck of vessel loading is in the yard. Therefore, in order to avoid efficiency bottleneck during vessel loading, it is better to designate two yard cranes to each quay crane, namely the yard cranes will deliver containers simultaneously. Therefore, it is necessary to fully consider the distribution of the blocks when making the block plan, that is, to distribute the export containers to multiple blocks, so as to improve the probability of concurrent operation of the yard cranes.

5.4.4.5 General Principles of Location Plan

Location plan in the yard refers to designating the specific stacking area for each group of export containers in the planned blocks, including two modes, namely row plan and bay plan, among which row plan is also known as slot serial plan. The principles below must be followed either in bay plan or row plan.

(1) **Principle of separate stacking of containers with different sizes**
20 and 40 ft containers cannot be stacked at the same bay, because it will bring great trouble to the operation scheduling of the container terminal, such as the increase of container reshuffle rate, the increase of stowage difficulty, and the frequent movement of yard cranes during operation. As shown in Fig. 5.13, the bays in the red box are stacked with both 20 ft and 40 ft containers, which is a violation operation.

(2) **Principle of concentration in bay plan**
When selecting bays in the block, it is necessary to give full consideration to how to reduce the shifting distance of the yard crane during container collection and vessel loading. In other words, the selected bays should be concentrated in the block as much as possible. When the distances between the selected bays are short, the yard crane always moves within a small range and its shifting distance is relatively short. As shown in Fig. 5.14, if the vessel needs to plan three bays in the block, the sum of the distances between the three bays selected in plan 1 is two bays, while the sum of the distances between the three bays selected in plan 2 is five bays, so plan 1 is more reasonable than plan 2.

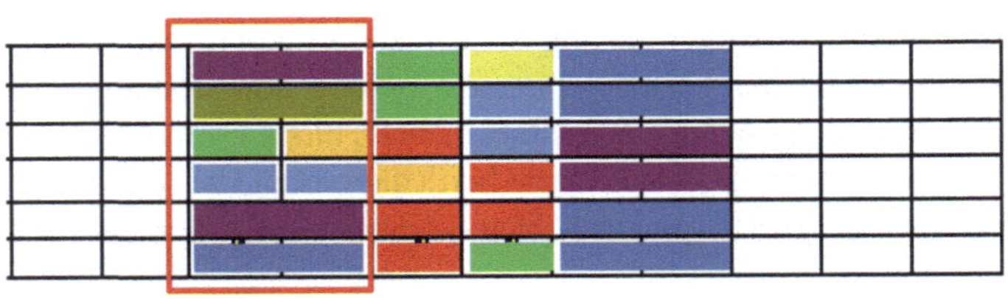

Fig. 5.13 Mixed stacking of containers with different sizes

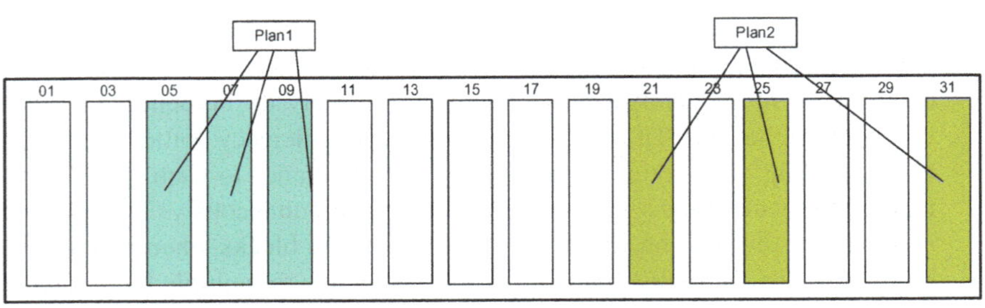

Fig. 5.14 Concentrating and distributing of bays

(3) **Principle of alternate stacking for small and large containers**
The small containers refer to 20 ft containers, while the large containers refer to 40 ft containers. Alternate stacking principle refers to stacking the large and small containers alternately in the mode of "small, large, large, small" when making the yard plan for the group with large container amount, so as to make it more convenient to shift from small containers to large containers in the future vessel loading in the cabin, and ensure two yard cranes to operate concurrently. As shown in Fig. 5.15, when making plan for 40 ft containers, it is better to select bay 10 and bay 18.

(4) **Principle of distributing and concentrating**
The principle means that the containers of the same discharging port should be distributed in several blocks, and concentrated at bays in one block.
If the container amount of a certain discharging port is relatively large, it can be planned to distribute the containers in two or three blocks to ensure that the concurrent operation of multiple yard cranes for one operation line during vessel loading, which requires that the export containers of the same discharging port should be distributed in different blocks as much as possible. As shown in Fig. 5.16, the containers of discharging port 1 are distributed in blocks 1 and 3, the containers of discharging port 2 are distributed in blocks 1 and 3, the containers of discharging port 3 are distributed in blocks 1, 2 and 4, and the containers of discharging port 4 are distributed in blocks 2 and 4. Such container distribution can ensure concurrent operation of multiple yard cranes for one operation line during vessel loading to the greatest extent. And location

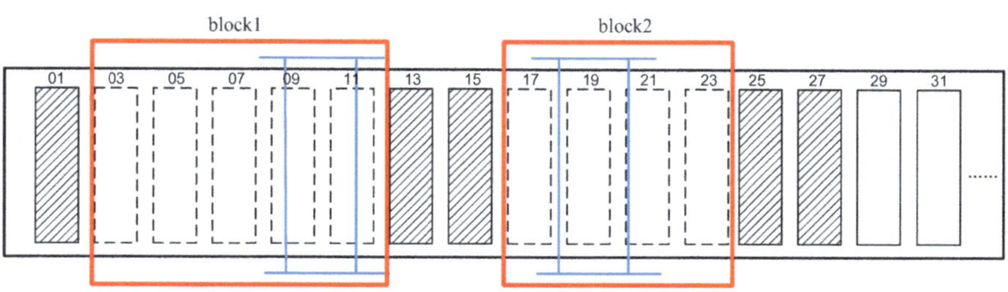

Fig. 5.15 Alternate stacking for small and large containers

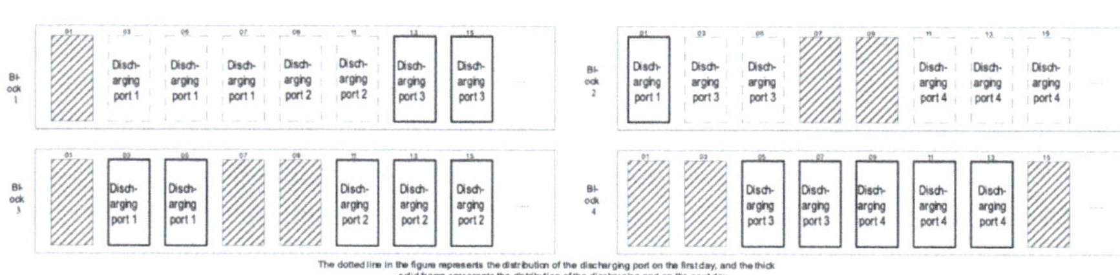

Fig. 5.16 Distribution of containers of different discharging ports

concentration at bays refers to that the bays nearby in one block should be stacked with the containers of the same discharging port as much as possible.

(5) **Principle of row plan**
This principle refers to that in the same row of the same bay, the containers of the same group, namely the export containers with the same voyage, same size, same discharging port, same tonnage level, and same type, should be stacked together. This principle can avoid the mixed stacking of containers with different tonnage levels or even different discharging ports in the same row.

5.5 Operations of Container Entry Through the Gatehouse

5.5.1 Introduction of Operations of Container Entry Through the Gatehouse

According to the different means of transportation, the export container collection can be divided into several modes, such as entry through the gatehouse (land transportation), entry with the barge (water transportation), and entry with the train (rail transportation), among which the entry through the gatehouse is the most typical. In this section, the operations of container entry through the gatehouse in the container terminal are taken as the main analysis target to elaborate the actual operation phases of container collection.

If the container collection operations of the port are divided into three phases, namely information collecting, planning, and scheduling, and actual operations, then the actual operation phase of the container collection can begin after the completion of the export container yard plan (the second phase). This phase can be summarized into four steps (taking the container entry through the gatehouse as the example) in the terminal operation system (TOS).

Process 1: information collection;
Process 2: location selection in the yard;
Process 3: container collection by the yard crane;
Process 4: exit of the empty container truck.

It is shown in Table 5.2 where the four steps occur and by whom they are performed or operated.

As can be seen from Table 5.2, step 2 is a virtual process completed by a computer server. For the TOS system, it is an important module for information flow processing. However, in terms of physical significance, there are only three locations involved in the operations of container entry through the gatehouse, as shown in Fig. 5.17.

Table 5.2 List of steps in container entry through the gatehouse

No.	Step	Location	Operator
1	Information collection	Gatehouse (in)	Gatehouse (in) clerk or intelligent gatehouse
2	Location selection in the yard	Gatehouse (in)	Server
3	Container collection by the yard crane	Yard	Yard crane driver
4	Exit of the empty container truck	Gatehouse (out)	Gatehouse (out) clerk

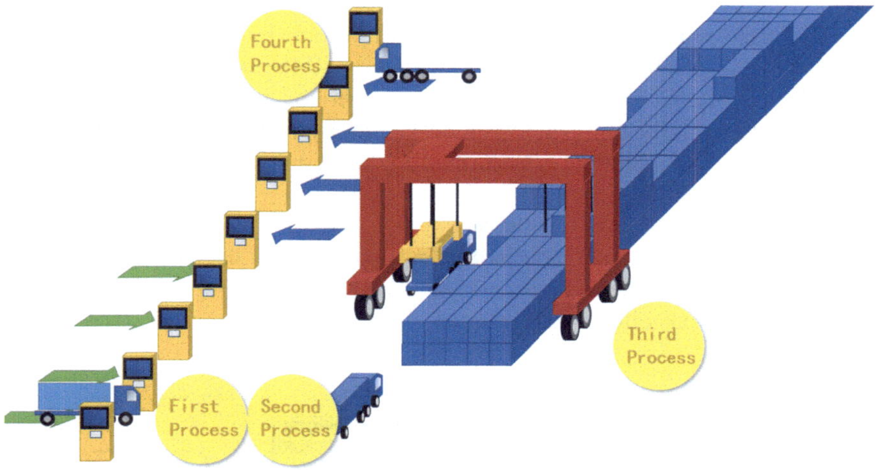

Fig. 5.17 Operation steps of container entry through the gatehouse

5.5.2 *Information Collection in Container Entry Through the Gatehouse*

Upon completion of the packing of export containers, the consignor/freight forwarder can entrust a truck fleet to transport the export containers to the port according to the operation schedule of the container collection. After arriving at the entrance gatehouse of the container terminal, the information collection of the container entry operation can be completed there.

(1) **Content of information collection**

- License plate no. of the container truck
- Container no. of the export container

(2) **Means of information collection**

There are two kinds of gatehouses in the container terminal, namely conventional gatehouse with clerk and intelligent gatehouse, so the information collection can be accomplished by manual collection or system automatic collection.

- Manual collection

Manual collection means that the clerk at the gatehouse inputs the license plate no. of the truck and export container no. into the TOS, and submits the collected information according to the system prompts, as shown in Fig. 5.18.

- System automatic collection

The intelligent gatehouse has been adopted in some terminals, and the system applies the graphic recognition technology or RFID technology to automatically identify the license plate no. and container no., and displays the collected information on the terminal screen of the intelligent gatehouse, and the information of entry operation is confirmed by the container truck driver, as shown in Fig. 5.19.

(3) **Function of information collection**
It is to provide data preparation for location selection decision of the export container entry. If the container entry though gatehouse is regarded as a system, the information collection is the input of the system.

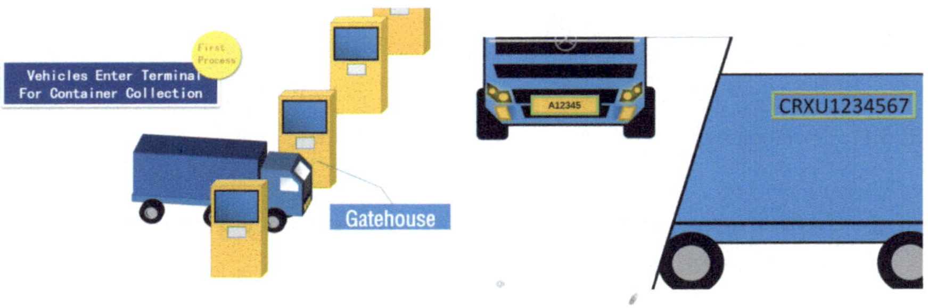

Fig. 5.18 Manual information collection at the conventional gatehouse

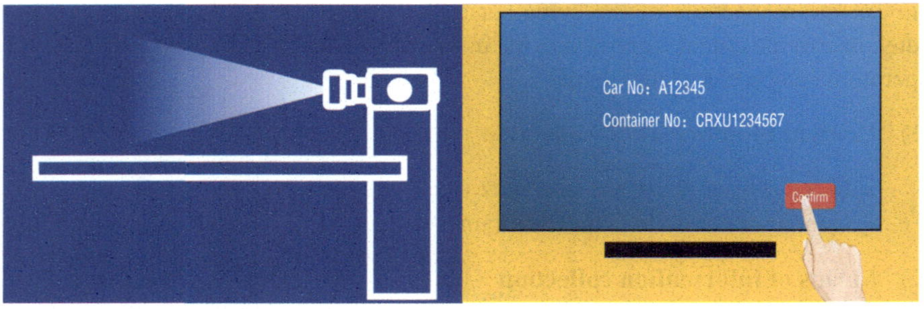

Fig. 5.19 Information collection at the intelligent gatehouse

5.5.3 Location Selection Algorithm

No matter it is the conventional gatehouse or the intelligent gatehouse, the collected information will be transmitted to the remote computer server (location selection algorithm server) through the network, and then the server will arrange the specific location in the storage yard according to the input information.

(1) **The decision content of location selection algorithm**
According to the container no. of the container at the gatehouse, the location selection system automatically retrieves the pre-input information of the export container corresponding to the container no., including name of vessel, voyage code, container size, discharging port, container weight, and other important parameters in details. According to these parameters, the export container group can be distinguished, which can help find the planned area in the yard for such container group. By combining with the current container storage and facility state in the yard, the location selection system can work out the final result through certain rules in location selection algorithm, namely the container location in the yard with information of block, bay, row, and tier.

(2) **Principles of location selection algorithm**
In the location selection system, different export containers will have different conditions in location selection when entering the yard, so they follow different rules in location selection. But generally speaking, all the containers waiting for location selection should follow the principles below when entering the yard.

- Principle of the light under the heavy: for the containers of the same voyage, size, discharging port, and tonnage level, the light one should be stacked under the heavy one.
- Principle of similarity: if the heavy container is to be stacked on the light one, the light one with the similar weight to the target container is preferred.
- Principle of concentration: the containers of the same size, discharging port, and tonnage level should be stacked concentratedly at the same bay to facilitate the future stowage.
- Principle of separate stacking of different container types: the containers of different types should be stacked separately at bays of different blocks if the storage space is adequate when making the plan.
- Principle of distributing: if the container amount is relatively large for a certain group of the same voyage, size, discharging port, and tonnage level, these containers should be distributed at different bays in the block as much as possible.
- Principle of affinity: containers with the same discharging port but different tonnage levels should be stacked in adjacent rows.
- Principle of mixer: if the containers have to be mixed, the tonnage level is the first to consider, then the discharging port. The mixed stacking containers of different vessels have the lowest priority. If the containers of different tonnage levels have to be mixed, the heavy should be above the light. If the containers

Fig. 5.20 Feedback information of location selection

of different discharging ports have to be mixed, the far should be above the near.

(3) **Data processing in location selection algorithm**
This process is not only to calculate a specific location in the yard, but also to process the following data.

- Calculating a specific location in the yard for the export container as mentioned above
- Generating operation task/instruction for the yard crane
In order to effectively connect to the next process (actual container collection by the yard crane), the system will generate the task/instruction of container collection for the yard crane according to the information of container no., license plate no. of the truck, selected location in the yard, operation time at the gatehouse, and operator ID.
- Sending the selected location information to the gatehouse
Information feedback is required after the location selection. The location selection information is sent to the gatehouse, and the entry receipt is printed for the container truck driver, which indicates the block and bay location that the driver should go to (Fig. 5.20).

5.5.4 Container Collection by the Yard Crane

As the container truck driver enters the yard with the receipt, the yard crane driver also receives the container collection instruction from the terminal installed in the crane, which indicates the information of the container truck and the container carried by it as well as the location for stacking including specific block and bay. Then, the yard crane driver will stack the container in the designated row and tier according to the actual stacking situation in the yard, and confirm the final location where the container is stacked and the completion of the operation in the system (Fig. 5.21).

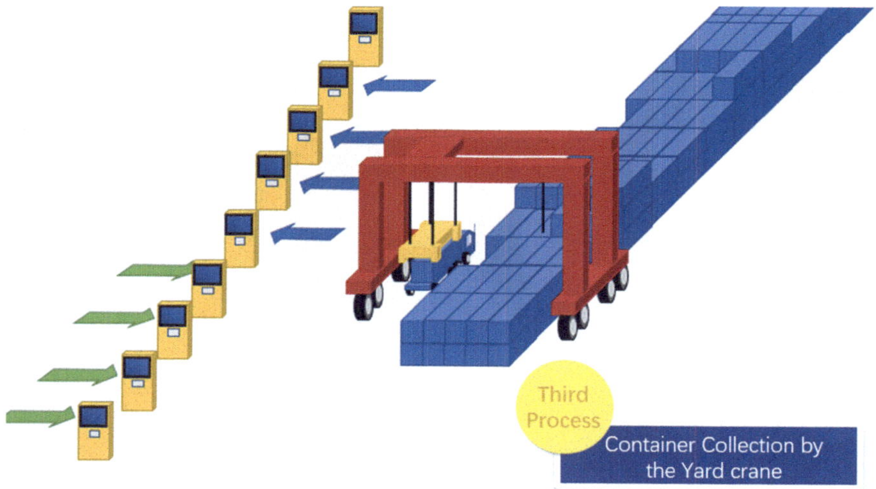

Fig. 5.21 Container collection by the yard crane

5.5.5 *Exit of the Empty Vehicle*

Since the external vehicles are not allowed to stay in the container terminal for a long time, after the completion of the yard crane operation, it is also necessary for the container truck driver to go to the exit gatehouse for vehicle elimination in the system to confirm the exit of vehicle and record the exit time. The purposes are:

(1) The terminal verifies the vehicle entry and exit information to ensure the integrity of terminal operation information.
(2) The service time of the vehicle in the terminal is recorded according to the time points of the entrance and exit at the gatehouse, which can be used as the reference for the terminal to analyze and optimize its operation process (Fig. 5.22).

5.6 A Comprehensive Case of Container Collection Operations

In this chapter, the whole process of export container collection operations is mainly introduced. In order to help the readers better understand and grasp the operation process of export container collection in the container terminal, the voyage E03 of vessel GUOTA in a container terminal is taken as an example to explain the export container collection operations and the corresponding steps in detail are in

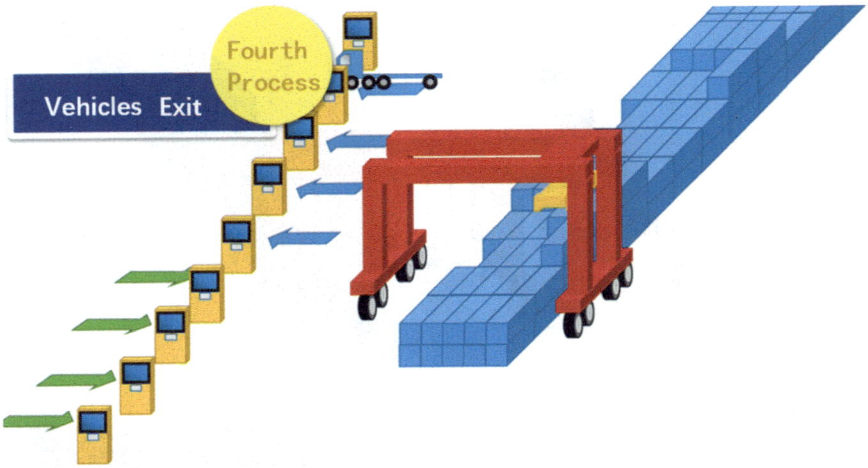

Fig. 5.22 Exit of the empty vehicle

the following section, so as to present to the readers each detailed operation and probable misoperations completely and clearly.

5.6.1 Basic Data

The basic data in the process of container collection is mainly the general information of the export containers, including the container size, discharging port, empty or full, container type, requirements of special goods, etc. Since the special containers such as reefer containers are stacked in special blocks, their container collection operation is relatively simple. In this case, the container collection process of general full containers is mainly explained. The basic data of the export containers of the vessel GUOTA, E03, are shown in Table 5.3.

Table 5.3 Amount of export full containers of the vessel GUOTA, E03

Discharging port	Amount of 20 ft containers	Amount of 40 ft containers	Amount of 45 ft containers
CNEVG	8	13	0
CNFES	0	7	0
CNLT8	5	26	0
CNRSH	0	9	0
CNSHA	237	20	0
CNYM8	24	8	0

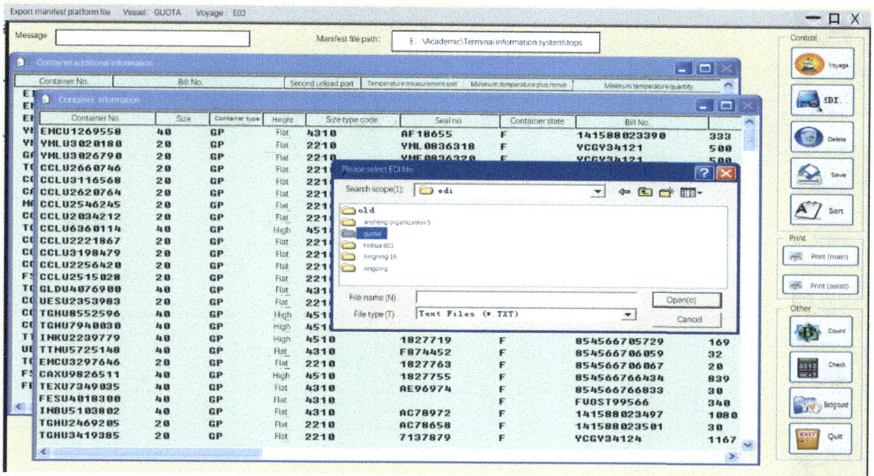

Fig. 5.23 EDI file reception of the export manifest

5.6.2 Input of the Export Manifest

In the export manifest input module of the TOS, EDI file of the export manifest of the vessel GUOTA, E03, is received and converted into the TOS system. Then, through the interface of manual input of export manifest, the export manifest data are verified, conducted statistics, and modified appropriately. Figure 5.23 shows the EDI file reception of the export manifest.

If it is necessary to manually add or modify records, select the voyage E03 in the export manifest input module to enter the interface for manual input of export manifest. As shown in Fig. 5.24, a record of export container no. "BHCU3039777" is manually added. Special attention should be paid to the container no. verification in manual input, to avoid input of illegal container no.

5.6.3 Port and Tonnage Grouping for the Export Containers

For the convenience of stacking and loading, when making the yard plan of export containers for each export voyage, the export containers of the voyage must be sorted according to the discharging port and tonnage. In the corresponding discharging port and tonnage grouping module for export containers, select the vessel GUOTA, E03 to enter the port and tonnage grouping interface. The system will automatically generate a detailed list of discharging ports and sizes of this voyage according to the export container sizes and the discharging ports of the voyage. One record is of the same size and discharging port. In each record, the operator designates the container weight range for each tonnage level. As shown in Fig. 5.25, the container weight range of

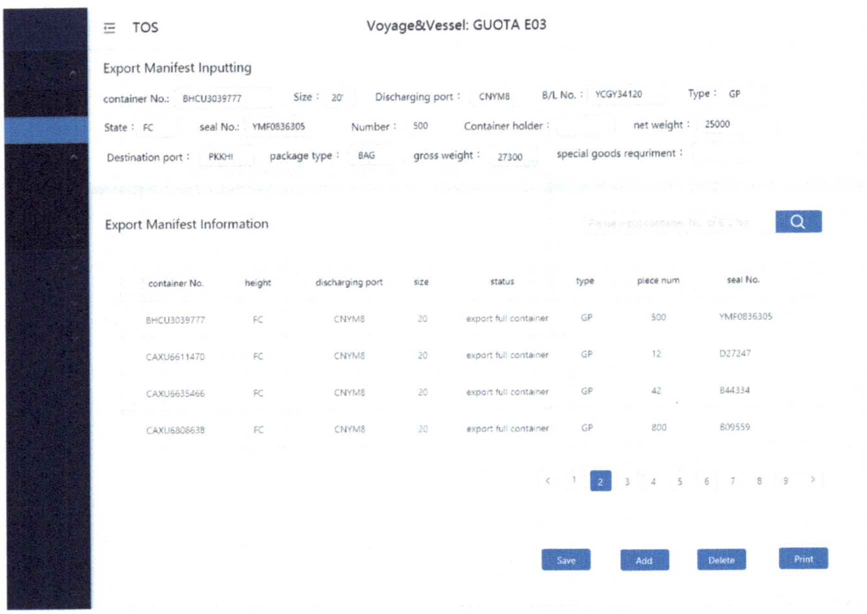

Fig. 5.24 Manual input of the export manifest

Fig. 5.25 Operation of port and tonnage grouping

tonnage level 1 is 0–10 t for the container group of size 20 ft and discharging port CNEVG, the container weight range of tonnage level 2 is 10–20 t, and the container weight range of tonnage level 3 is 20–40 t. Tonnage 99 is the default maximum container weight for the highest level.

To modify the maximum container weight for a tonnage level (except the highest level), a new value can be directly input in the corresponding position. If the level

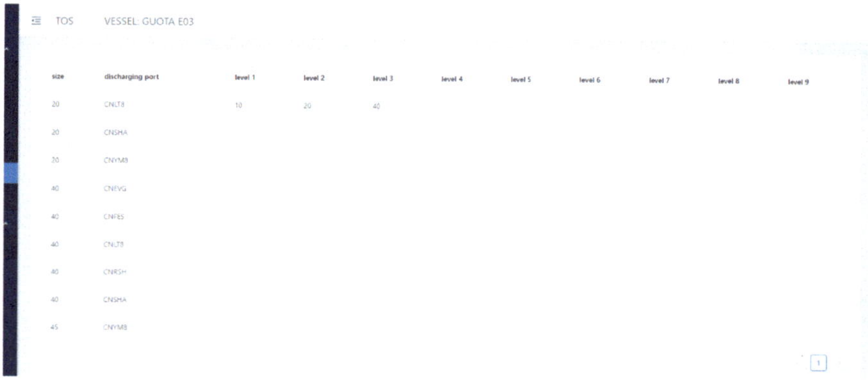

Fig. 5.26 Results of port and tonnage grouping

definition is too fine, a level can be deleted. When the grouping of all records is completed, the results of discharging port and tonnage grouping for all records are displayed, as shown in Fig. 5.26.

5.6.4 Yard Plan for Export Containers

The yard plan for export containers includes export container grouping, block plan, ay plan or row plan in the block.

1. **Export container grouping**

According to the discharging port, tonnage level, size, and other information of the vessel GUOTA, E03, the export full containers are sorted into 32 groups, and the specific information is shown in the following table. In container grouping, the size of export containers shall be selected first, then the discharging port, and then the tonnage level and container type. The grouping interface is shown in Fig. 5.27 (Table 5.4).

2. **Block plan**

When selecting the blocks, the no. of blocks should be determined first. Two vessel loading operation lines are planned for the vessel GUOTA, E03, and the ratio of quay cranes to yard cranes is usually 1:2. According to the principle of concurrent operation of multiple yard cranes, the planned number of blocks should be four. Then, the specific blocks should be selected and container amount should be designated to the corresponding blocks. When determining the specific block, the distance between the berth and the block should be minimized as much as possible and the container amount in each block should be balanced. Meanwhile, the principle of matching the container amount and the number of bays in the block and the principle of no-time crossing of vessel loading and unloading in the block should be satisfied. As

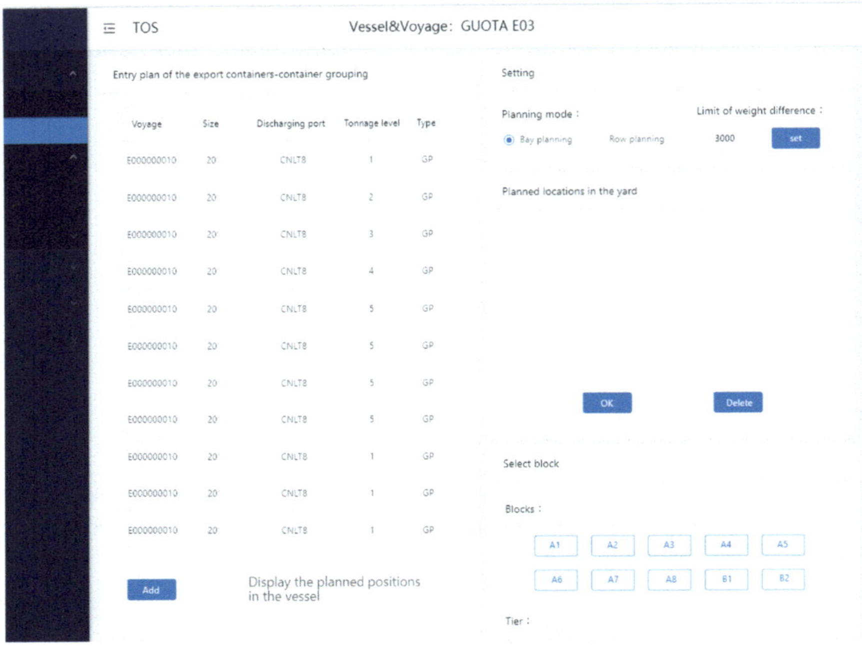

Fig. 5.27 Interface of export container grouping

shown in Fig. 5.28, the anticipated berth of the vessel GUOTA is close to blocks A1, B1, A2, and B2, and the number of unoccupied bays in the four blocks is adequate enough to ensure uniform container amount distribution. The loading time of the vessel GUOTA does not cross with loading or unloading time of other vessels in the four blocks, so the blocks selected for the block plan of the vessel are blocks A1, B1, A2, and B2. In order to satisfy the principle of minimizing the container amount difference among the blocks, 20 and 40 ft containers should be distributed to each block evenly. There are 284 20 ft containers for the vessel GUOTA, E03, so 71 20 ft containers are designated, respectively, to the blocks A1, B1, A2, and B2. There are 83 40 ft containers for this voyage, so the container amount designated to blocks A1, B1, A2, and B2 is 21, 21, 20, and 21, respectively.

3. **Planning accuracy control**

There are two ways for planning accuracy control, namely bay planning and row planning. The contents of planning accuracy control include the bay selection within the block, the distribution of loading ports in different blocks, and the designation of container groups by row, etc., which will be described in the following case.

(1) Bay selection in the block
When selecting bays in the block, the number of bays should be determined first. Since in this actual case, the row planning is adopted, that is, export containers

Table 5.4 Export full containers grouping of the vessel GUOTA, E03

Group No.	Size	Discharging port	Tonnage level	Container amount
1	20	CNEVG	1	2
2	20	CNEVG	2	3
3	20	CNEVG	3	3
4	40	CNEVG	1	4
5	40	CNEVG	2	6
6	40	CNEVG	3	3
7	20	CNLT8	1	2
8	20	CNLT8	2	2
9	20	CNLT8	3	1
10	40	CNLT8	1	2
11	40	CNLT8	2	4
12	40	CNLT8	3	2
13	20	CNSHA	1	82
14	20	CNSHA	2	65
15	20	CNSHA	3	52
16	20	CNSHA	4	30
17	20	CNSHA	5	8
18	40	CNSHA	1	8
19	40	CNSHA	2	7
20	40	CNSHA	3	5
21	20	CNYM8	1	12
22	20	CNYM8	2	7
23	20	CNYM8	3	5
24	40	CNYM8	1	4
25	40	CNYM8	2	3
26	40	CNYM8	3	1
27	40	CNRSH	1	4
28	40	CNRSH	2	3
29	40	CNRSH	3	2
30	40	CNFES	1	5
31	40	CNFES	2	3
32	40	CNFES	3	2

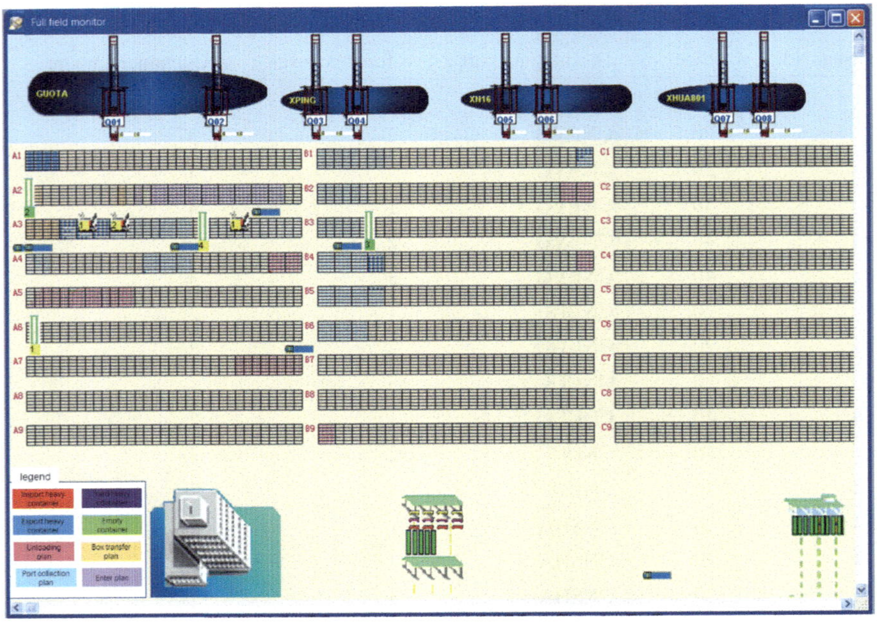

Fig. 5.28 Layout of the container terminal

of the same group are planned in the same row. The equation for calculating the number of rows required for each export container group is listed below.

$$\mathrm{RNum} = \mathrm{ceil}\left(\frac{\mathrm{CTNNum}}{\mathrm{Tier}}\right)$$

where Rnum is the number of rows required, CTNNum is the container amount of the group, and Tier is maximum stacking tier allowed.

The equation for calculating the number of bays required for each export container group is listed below.

$$\mathrm{BayNum} = \mathrm{ceil}\left(\frac{\mathrm{RNum}}{\mathrm{Row}}\right)$$

where BayNum is the number of bays required, RNum is the total number of rows required, and Row is the number of rows allowed to stack at one bay.

According to the container amount of each container group and the allowable stacking tiers (in this case are four tiers) for the export container grouping of the vessel GUOTA, E03 (Table 5.3), it can be calculated that the number of rows required for 20 ft and 40 ft export containers is 74 and 23, respectively, which are 13 and 4 for bays, respectively. Since a 40 ft container holds two adjacent bays, all the 40 ft containers need eight small bays. Distributed evenly

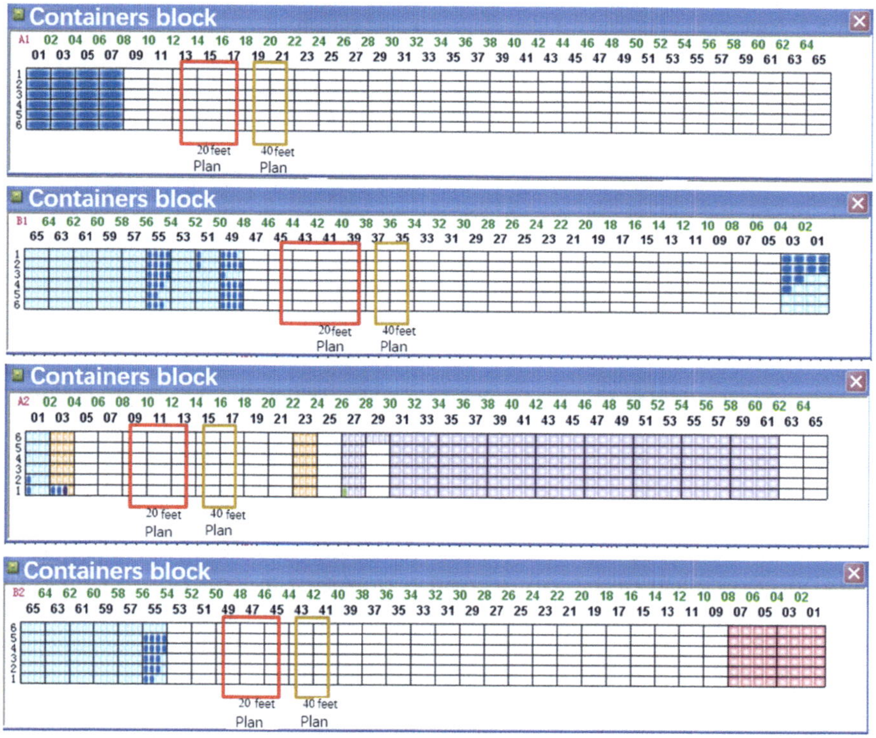

Fig. 5.29 Bay selection in the blocks

in the planned blocks A1, B1, A2, and B2, the number of bays required for 20 ft containers in each block is 3, while number of bays required for 40 ft containers in each block is one big bay, namely two adjacent bays.

According to the principle of concentration for bay selection, bays selected in the planned blocks are as shown in Fig. 5.29. The selected bays for 20 ft containers in block A1 are 13, 15, and 17, while the selected bay for 40 ft containers is 20; the bays for 20 ft containers in block B1 are 39, 41, 43, 45, while the bay for 40 ft containers is 36; the bays for 20 ft containers in block A2 are 9, 11, 13, while the bay for 40 ft containers is 16; the bays for 20 ft containers in block B2 are 45, 47, 49, while the bay for 40 ft containers is 42.

(2) Distributing containers of the same discharging port

When determining whether to distribute containers of the same discharging port in multiple blocks, the principle of concurrent operation of multiple yard cranes should be mainly considered. First, the number of rows required for each discharging port group is calculated according to the export container grouping table. If the number of rows is small, this group is planned in one block; otherwise, the containers of this group should be distributed in multiple

Table 5.5 Number of rows required for each discharging port group

Discharging port	Number of rows required for 20 ft containers	Number of rows required for 40 ft containers
CNEVG	3	4
CNFES	0	4
CNLT8	3	3
CNRSH	0	3
CNSHA	61	6
CNYM8	7	3

blocks. The number of rows required for each discharging port group is shown in Table 5.5.

According to the number of rows required for each discharging port group, for the vessel GUOTA, E03, the containers with the discharging port of CNEVG are planned to blocks A1 and A2; the containers with the discharging port of CNFES are planned to block A1; the containers with the discharging port of CNLT8 are planned to block A2; the containers with the discharging port of CNRSH are planned to block B1; the containers with the discharging port of CNYM8 are planned to block B2; and the containers with the discharging port of CNSHA are planned to blocks A1, A2, B1, and B2.

(3) Row planning for the container group

When the container groups are planned in rows, the container groups shall be sorted first. Generally, the discharging ports shall be sorted from far to near, container sizes from 20 to 40, and tonnage levels from low to high, and the container group shall be numbered according to the sequence. The farther the discharging port, the higher the tonnage level, and the smaller the group no. will be. The order of discharging ports of the vessel GUOTA, E03 from near to far is shown in Table 5.6, and the order of each container group is shown in Table 5.7. After sorting the container groups, it is necessary to plan each container group into specific rows. At this moment, the main factor to be considered is to reduce the shifting distance of the yard crane during vessel loading. Therefore, the container groups should be planned to rows according to the principle of stacking the containers of the same tonnage level and discharging port in the same rows

Table 5.6 Order of discharging ports of the vessel GUOTA, E03

Discharging port	Port order no.
CNSHA	1
CNEVG	2
CNFES	3
CNLT8	4
CNRSH	5
CNYM8	6

Table 5.7 Order of export container group of the vessel GUOTA, E03

No.	Group	Size	Discharging port	Tonnage level
1	23	20	CNYM8	3
2	22	20	CNYM8	2
3	21	20	CNYM8	1
4	26	40	CNYM8	3
5	25	40	CNYM8	2
6	24	40	CNYM8	1
7	29	40	CNRSH	3
8	28	40	CNRSH	2
9	27	40	CNRSH	1
10	9	20	CNLT8	3
11	8	20	CNLT8	2
12	7	20	CNLT8	1
13	12	40	CNLT8	3
14	11	40	CNLT8	2
15	10	40	CNLT8	1
16	32	40	CNFES	3
17	31	40	CNFES	2
18	30	40	CNFES	1
19	3	20	CNEVG	3
20	2	20	CNEVG	2
21	1	20	CNEVG	1
22	6	40	CNEVG	3
23	5	40	CNEVG	2
24	4	40	CNEVG	1
25	17	20	CNSHA	5
26	16	20	CNSHA	4
27	15	20	CNSHA	3
28	14	20	CNSHA	2
29	13	20	CNSHA	1
30	20	40	CNSHA	3
31	19	40	CNSHA	2
32	18	40	CNSHA	1

and the principle of stacking the containers of farther discharging port and higher tonnage level in the rows near to vehicle lanes.

When planning container groups to rows, the number of rows required for each container group should be first calculated. For example, number of rows required for container group 1 in Table 5.7 is 2. According to the discharging port grouping, this container group is planned to block B2, and the bays planned for the 20 ft containers of the vessel GUOTA, E03, in block B2 are 45, 47, and 49; the rows planned for this container group are row 5 and row 6 of bay 45, which are referred to as "B2456" and "B2455" in the row plan as shown in Fig. 5.30.

The planned rows at the profile view of the bay in the block are shown in Fig. 5.31.

If bay planning is adopted, the whole bay will be planned for the container group, as shown in Fig. 5.32.

The results of row planning for all the container groups of the vessel GUOTA, E03, are shown in Table 5.8.

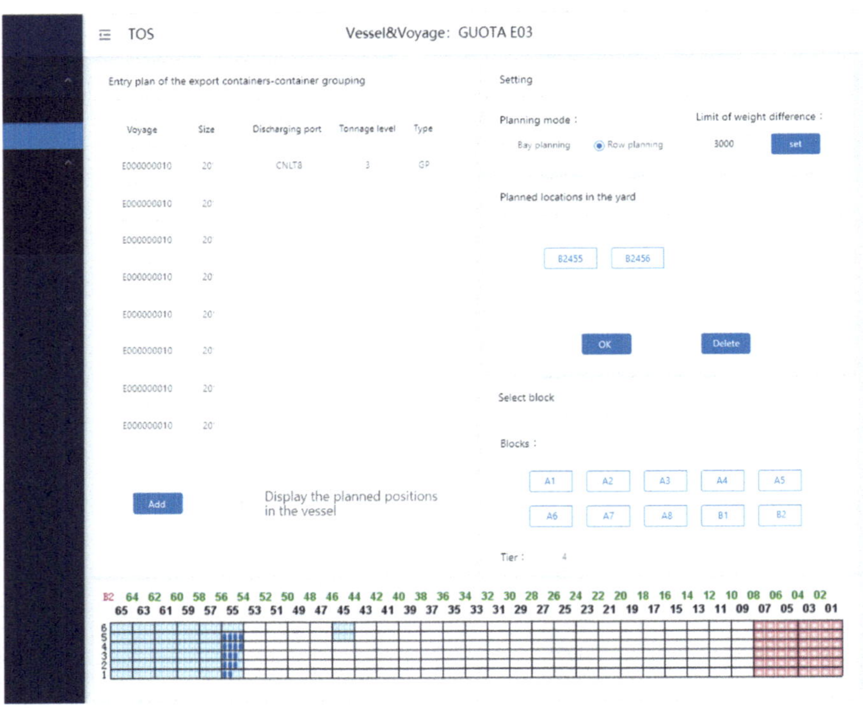

Fig. 5.30 Row planning for the container group

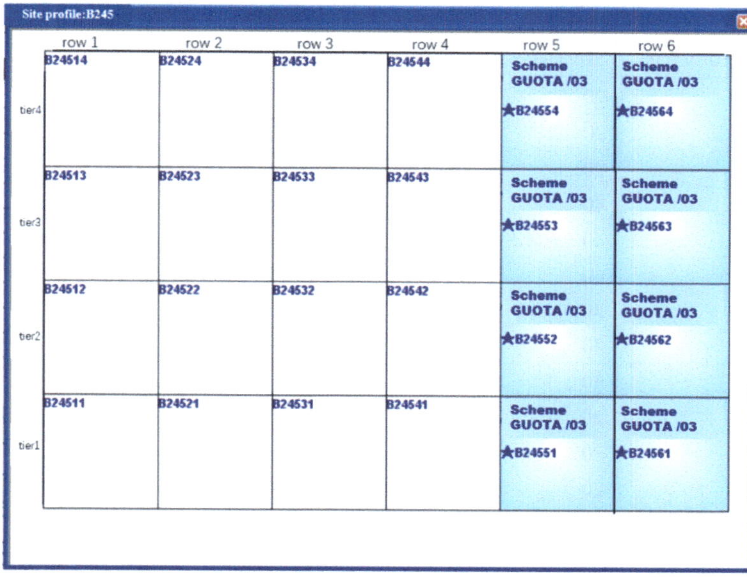

Fig. 5.31 Row plan at the profile view

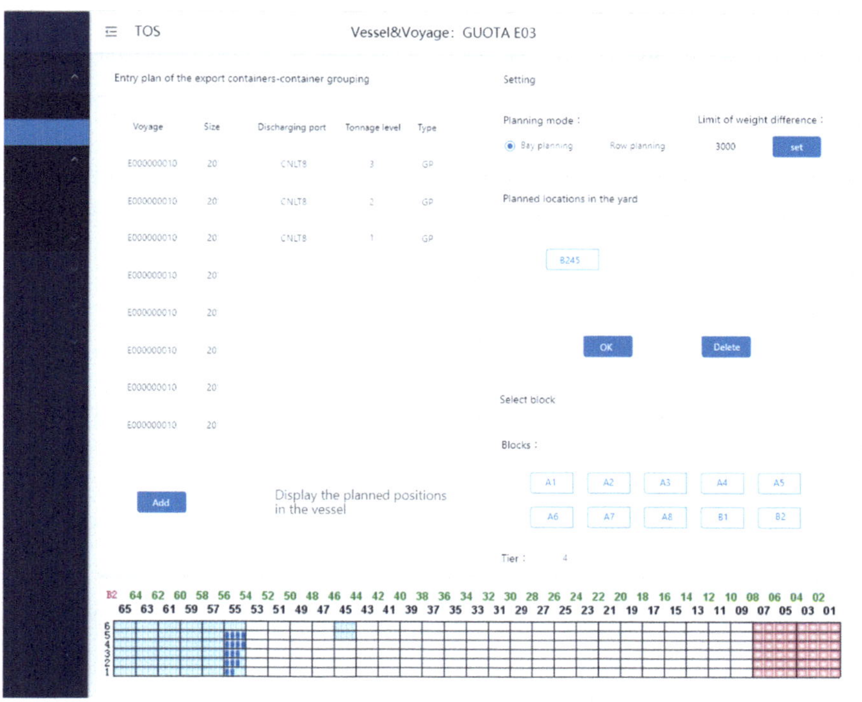

Fig. 5.32 Bay planning for the container group

Table 5.8 Results of yard planning for the export containers of the vessel GUOTA, E03

Container group no.	Size (ft)	Result of row planning
1	20	B2456–B2455
2	20	B2454–B2453
3	20	B2452–B2451, B2476
4	40	B2426
5	40	B2425
6	40	B2424
7	40	B1366
8	40	B1365
9	40	B1364
10	20	A2096
11	20	A2095
12	20	A2094
13	40	A2166
14	40	A2165
15	40	A2164
16	40	A1206
17	40	A1205
18	40	A1204–A1203
19	20	A1136
20	20	A1135
21	20	A1134
22	40	A1202
23	40	A1201, A2163
24	40	A2162
25	20	A1133–A1132
26	20	A1131, A1156–A1151, A1176
27	20	A1175–A1171, A2093–A2091, A2116–A2112
28	20	A2111, A2136–A2131, B1396–B1391, B1416–B1413
29	20	B1412–B1411, B1436–B1431, B1456–B1455, B2475–B2471, B2496–B2491
30	40	B1363–B1362
31	40	B1361, B2423
32	40	A2163–A2162

5.6.5 *Location Selection at Gatehouse for Export Containers*

After the yard plan is made for the export containers, the next step is the actual entry of the containers. In order to facilitate location guideline after the container enters the terminal, it is necessary to select the container location at the gatehouse before entering the terminal. At present, many advanced container terminals have applied RFID to automatically identify the vehicle and container information, but for some terminals, the relevant information still has to be input manually by the tally clerk at the gatehouse.

The entry plan is made according to different container groups. The purpose is that after the "automatic location selection" is selected, the system will select location for the container according to the specified rules. For the container terminal with RFID, if the check box of "card reading" is selected, the information will be read automatically when the vehicle enters. The manual operation mode is mainly introduced in this section, which is divided into two steps: vehicle selection and data input of the export container.

1. **Vehicle selection**

In the module of automatic location selection, it is necessary to firstly select the vehicle according to the vehicle information in the database. If it is the first visit for the vehicle, the relevant information of the vehicle should be input in the module of "vehicle selection." Then the system will automatically prompt whether to input the vehicle information. The operation after manual input will be the same as above. The vehicle selection operation is shown in Fig. 5.33. It is the first visit for the container truck "J03568," which is carrying an export container of the vessel GUOTA, E03, so the vehicle information should be input manually.

2. **Information input of the export container**

There are two different ways to input export container information. The first is that after selecting the vehicle, the container no. is directly input, and an empty record will be automatically added in the window of "storage information of export containers." When the clerk selects the voyage to which the container belongs, the system will automatically generate the corresponding information according to the previously entered export manifest. For individual default information, it should be input according to paper documents at hand. As shown in Fig. 5.34, the export container carried by the container truck "J03568" is with the container "BHCU3039777." After the container no. is entered, the detailed information of the container, including voyage, size, destination port, bill of lading no., type, weight, status, discharging port, and other information will be automatically generated.

3. **Location selection at entrance**

The location selection at entrance shall be based on the container group to which the export container belongs and the planned rows for the container group to select

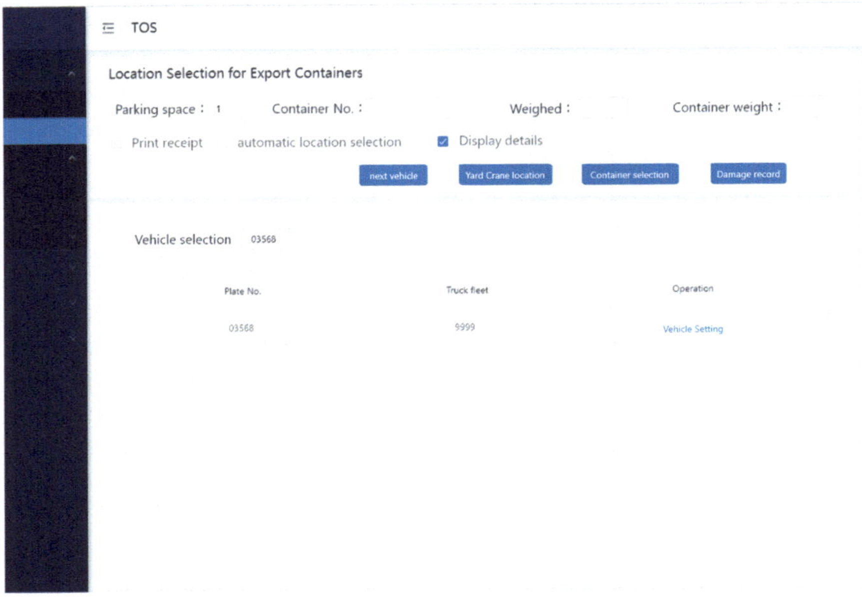

Fig. 5.33 Vehicle selection interface

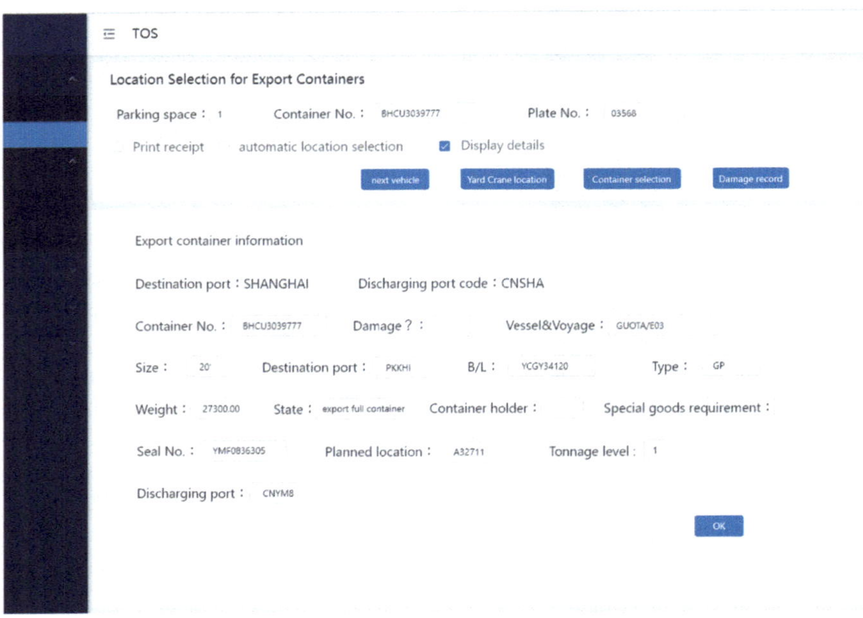

Fig. 5.34 Information input of the export container at entrance

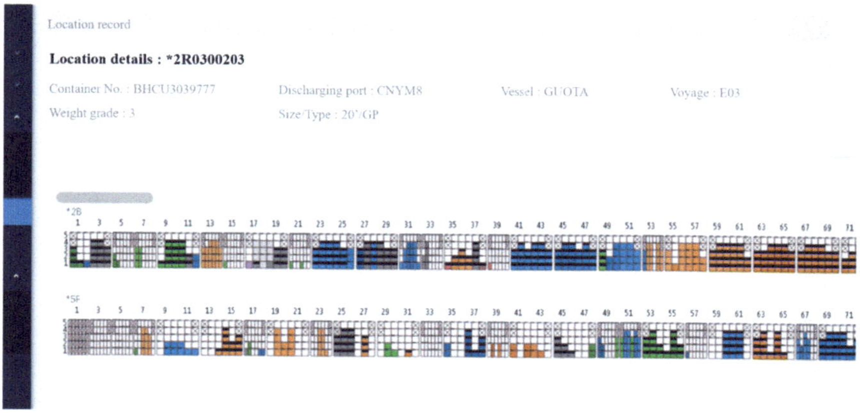

Fig. 5.35 Location selection of export container

a reasonable location. The row planning is adopted in this case; therefore, the principles of row planning should be followed to select location. The export container "BHCU3039777" has the attribute of "size: 20; discharging port: CNYM8; tonnage level: 3." According to the sorting table of export containers, the container group to which the container belongs is sorted as 1, and its planned rows are B2456–B2455. Therefore, the selected location for the container is B24561, namely in block B2, bay 45, row 6, and tier 1. The location selection interface is shown in Fig. 5.35.

5.6.6 Container Collection in the Yard

After the location selection is finished and the container passes through the gatehouse, it will enter the storage yard designated by the system according to automatic location selection. If the yard control scheduler has arranged a yard crane for the block, the container operation in the yard can continue. The function of the operation is to save the actual stacking location of the export container into the database, and feed it to the exit gatehouse, so as to release the vehicle when the empty truck exits and calculate the duration of the external container truck in the yard. When collecting the container in the yard, the yard crane driver should select the collection task, and meanwhile, the stacking location of the container can also be adjusted. The collection of the export container "BHCU3039777" carried by the container truck "J03568" is shown in Fig. 5.36.

When the container collection in the yard is completed, the actual arrival of this container will be displayed in the yard information, as shown in Fig. 5.37.

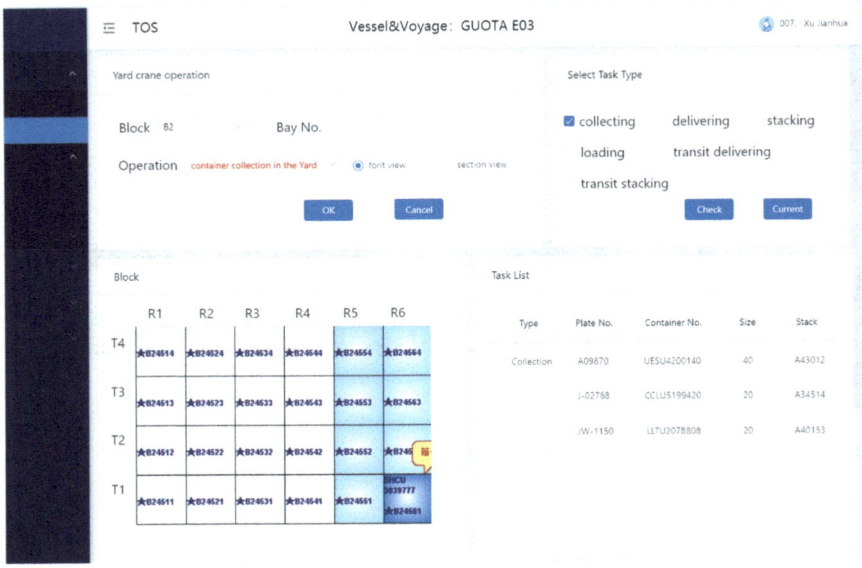

Fig. 5.36 Container collection in the yard

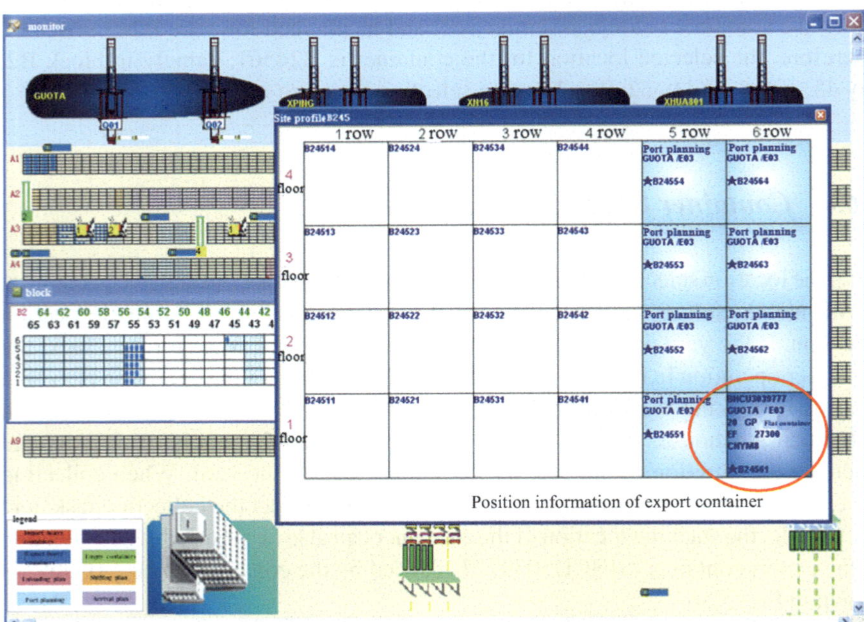

Fig. 5.37 Export container information indicating its actual arrival

5.7 Intelligent Container Collection System

5.7.1 Definition of Intelligent Container Collection

Intelligent container collection is based on the traditional container collection method to construct probability graph model according to the two-way historical data of container stowage and container entry, for representing, reasoning, learning, and predicting container entry state of the export containers, and then dynamically designate location in the yard by computer for the container entering the terminal, and dynamically determine the container collection mode by calculation according to the port and tonnage grouping as well as the yard plan in the background. The "unplanned" feature makes the decision-making of location selection for container collection more real time and dynamic, which can significantly improve the space utilization rate of resources, optimize the container stacking state, and improve the operating efficiency by refining the operating technique mode and developing targeted container collection rules according to the different technique modes of vessel loading for different voyages.

5.7.2 Major Features

- No manual division of the container groups (the port and tonnage grouping) is required.
- No manual preparation of yard plan for export containers is required.
- Container location can be dynamically selected, and plan be dynamically adjusted.
- Based on the historical data of container entry and container stowage, the decision is made through the real-time dynamic rule pool without yard planning.

5.7.3 Comparison with Traditional Mode

The main target of the intelligent container collection is to guide operation and improve operating efficiency in the terminal. Reasonable utilization of historical data and stowage demand is the key factor to achieve this goal, so as to realize data driven. For the traditional two-staged mode, which has been adopted by most container terminals at present, the yard plan is made by the planner in advance and the specific container location is then determined by computer. The traditional mode has the following disadvantages.

(1) The quality of yard planning depends on the skill level of the planner.
(2) Yard planning only relies on human experience referring to a few historical data, and cannot discover more detailed rules from a large number of historical data of container entry to assist decision-making.
(3) The planner does not consider the historical stowage plan of the voyage when making the plan and conducts no targeted yard planning (by considering the vessel structure, the container amount of each container group, and different handling techniques).
(4) The static planning requires the planner to constantly update the yard plan according to the actual situation; otherwise, it may lead to the situation of an error that no location is selected for the container.

5.7.4 Significance of Intelligent Container Collection

5.7.4.1 Application Significance

The container stacking status in the yard will directly affect the efficiency of the four operational sections of "vessel loading, vessel unloading, container collection, and container delivery" in the terminal. How to select a suitable location for the container is the most important problem to improve the operating efficiency of the terminal. Therefore, it is the most critical technology in the process of intelligent transformation from the traditional terminal of manual operation to realize intelligent location selection for the export containers.

At present, the intelligent location selection for the container terminal is still at the initial stage, and most of the location selection systems have only reduced the work intensity of employees, which has barely improved the operating efficiency of the terminals. The intelligent location selection system for the next generation of container terminal should aim at guiding operations and improving terminal operating efficiency. The rational utilization of historical data and stowage requirements is the key factor to achieve this goal. The significance of realizing the next generation of data-driven intelligent location selection system for the container terminal lies in the following aspects.

(1) The value of the historical data will be explored to guide the decision-making of location selection.
In the era of data explosion, the intelligent location selection system for the next generation of container terminal can make full use of the historical data accumulated for years, by mining potential objective rules as prior knowledge from massive historical data through analysis, so as to provide more reliable and detailed reference for the decision-making of intelligent location selection.
(2) A pulling strategy of container collection will be established for the purpose of stowage plan.

Combined with the intelligent stowage system, the intelligent location selection system is pulled to make decisions based on the requirements of stowage plan for container stacking in the yard so as to facilitate the stowage plan and vessel loading. The final purpose of intelligent location selection of export containers is to coordinate with the stowage plan and actual container loading. Therefore, the requirements of stowage are of great significance to the guidance of intelligent location decision.

(3) The dynamic planning mode will improve the utilization of yard resources.
One of the advantages of the intelligent location selection system for the next generation of container terminal is that the manual yard planning is not necessary. All location selection decisions are dynamically made by the system at the moment when the container enters the terminal through the gatehouse. These decisions are all referred to historical data, and can analyze their location selection rules in a more detailed way without the support of yard planning. Meanwhile, the phenomenon that yard planning takes up yard resources in advance can be avoided, so as to improve the utilization rate of yard resources.

(4) The process modes will be refined.
According to the different vessel loading technique modes of different voyages, the targeted location selection rules have been developed. In recent years, the continuous emergence of ultra-large vessels has brought more challenges to the operations of the terminal. In order to better serve the ultra-large vessels, the operators of the terminal have also put forward many new handling techniques, such as dual cycling, reverse stowage in the cabin, and centralized storage for ultra-light containers. These new handling techniques all put forward new requirements for container stacking. By refining the operation technique modes, the configurable location selection rules pool can be realized to make the results more consistent with the actual vessel loading requirements and improve the vessel loading efficiency indirectly.

(5) The work intensity of employees will be lowered.
Another great advantage of the intelligent location selection is that it eliminates the manual yard planning, port and tonnage grouping, and location supplement of containers with no location assigned, etc. These operations are limited by the human experience and the working state of employees. The human experience will be solidified through intelligent calculation, and the computer will automatically take over the related work. Terminal employees can be liberated from the trivial and complicated work, and have more time to find problems, summarize the rules, and further put forward the optimization plan to help the intelligent system to further upgrade the location selection results.

It can be seen that the intelligent location selection system for the next generation of container terminal is one of the key technologies to realize the intelligent transformation of the traditional container terminal, which is of great significance for the improvement of operating efficiency and the embodiment of the intelligent level.

5.7.4.2 Social Benefits

(1) The research of intelligent container collection is helpful to promote the integration of intelligence and industrialization.
(2) It will help speed up the intelligent transformation for domestic traditional container terminals. The cutting-edge artificial intelligence technology will be applied to optimize the decision-making problems of location selection of export containers in the terminal, and continuously optimize the general stacking state of the yard, so as to improve the overall efficiency of the terminal, greatly reduce labor cost and operation cost of the port, showing the advantages of intelligence, reliability, stability, equipment utilization and capacity, operating costs, and so on.
(3) It is conducive to promoting the industrial upgrading of the high-end manufacturing industry.
(4) It is beneficial to improve the business development capacity and market competitiveness of large port terminals.

5.7.5 Development Status and Case Study

5.7.5.1 Development Status

At present, most container terminals have adopted two main modes for location selection: two-staged mode and no-plan mode. The features of the two decision-making modes for location selection are shown in Table 5.9.

The two-staged mode is adopted by most container terminals. In this mode, the yard plan is made by the planner in advance, and then the specific container location is determined by the computer.

At present, the no-plan mode is only adopted by a few terminals, such as the SMCT. In this mode, there is no need for a planner to make the yard plan, and the location selection decision is directly made in real time by the computer when the container enters the terminal through the gatehouse. The advantage of this mode lies

Table 5.9 Comparison of two decision-making modes for location selection

No.	Description	Adoption rate	Timing for decision-making	Decision-making subject
1	Two-staged mode	Majority	Pre-entry planning; Location selection at the gatehouse (in)	Stage 1: human decision Stage 2: computer decision
2	No-plan mode	Minority	At the gatehouse (in)	Computer decision

in real-time dynamic decision, with which the results of location selection will not be affected by human experience.

In general, no-plan mode is the inevitable trend of intelligent location selection of export containers in the future. On the basis of no-plan location selection, intelligent container collection aims to realize the next generation of intelligent location selection system by means of data driving and stowage pulling to overcome the drawbacks of the existing decision mode, and further improve the quality of location selection results, thus to optimize the container stacking state in the yard and improve the operating efficiency of the terminal.

5.7.5.2 Case Study

Some terminals in China are pioneers of intelligent reform, for example, CMICT, Ningbo, has adopted the TOS with automatic stowage function, and actively promoted the research of intelligent container collection system, which was officially put into use in CMICT in September 2018. By January 2019, the intelligent container collection system has covered 78% of the total volume of container collection in the terminal and over 110 voyages. The container reshuffle rate and facility shifting are greatly reduced. The following are the main functions of the intelligent container collection system of CMICT.

- Function of rule engine
  The rules of location selection calculation and the priority relationship among the rules can be customized by the rule engine to realize the customized container collection rule management of a specific route (Fig. 5.38).
- Function of intelligent location selection
  The system dynamically selects a reasonable location for the export container waiting for entry at the gatehouse by taking into account facility status, stacking state in the yard, container attributes, and other factors, as shown in Fig. 5.39.
- Function of dynamic planning
  Similar with the traditional location selection mode based on yard plan, the intelligent location selection system also needs a general short-time plan for the use of space resources. However, in the intelligent location selection mode, the yard plan is made by a computer according to the container amount of each discharging port automatically estimated based on the situation of the collected containers and historical data. The detail level of the yard plan is also independently controlled by the computer, as shown in Fig. 5.40.
- Function of real-time monitoring
  The dashboard screen is adopted in the system to monitor various dynamic operations of container collection and the busy level of each block in the yard, as shown in Fig. 5.41.

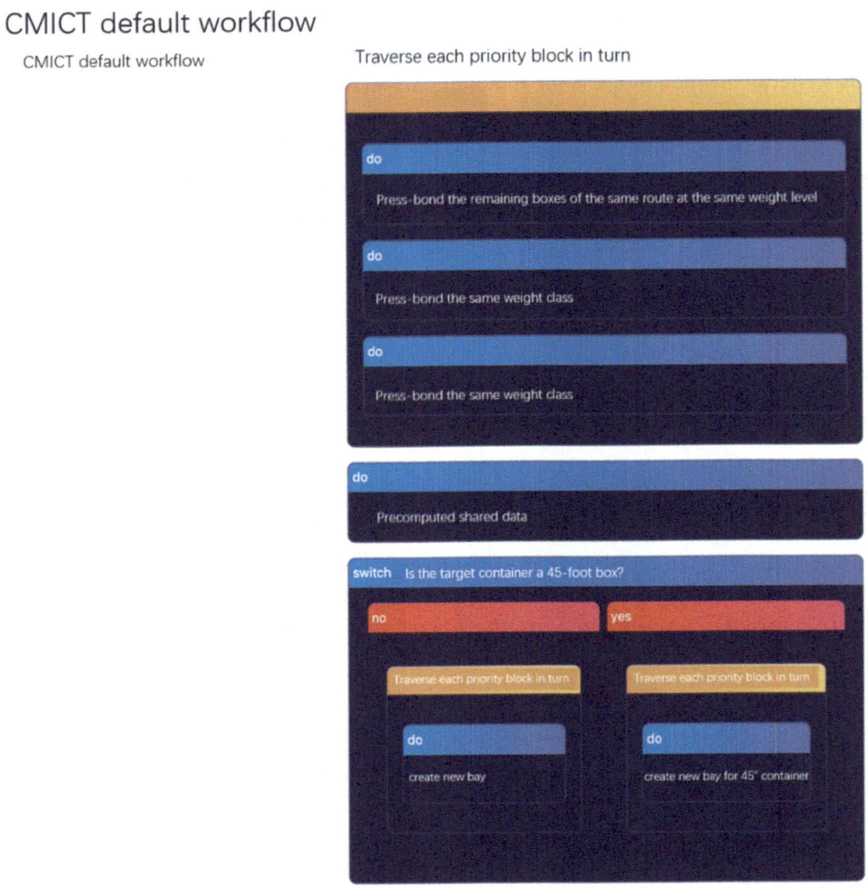

Fig. 5.38 Customerization of rule engine

- Function of autonomous learning
  The system will automatically retrieve and mine the stowage data and other historical data in the background, on one hand to improve the accuracy of container volume estimation, on the other hand to learn and adjust relevant parameters of the algorithm.

Fig. 5.39 Intelligent location selection triggered at the gatehouse

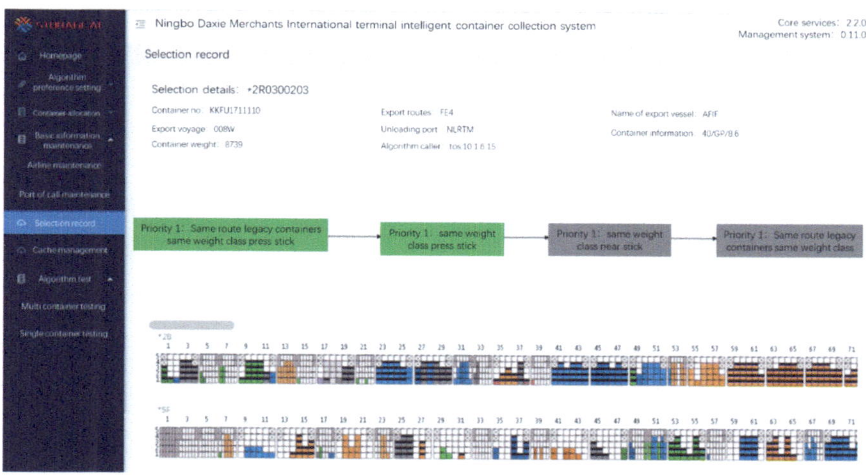

Fig. 5.40 System function of dynamic planning

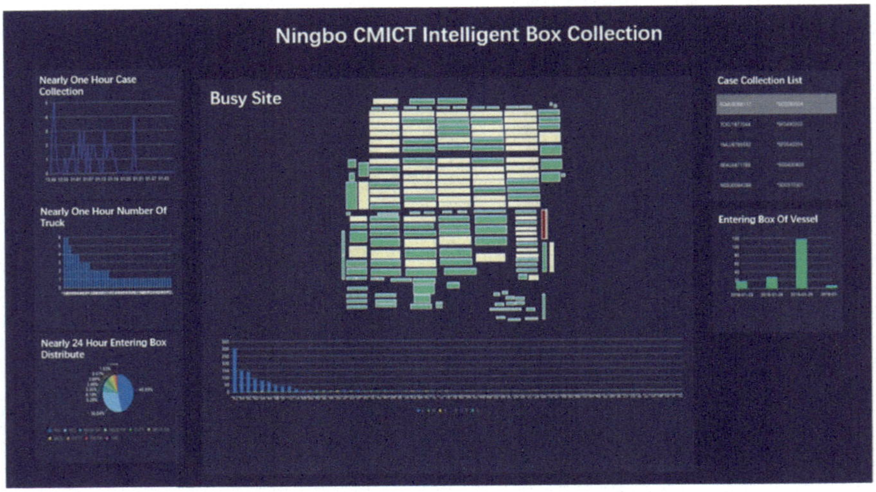

Fig. 5.41 Dashboard screen for monitoring

Chapter 6
Management of Vessel Loading Operations in the Container Terminal

6.1 Introduction of Vessel Loading Operation Process

Vessel loading is the second phase of the export operations (the first phase is the container collection). The main task of this phase is to verify, release and confirm, stow, and finally load the export containers that have been collected at the port.

As shown in Fig. 6.1, the entire export process from container collection to vessel loading is a process from distribution to concentration, which is the opposite process to import process. Similar to the vessel unloading operations, the vessel loading operations also need to go through three stages, namely ① information collecting and processing, ② resource planning and scheduling, and ③ actual vessel loading operations. After the export containers enter the yard of the terminal by container collection, the preparation and control of the vessel loading operations generally go through the following steps.

Shown in Fig. 6.2 is the decomposition of the vessel loading process. In the stage of information collecting and processing, there are two main aspects of preparation. First, the information of the collected containers is checked against the information of the containers in the export manifest, so as to verify the information of the collected containers and ensure the effectiveness and continuity of the subsequent operations. Meanwhile, the customs will release the export containers. The port needs to release the corresponding containers in the system after getting the dock receipt from the customs. After information verification and customs clearance, containers can be stowed. Stowage is equivalent to a plan, whose main task is to make one-to-one correspondence between the containers to be loaded in the yard and the designated container positions in the corresponding vessel, that is, to plan where each container will be loaded in the vessel in the future. Only after the completion of stowage can a complete distribution diagram of container amount be obtained, the vessel

The original version of this chapter was revised: Figures have been corrected. The correction to this chapter is available at https://doi.org/10.1007/978-981-15-2937-5_7

N. Zhao et al., *Digital Management of Container Terminal Operations*, https://doi.org/10.1007/978-981-15-2937-5_6

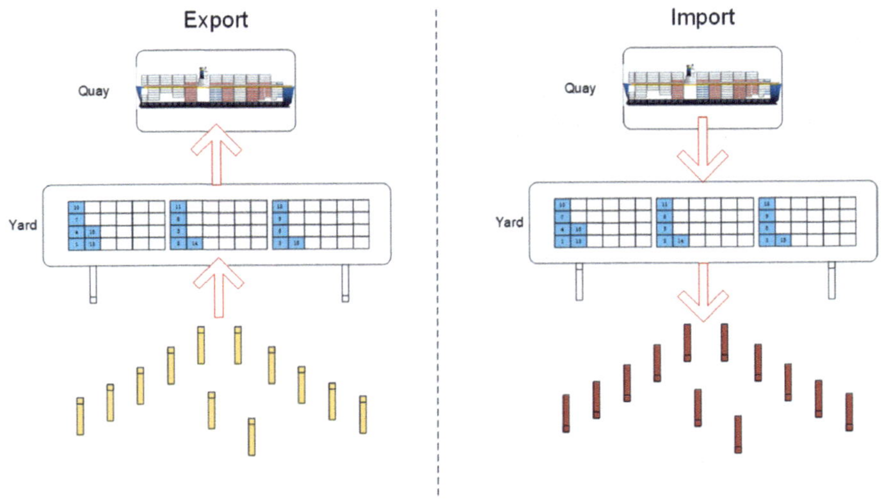

Fig. 6.1 Container collecting and delivering of export and import operations

control scheduler can carry out specific operation line dispatching (i.e., quay crane dispatching) based on this, as well as the block coverage of the yard facilities and the configuration of container trucks. After arranging the quay cranes, yard cranes, container trucks, and other operating facilities, the vessel loading will begin. In order to control the order and pace of vessel loading, vessel control scheduler should first make the operation instruction sequence for the current operating bay; that is, he should decide which containers are loaded first and what is the specific sequence. Since then, the containers with the loading sequence no. will be selected by the server successively and corresponding operation task generated for each container. Meanwhile, the empty container truck will be designated to collect the container in the yard. After receiving the task, the container truck driver will drive to a certain bay of the specific block, waiting for the container delivery of the yard crane. Once the yard crane has confirmed the operation, the container truck driver will receive a vessel loading task and drive to the quay crane specified in the task. The quay crane driver will lift the container from the container truck to the position in the vessel according to stowage plan, and the tally clerk needs to update the container status and submit it to the server through wireless handsets, by which the loading of a single container is completed.

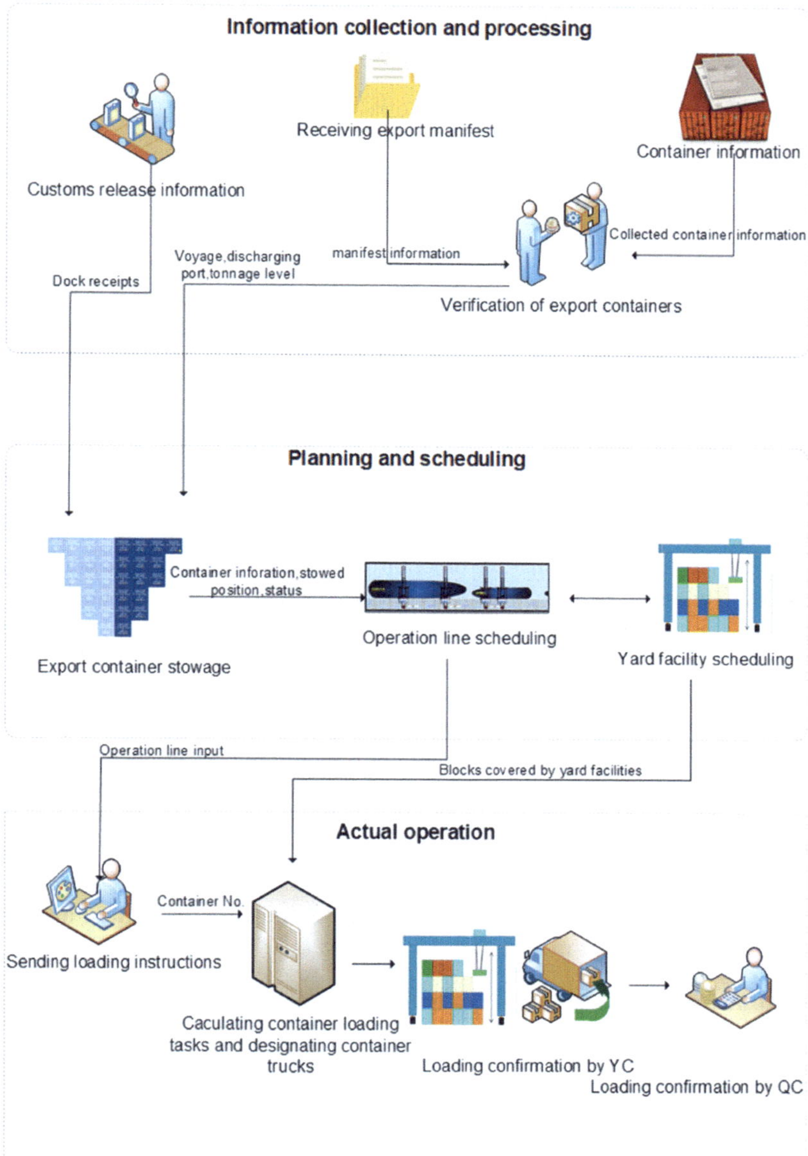

Fig. 6.2 Decomposition of vessel loading process

6.2 Information Collection and Data Preparation for Vessel Loading Operations

6.2.1 Information Verification of Export Containers

1. **Targets of information verification**

Information verification refers to the collation of information from different sources. The targets of information verification of export containers are from two aspects: (1) the information of the collected containers in the yard and (2) the information of containers in the export manifest.

(1) **Information sources of the collected containers**
The information sources of the collected containers in the yard at different ports may be different, but there are two main sources. One is the pre-input information from the carrier outside the gatehouse, which is adopted by many large terminals at seaports. Some ports directly use the export manifest as pre-input information, which is the other information source. Since the ports get the complete export manifest relatively late, they can only begin container collection operations in 2–3 days in advance. In addition, if the manifest information of the corresponding container has not been received when the vehicle loaded with the container enters the gatehouse, then the information of the container needs to be manually input by the tally clerk at the gatehouse according to the container load plan. In this mode, the information of the collected containers in the yard comes either from the manifest information or manually input by the tally clerk at the gatehouse.

(2) **Information source of the export manifest**
The export manifest comes from the shipping company's booking note, which is generated in the booking process. Therefore, it is late for the terminal to get the complete export manifest of the whole vessel.

2. **Significance of information verification of export containers**

The verification of container information from different information sources is to ultimately improve the accuracy of export container information and avoid huge property losses caused by information errors. For example, if a container is supposed to go to the USA, but Singapore is selected as the discharging port instead of the USA in information pre-input by mistake. The container can enter the port without information verification, and the mistake is difficult to be noticed. Then, the container in the process of stowage and vessel loading will be considered as a container with Singapore as the discharging port. The lesser consequence of such mistake will be container reshuffle in the vessel, while the severer consequence will be wrong container unloading in Singapore.

3. **Main tasks in information verification**

The main information of export containers for verification includes discharging port, container size, container type, special requirements of goods, status, bill of lading no., etc., among which the most important is the discharging port. In addition, it is necessary to check whether there is over-loading or short-loading of containers. For example, the container has been collected to the port, but there is no information of this container in the export manifest; that is, the container at the port is over-loaded, which needs to be confirmed by contacting the shipping company in time.

4. **Means of information verification**

The means of information verification include manual verification and automatic verification. Manual verification can be divided into single container verification and batch verification, as shown in Fig. 6.3. If it is single container verification, the information can be verified according to the container no. or the yard location. In other words, after inputting the container no. or the yard location, the corresponding container information will be displayed in the system and then checked with paper information manually. Manual batch verification is mainly to verify the discharging port. By inputting the discharging port for verification, and then according to paper

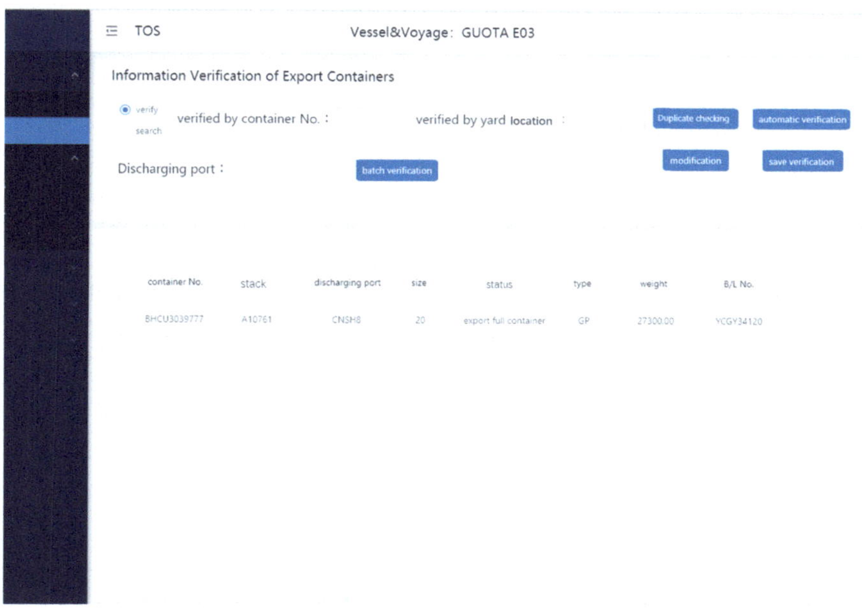

Fig. 6.3 Manual information verification of export containers

Fig. 6.4 Automatic information verification of export containers

documents, inputting the containers with this discharging port into the system successively, the system will check the attribute of discharging port of the corresponding containers.

There is an automatic verification button at the lower right corner of the figure above. By clicking it, the automatic verification module of export containers can be logged in as shown in Fig. 6.4.

As shown in the figure above, the export container information for automatic verification is divided into two groups, the upper "master file" is the information of the collected containers, while the lower "export manifest" is the information of the vessel's export manifest. By clicking the "verification" button, the system will automatically match container information in the two groups and check the information of the discharging port for each container. If there is any inconsistency, the corresponding container information will be displayed in a blue background.

6.2.2 *Dock Receipts and Customs Release of Export Containers*

6.2.2.1 Contents and Functions of Dock Receipts

1. **Definition of dock receipt**

Dock receipt is a certificate issued by the carrier to certify that the consignment has been received and responsibility for the cargo has begun. Compared with the shipping documents used in the traditional general cargo transportation, the dock receipt is a comprehensive document, which has integrated booking form, customs declaration, mate's receipt, freight notice, freight allocation receipt, etc. This is of great significance to improve the efficiency of container goods shipment.

When booking space orally or in written form, the shipper and the shipping company or its agent agrees on the carriage of goods. After the shipping agency confirms the booking, it will deliver the dock receipt to the shipper or the forwarder to fill in. The dock receipt will be issued and come into effect after the FCL or LCL cargos have been received in the carrier's commissioned terminal yard or container freight station or inland station. The shipper or the forwarder may obtain the bill of lading of the vessel already loaded or to be loaded from the shipping agency with the dock receipt.

2. **Dock receipt formats of ten, seven, and 12 copies**

The dock receipt is a special export document for container transportation, and its composition varies from many sources. Its usage at different ports and docks is also different. The number of copies could be ten, 12, and seven. Here, the ten-copied dock receipt is taken as an example to illustrate its composition (Table 6.1).

The no. 1 l and 12 copies of the standard 12-copied dock receipt are for warehouse receiving and counting. The standard seven-copied dock receipt is without the first, third, fourth, and tenth copies, but the copy for the container tally is added (the fifth copy). The numbers in brackets after each copy above indicate the serial no. of each copy in this format (the same below).

3. **Functions of dock receipt**

(1) The shipping company or shipping agency confirms the booking of shipping space and stamps showing customs declaration qualification on the dock receipt and then delivers it to the shipper or his agent, which means that the contract of carriage begins to implement.

Table 6.1 Ten-copied dock receipt

Copy no.	Direction and function	Color	Page number of the seven-copied dock receipt
The first copy	Certificate for the shipper		
The second copy	Booking form (certificate for the shipping agency)	White	(1)
The third and fourth copies	Freight notice (1) and (2)	White	
The fifth copy	Customs declaration	White	(2)
The sixth copy	Copy of mate's receipt	Pink	(3)
The seventh copy	Dock receipt (original)	Light yellow	(4)
The eighth copy	Certificate for the freight forwarder	White	(6)
The ninth and tenth copies	Freight allocation receipt (1) and (2)	White	(7)

(2) It is one of the export goods customs declaration documents.
(3) It is a proof that the carrier has received the consignment and taken responsibility for it.
(4) It is a certificate in exchange for a shipping or combined transport bill of lading.
(5) It is the reference of vessel loading/unloading, tallying and stowage organized by shipping companies and ports.
(6) It is the reference of freight settlement.
(7) If stipulated in the letter of credit, it can be used as a document to settle exchange with a bank.

4. **Dock receipt circulation process and procedures**

In the export consignment process of container goods, the dock receipt should be circulated among multiple institutions and departments. In the process of circulation, the shipper, freight forwarder, shipping agency, customs, storage yard, tally company, captain, or mate are involved. The following shows the functions of ten-copied format and seven-copied format, respectively.

(1) After filling, the shipper (or freight forwarder) shall keep one copy as certificate (the first copy) and send the second to the tenth copies (or the first to the seventh copies of the seven-copied dock receipt) to the shipping agency (booking certificate) for numbering.
(2) After numbering, the shipping agency keeps the second to the fourth copies (or the first copy of the seven-copied dock receipt) and return the fifth to the tenth copies (or the second to the seventh copies of the seven-copied dock receipt) to the forwarder after stamping for booking confirmation and customs declaration on the fifth copy (or the second copy of the seven-copied dock receipt). The forwarder keeps the eighth copy (or the sixth copy of the seven-copied dock receipt) and sends the nineth and tenth copies (or the seventh copy of the seven-copied dock receipt) to the shipper as the freight allocation receipt.
(3) The fifth to the seventh copies (or the second to the fifth copies of the seven-copied dock receipt) are for the customs declaration.
(4) After customs approval, the release seal is stamped on the fifth copy (or the second copy of the seven-copied dock receipt) and all the copies are returned to the applier or forwarder or shipper.
(5) The forwarder is responsible for filling in the fifth to the seventh copies (or the second to the fifth copies of the seven-copied dock receipt) the container no. and seal no., and delivering the container goods together with these copies to the storage yard at the specified time.
(6) After accepting the containers, the dock clerk shall fill in the number of containers actually received and the date of completion of entering, and sign for the containers and stamp the official seal of the dock on the fifth to the seventh copies (or the second to the fifth copies of the seven-copied dock receipt). The fifth copy (or the second copy of the seven-copied dock receipt) shall be retained by the dock, and the sixth (or the third and the fifth copies of the seven-copied dock receipt), i.e., the copy for the mate, shall be delivered by

the dock clerk to the external vessel tallyman in batches in 24 h before vessel loading, and the last batch must be delivered in no less than 4 h before vessel loading. The tallyman shall deliver the third copy to the mate during vessel loading and return to the forwarder the seventh copy (or the third copy of the seven-copied dock receipt), i.e., the original copy signed by both parties.

5. **Main information composition of dock receipt**

According to the circulation procedures of each copy of the dock receipt, the fifth to the seventh copies are used for customs declaration. If the containers are released by the customs, the customs release seal shall be stamped to the container load plan on the fifth copy. Therefore, the dock receipt is the most important reference for the terminal to confirm the release of export containers and make the subsequent stowage plan. In the operation system of the terminal, the dock operators need to input the dock receipt with customs release seal into the system, so as to facilitate the operators to confirm customs release.

The main contents of the dock receipt can be divided into two layers: one layer for the shipment and the other for the containers. At the first layer, a shipment simply corresponds to a shipping agreement. So the main contents of the first level include receipt no., consignor, consignee, notifier, vessel name and voyage, loading port, discharging port, receiving place, delivery place, and so on. The second layer is about the container information for carrying and transporting the goods of the shipment, mainly including the container no. (as core), lead seal no., shipping mark, quantity, weight, and volume.

Figure 6.5 shows the module of the dock receipt information input for customs release. The original information of the dock receipt does not need to be all input into the operation system, but the information that affects the subsequent release judgment must be input accurately and timely, including the most important information at the

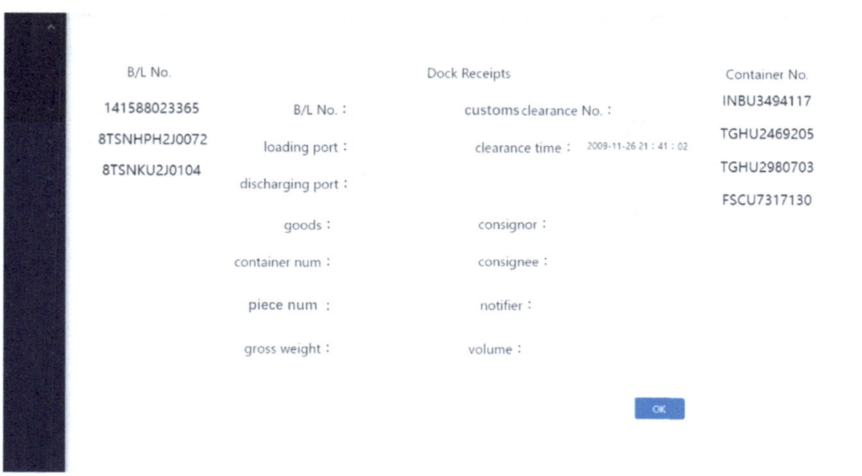

Fig. 6.5 Dock receipt input for customs release

first layer, namely the bill of lading no., customs clearance no., customs clearance time, as well as the container no. at the second layer. The bill of lading no. and container no. are the main parameters for release calculation.

6. **Paperless customs clearance**

The Chinese government has steadily promoted the electronic mode of customs clearance, namely paperless customs clearance system, on the basis of paper dock receipts.

Paperless customs clearance is by making use of the functions the Chinese electronic port and the modern customs business information system, to transform the traditional clearance procedures via verifying and auditing the written declaration form and the attached documents submitted by import and export enterprises into the customs clearance mode of paperless examination and approval directly according to the electronic data of the import and export goods declared through the enterprise network.

Since the pilot reform of paperless customs clearance was launched on August 1st, 2012, all the work has been carried out in an orderly manner, and the system is running smoothly. In order to further improve the efficiency of customs clearance, enhance the efficiency of supervision, and expand the effectiveness of the reform, the Central Administration of Customs decided to deepen the pilot work of paperless customs clearance throughout the whole country on the basis of the pilot work of paperless customs clearance at the early stage. On April 23rd, 2013, Shanghai entry-exit inspection and quarantine bureau and Shanghai customs jointly announced establishing the pilot of paperless customs clearance at Shanghai ports. The pilot program covers all export inspection commodities specified in the law and some import legal inspection commodities of some enterprises. The program will further simplify import and export declaration procedures and improve clearance efficiency at Shanghai ports.

The goal of paperless customs clearance is to change the audit reference from traditional paper documents to electronic data and realize paperless customs clearance. In the past, before customs declaration, enterprises should apply for and get the paper customs clearance form on site in the inspection and quarantine department. After the realization of paperless customs clearance, the application for paper customs clearance form has been cancelled, the import and export customs clearance of legal inspection commodities simplified, the visits of enterprises to the customs clearance reduced, the customs clearance efficiency further improved, and the customs clearance cost of enterprises reduced accordingly.

6.2.2.2 Customs Release of Export Containers

1. **Difference and connection between customs release confirmation and dock receipt release**

The customs release confirmation means confirming the release of the corresponding export containers according to the release stamp on the dock receipt and the export manifest, and only after the release can the corresponding containers be stowed. Although dock receipt is the most important reference for release, containers released with dock receipt may not be released for stowage, because dock receipt is related to B/L instead of container, while customs release confirmation is for containers, which is mainly for LCL cargos. LCL is one of the packing modes of container goods, which can be divided into FCL and LCL according to the number and mode of packing.

(1) Full container load (FCL) refers to the container which is filled with goods by the shipper and shipped as the basic unit. This situation is usually used when the shipper has enough consignment to load one or several FCL containers. Except some large shippers owning containers, the shippers generally rent certain containers from the carriers or container leasing companies. After empty containers are delivered to the factory or warehouse, under the supervision of the customs personnel, the owner of the goods will put the goods into the containers, lock and seal the containers with lead seal and deliver them to the carrier and obtain the dock receipt, and later exchange the receipt for the bill of lading or waybill.

(2) Less than container load (LCL) refers to that the carrier (or agent) accepts the shippers consignment of goods less than FCL and sorts the goods according to the nature of goods and destination. The carrier consolidates the goods with the same destination into a certain quantity and packs them into containers. Since there are goods of different owners together in one container, it is called LCL. This situation is used when the shipper's consignment is insufficient to fill the full container. The sorting, organizing, centralization, packing (unpacking), and delivery of LCL cargos are all carried out in the container terminal or inland container transfer station of the carrier. For the terminal operators, the packing of goods of different bills of lading in a container is called LCL.

According to the regulations, for LCL cargos, the dock receipt should be signed according to containers with one copy for one shipment. However, if any goods of LCL cargo are not released by the customs, the whole container shall not be loaded. For example, if a container contains the goods of five shippers, namely five shipments of goods, then the corresponding dock receipt should have five copies. The first layer of each dock receipt has the information of all shipments, while the second layer has the information of the container and the goods packed in it. If four of the five dock receipt copies have been stamped with the release seal, and the other one has not been released, the container still cannot be confirmed for customs release, so it cannot be stowed and loaded.

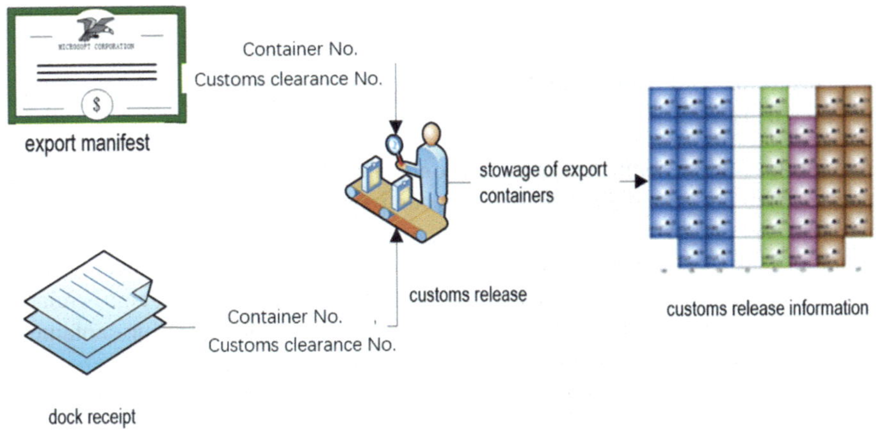

Fig. 6.6 Information sources of customs release confirmation

2. **Customs release confirmation**

In order to ensure the timely and smooth stowage of export containers, as shown in Fig. 6.6, after receiving the export manifest, it is possible to confirm the release of the containers that have received the customs release copy of the dock receipt.

Figure 6.7 is the function module for customs release confirmation. With the continuous input of dock receipts, the operator can constantly update the export containers recommended by the system for release.

In the list of export containers shown in the figure above, there are two types of containers recommended for release by the system. The first is FCL which has received the release stamp on the dock receipt; the second is LCL which has received the release stamp on the dock receipt for all the clearance no. (all the shipments) of the corresponding container no. in the export manifest.

6.3 Stowage of Export Containers

With the continuous development of computer technology and network technology, the implementation of informatization in various fields is deepening constantly in enterprises. At present, all container terminals have adopted the vessel stowage management system, which has realized the high-speed operation of containers and greatly improved the economic benefits of enterprises.

Vessel stowage in container terminals is an important part of the whole operating system. The quality of the stowage directly affects the efficiency of loading and unloading operations, the berth date of the container liners, and the reputation of port enterprises. Meanwhile, stowage has a highly technical content. Under the same conditions of vessel, containers, and storage yards, the stowage quality achieved by

With dock receipt

Container No.	Size	Weight	Height	Type
FSCU7317130	20'	19949	GP	Flat container
FSCU9003479	20'	20721	GP	Flat container
INBU3494117	20'	19949	GP	Flat container
TGHU2469205	20'	19949	GP	Flat container
TGHU2980703	20'	22600	GP	Flat container

Unreleased：6

Release

Fig. 6.7 Customs release confirmation

planners with different technical abilities and professional quality also varies greatly. Due to the large amount of stowage data, there are many factors to be considered. At present, the stowage in container terminals in China mostly relies on personnel experience.

6.3.1 Meaning and Functions of Stowage

1. **Meaning of stowage**

Stowage refers to the specific loading plan made according to the transportation requirements of the vessel and the operational requirements of the terminal for the containers scheduled to be loaded in the established vessel.

Stowage must meet the requirements of shipping, namely the safety of shipping, the shipping quality of containers and goods inside, and the economic benefits of shipping operations. Meanwhile, the handling techniques and operation modes of the container terminal should also be taken into account, so that the terminal can organize production reasonably, orderly, and efficiently.

Stowage must be clear, scientific, and reasonable. Each container corresponds to a container position in the vessel, and the container must conform to the vessel specifications. The whole stowage should be scientific, reasonable, and highly operational, so as to ensure the continuous and efficient vessel loading operations and meet the berth date requirements.

2. **Functions of stowage**

Stowage is a very important operation and plays a very crucial role in the vessel loading operations in the container terminal.

(1) The storage area in the blocks can be arranged reasonably. According to the requirements of loading, reasonable arrangement should be made for the collection operations of export containers in advance, such as grouping by discharging port, tonnage, size, and type of containers, so as to improve the utilization rate of the storage yard as much as possible on the premise of following the loading sequence.
(2) The shipping requirements can be met. Through careful stowage, it can meet the requirements of stability, draft difference, load intensity, and other navigation safety requirements of the vessel, as well as the utilization of shipping space requirements and transportation quality requirements of containers and cargos. It can also avoid frequent transfer of containers, causing confusion or even paralysis of loading operations when the transporting requirements of the vessel are not met.
(3) Operations in the terminal will be organized effectively. The stowage plan is a specific and detailed plan for the vessel loading operations, which can help the central control office effectively monitor and coordinate the loading operations, the storage yard control scheduler deliver containers in turn, and the vessel commander load the vessel in sequence, so that all the loading operations are organized in an orderly manner.
(4) It is the original information of vessel loading operations. After loading, the terminal will check the stowage plan with the external tally clerk, and the confirmed stowage plan can be used as the original certificate of loading operations and the statistical data of actual throughput of the terminal.

6.3.2 Factors Affecting Stowage Quality

Stowage is an extremely complicated planning process. There are many factors affecting the stowage results, mainly including the following.

(1) Stowage planner's professional levels
The proficiency to the rules, the technique of stowage, the working experience, and other personal conditions are important factors that affect the stowage result.

(2) Container storage conditions in different blocks in the yard
The utilization rate of the yards in the terminal will change with the adjustment of marine transportation form and terminal routes. Different container storage conditions in the yard also have great influence on stowage.
(3) Time affluence for stowage
Under normal circumstances, the work quality is directly proportional to the limited time to complete the work, and the affluence of stowage time will also affect the stowage quality. With the upgrading of vessels and the postponement of the closing date, the time left for stowage planner is becoming increasingly tight.
(4) Uncertainty of operations in the container terminal
Weather changes, equipment failures, temporary operation requirements of shipping companies and port departments, accidents of vessels and containers, and other emergencies will all affect the stowage plan.
(5) Loading condition of the vessel
The loading condition of the vessel will change with the seasons and years. The world economic situation and the slack and busy seasons of shipping will cause the number of containers carried by the vessel to fluctuate, and the requirements for seaworthiness parameters of different loading capacity are not the same.

Improving the quality and stability of stowage results has become an important measure to improve the fine operation management of the container terminal. The intelligent stowage system can continuously collect and improve the stowage rules of all types of vessels, which will not be affected by factors such as the change of stowage planners and the difference in state, so as to ensure the stable and high-quality output results. The stowing speed of an ordinary stowage planner is 300 containers per hour; however, for a vessel with 1000 containers, it takes only about 10 min for the intelligent stowage completed by the computer, which has greatly increased the stowing speed. The intelligent stowage system can also adjust the stowage result in real time in case of emergency and reduce the impact of emergency on operations in the terminal.

6.3.3 General Principles of Stowage

There are various types of container vessels in service, from modified vessels with a capacity of 100 TEU to ultra-large container vessels with a capacity of 18,000 TEU. Different types of container vessels have different stowage principles and special requirements. In Table 6.2, taking CMICT as an example, the operation principles of stowage plan corresponding to various common vessel forms have been listed.

The performance parameters of the vessel, such as stability, shear force, torque, stack weight, and lashing force, should be considered by the vessel handler. The reasonable number of operation lines and loading sequence, the smooth scheduling of facilities, the control of reshuffle containers in the yard, the reduction of the

Table 6.2 Stowage principles for different vessel forms

No.	Vessel form	Loading capacity	Deck features	Cabin features	Percentage (%)
1	Small vessel	Below 2000 TEU	The light on top and the heavy at bottom	The light on top and the heavy at bottom	10
2	Medium vessel with poor stability	2000–4500 TEU	The heavier at bottom and the lighter above the 4th tier, considering the lashing force	The light on top and the heavy at bottom	15
3	Medium vessel with high loading weight	4500–7000 TEU	The light on top and the heavy at bottom strictly, considering the loading capacity of each row	All the heavy in the cabin, the light on top and the heavy at bottom, considering the loading capacity of each row	15
4	Ordinary vessel	7000–9000 TEU	The light on top and the heavy at bottom	The light on top and the heavy at bottom	50
5	Ultra-large vessel	Above 12,000 TEU	The heaviest at the 1st and 2nd tiers and the lighter above the 5th tier, considering the lashing force	The light at bottom and the heavy on top	10

operating difficulty in the yard and the vessel, and the control of operation cost should be considered by the terminal. To some extent, the two considerations are in conflict. Stowage planner of the terminal needs to balance the interests of both from the actual vessel form, so as to ensure that the final vessel loading plan can not only meet the requirements of various seaworthiness parameters of the vessel, but also reduce the operation cost and improve the operating efficiency in the terminal.

6.3.3.1 Principles of the Vessel

The precise stability, shear force, torque, stack weight, and lashing force of the vessel are the main reference parameters for the pre-stowage center of the shipping company and the vessel's mate to consider. This part of data does not need to be accurately calculated for stowage in the terminal, but the principles should be understood in the actual stowing process, to reasonably plan the distribution of export containers loaded from the port into the vessel, so as to ensure that the vessel still meets the seaworthiness parameters requirements after operation. The general principles are as follows.

(1) Weight distribution
For the container vessels sailing in the vast ocean, their safety must be ensured, and the stability is an important parameter to measure their safety. Stability refers to the ability of the vessel to tilt due to the action of external forces (such as wind and surge) and automatically returns to the original equilibrium position when the external forces disappear. There are a large number of containers stacking on deck, which usually accounts for more than 30% of the total loading capacity of the vessel. The proportion for modern container vessels, especially large ones, is even higher, almost a half. For example, the vessel "BH" of 1700 TEU has a deck container capacity of 828 TEU, and the deck container capacity accounts for 48.8% of the total container capacity of the vessel. The vessel "LH" of 5250 TEU has a deck container capacity of 2460 TEU, accounting for 46.8% of the total. As there are a large number of containers on deck, the stability will decrease if the gravity center rises, which should be thoroughly considered by the stowage planner.
The following rules should be noted during stowage.
In the longitudinal direction, the heavier should be in the middle and the lighter at both ends. The bays from the bow to about one-fourth of the hull and behind the living quarters of the vessel should be loaded with lighter containers and the middle of the vessel with heavier containers.
In the lateral direction, the heavier should be in the middle and the lighter at both sides. In the case of actual stowage of a single bay, the heaviest containers are loaded to the middle rows, and the lighter to the both sides.
The weight distribution within the bay in longitudinal direction should be rational. In most cases, it is necessary to ensure that the heavier containers at bottom and the lighter containers on top, so as to ensure the stability of the vessel and increase the GM value. Not all vessels need to be loaded with the heavier at bottom and the lighter on top. For ultra-large vessel with container loading capacity higher than 12,000 TEU, the loading inside the cabin needs to be reversed, that is, the heavier on top and the lighter at bottom. This is because the ultra-large vessel is too stable, and the GM value needs to be lowered to increase stability. The weight of containers at the upper tier of this bay is greater than that at the lower tier.
(2) Stack weight
In order to avoid the damage of rigid structure of the vessel, each bay of the container vessel is indicated with the maximum stack weight of each row. The containers carried on the Middle East route, India–Pakistan route, and other shipping routes are usually very heavy, so it is necessary to pay attention to the stack weight of each row in the actual stowage to avoid overweight.
(3) Lashing force
Lashing force is for the purpose of reducing or preventing the accidental drop or loss of cargo due to the rocking of the vessel's hull by waves, winds, and other

external forces during navigation. The deck of each vessel must be reasonably lashed.

In the actual stowage, heavy containers (generally weighing about 25 t) should be stowed at the first or even the second tier on deck. Containers below 15 t should be stowed above the third tier on deck. The higher the no. of tiers, the lighter the containers will be, and the lighter the containers will be as they are away from the center toward both sides. As shown in Fig. 6.8, the weight of containers stowed at the first deck tier is obviously larger than that of containers at other tiers, and the weight of the same tier gradually decreases toward both sides.

(4) Avoiding the operation difficulty at other ports along the route caused by improper stowage
In addition to loading and unloading consideration at local port, operating requirements at ports of call along the route should also be considered, especially at the port of departure. In general, the following situations should be avoided as much as possible.

① Container reshuffle at intermediate ports
Usually there are many ports of call along the route of a container vessel, and this is especially true for the ocean-going vessels. To this end, there should be an overall view of the whole route during stowage. Containers should be stowed according to the sequence of ports of call of the container vessel. The previous port should provide convenience for the latter port, and at the port of departure, a good foundation should be laid for the whole route. In particular, it is necessary to avoid container reshuffle caused by the containers of latter ports stacking on the containers of previous ports, thus reducing the loading and unloading speed, increasing costs and causing losses.

② Excessive concentration of containers of the same discharging port
Due to the structural characteristics of the container cranes, two quay cranes cannot operate together and should be separated by at least four small bays (20 ft container width) during loading and unloading. When the container amount of a certain discharging port is large and exceeds the capacity of one bay and has to be divided into two bays, generally it is not allowed to

Fig. 6.8 Container weight distribution on deck

8	10	11	12	12	11	10	8
13	14	14	15	15	14	14	13
15	15	16	18	18	16	15	15
20	22	25	30	30	25	22	20

be configured in two adjacent bays, but should be separated by at least one 40 ft bay for stowing. In this way, the two quay cranes can operate at the same time, avoiding the formation adjacent bays and ensuring the loading and unloading efficiency and berth date.

③ Skewed stowage

The skewed stowage of containers refers to stowing all containers of a port or several ports together on one side of the vessel. When unloading at the intermediate port, it will cause vessel rolling in a short time, leading to difficulties in loading and unloading operations and affecting the operating speed. Therefore, during stowage, the skewed stowage of containers should be avoided if possible. Instead, containers of the same port or several ports should be stowed symmetrically on the both sides of the vessel.

(5) Meeting the stowage requirements of special containers

Due to characteristics of structure, size, or weight of special containers, there are certain special requirements for stowage, such as that reefer containers must be stowed in the cabin or on top of the cabin and so on. In addition, for the containers without determined discharging port, it shall be ensured that the containers can be unloaded smoothly at all the selected ports. Stowage planner must understand the special requirements for stowage of all types of special containers, so as to avoid difficulties or losses caused by improper stowage.

6.3.3.2 Principles of the Terminal

For the terminal, the most important consideration is how to reasonably utilize the equipment resources, ensure the smooth and uninterrupted operations of the quay cranes, and reduce the energy consumption of the equipment. Therefore, under the premise of meeting certain seaworthiness parameters of the vessel, the operational principles of the terminal must be fully considered. The general principles are as follows.

Principle 1: Reshuffle of export containers in the yard should be controlled.
Reshuffle of export containers in the yard is due to the unreasonable container collection, timely container release failure, temporary port change, shut out and unreasonable stowage plan, which causes that, in the process of vessel loading, the yard crane has to reshuffle a number of containers stacking above to get the target container, and the reshuffle is an extra operation. The ideal sequence of container operations in the yard is shown in row 01 of bay 58 in Fig. 6.9a, while the sequence of delivering containers in row 01 in Fig. 6.9b generates three additional reshuffles in the yard when the target container is delivered.

Unnecessary reshuffle in the yard will cause a waste of the processing capacity of equipment, seriously affect the continuity of quay crane operations, and reduce the efficiency of the whole vessel. Reshuffle in the yard will also increase the energy consumption of the yard crane and even increase the amount of equipment and personnel configured for the single-vessel operations.

The stowage quality can be improved to some extent by housekeeping in advance.

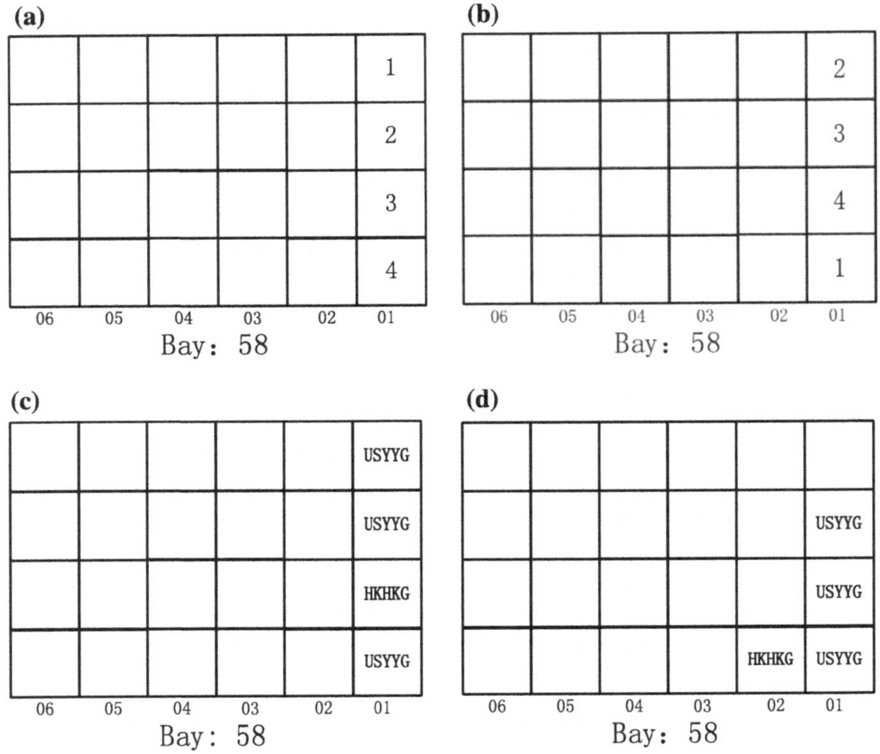

Fig. 6.9 Operating sequence of a yard crane

After the release and before the berthing of the vessel, the shipping company will adjust the shipped goods according to the actual release situation, namely changing discharging port, shutting out, and subsequent stowage. This will cause the storage of the containers of the vessel in the yard to become disorderly. As shown in Fig. 6.9c, the discharging port of a container in row 01 has been changed to 'HKHKG' (Hong Kong), and sandwiched by three USYYG (Los Angeles) containers. For general stowage, the containers of the same discharging port at the same bay in the yard will be stowed at the same bay in the vessel, and the transshipment order of containers of HKHKG and USYYG could not be determined in advance when loading the vessel, thus there is a high probability to cause extra reshuffle in the yard. In addition, if a container is not released in time (the same as being shut out), the container will remain in the yard and cannot be shipped. Therefore, when loading the two containers under the container into the vessel, the container has to be reshuffled first.

Due to the high utilization rate of the yard, the shortage of storage space in the yard, the faults of yard planners, and the large number of unplanned containers, etc., will lead to concentrated storage of export containers in the yard, which will make the blocks become congested when loading, and cause reduction of operating efficiency, even shut down of vessel loading operations due to the slack delivery of containers.

In order to relieve or eliminate such situations, stowage planners should check the storage situation of export containers of the vessel in 6–12 h before the vessel berthing, find abnormalities in the yard, and ask for rearrangement of the yard plan, separating the containers that have not been released or containers that have been changed to another port. The rearranged bay in the yard is shown in Fig. 6.9d. By paying a certain amount of container shifting fees in advance, the number of containers reshuffled during loading can be effectively reduced and the operating efficiency of the whole vessel can be improved.

Principle 2: container delivery sequence in the yard

Generally speaking, the export full containers are stored at the bays with a gantry crane, while the export empty containers stored at the bays with a stacker. Due to different operation techniques, the stowage for these two kinds of situations is slightly different.

(1) Bay with a stacker

The stacker cannot operate over containers, so it must operate from the outer row near to the lane to the inner row. As shown in Fig. 6.10, the operating sequence of containers stacking in the bay must be from the outer row to the inner row. If the container stored in the inner row needs to be delivered first, all the containers in the outer rows must be reshuffled to other bay before the target container can be delivered. Generally, there is a large amount of containers stacked in the block with the stacker, sometimes over a hundred. If it is not handled properly, it will cause unnecessary reshuffle of containers, which must be considered when making stowage.

(a) Rule of bay

It is ideal to designate all the containers at one bay to one operation line of the vessel when making stowage of the bay with a stacker.

(b) Rule of row

Fig. 6.10 Operating sequence of a stacker at one bay

21	17	13	9	5	1
22	18	14	10	6	2
23	19	15	11	7	3
24	20	16	12	8	4

Location in the yard with a stacker

The pre-stowage results of the shipping companies often cannot match with the storage situation in the yard of the terminal, and there will be the situation that the containers at one bay must be split and designated to two or more operation lines. In this case, the stowage planner needs to carefully observe the operation line plan of the vessel, estimate roughly the operation sequence of the vessel, and designate the containers in the middle and outer rows at the bay with a stacker to the early operation line and the containers in the inner rows to the later operation line.

(c) Rule of operation line

When the operation sequence cannot be completely determined, the containers in the outer rows at the bay with a stacker should be designated to operation lines with small container amount, so as to minimize the possible reshuffle. For example, three containers of port A at the bay with a stacker need to be delivered, and two stacking situations are shown in Fig. 6.11. If the containers in the outer rows need to be reshuffled during vessel loading, the container reshuffle in (a) is 0 and 21 in (b), and (a) is definitely better than (b).

(2) Bay with a gantry crane

The gantry crane can lift vertically at a high point and operate across containers. It is very flexible when delivering containers. When making stowage, only attention should be paid to the operation techniques to reduce the container reshuffle.

From the perspective of operation techniques, the gantry crane needs to avoid the operation of container insertion, reduce the operation over high tiers, and shorten the running distance of the spreader (shorten the operation time of one container). Shown in Fig. 6.12a is the operation of container insertion. High tiers of containers are stacked on both sides of the target container, which increases the difficulty of the operation of the target container. During the operation, the containers on both sides may be rubbed against, resulting in the damage or fall of the containers

(a)

B	B	B	B	B	A
B	B	B	B	B	A
B	B	B	B	B	A
B	B	B	B	B	B
06	05	04	03	02	01

Bay with a stacker

(b)

B	B	B	B	B	B
A	B	B	B	B	B
A	B	B	B	B	B
A	B	B	B	B	B
06	05	04	03	02	01

Bay with a stacker

Fig. 6.11 Reshuffle operation at the bay with a stacker

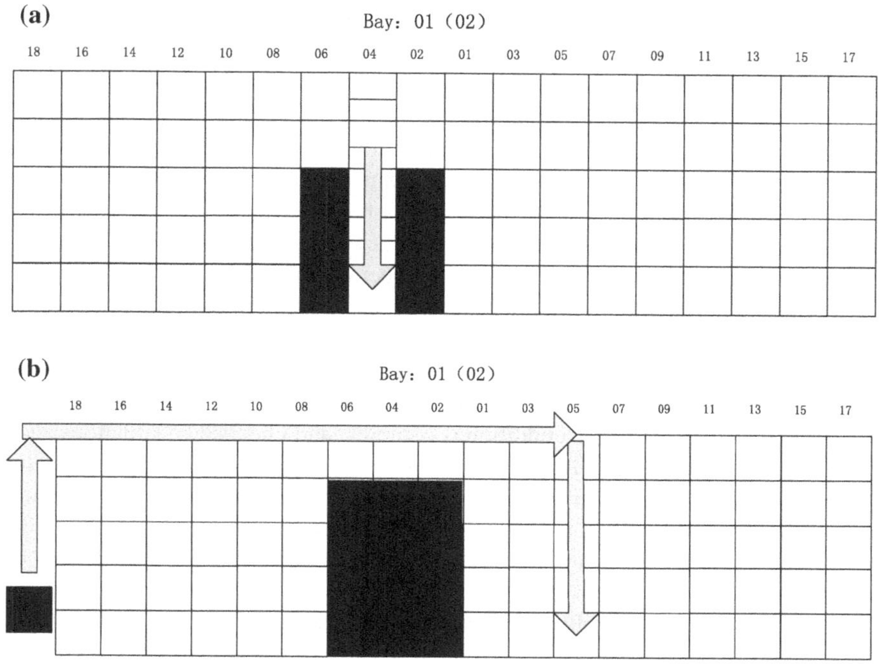

Fig. 6.12 Operation techniques of the gantry crane

on both sides. In this case, the operator will subconsciously reduce the operation speed to ensure safety. Shown in Fig. 6.12b is the operation over high tiers. When the gantry crane is delivering containers, the spreader should be lifted to a safety height. However, unreasonable container delivery order will cause the situation that when delivering the target container, the spreader has to be lifted up and down more frequently to avoid collision and rub. Unnecessary operations over high tiers in large amount also have a negative effect on the efficiency, which should also be avoided as insertion operation.

There are few stacking tiers at the bay with a gantry crane, usually 4–5 tiers high, and the gantry crane can operate across containers. Therefore, even if there are container reshuffles, the number will not be too high, and the stowage requirements are not as strict as those at the bay with a stacker, but certain rules must be followed. When the containers at a certain bay must be split and designated to two or more operation lines of the vessel, the rules to be followed are similar to those for bay with a stacker, namely the rule of one bay to one operation line, the rule of prior stowage for containers in the middle and outer rows, and the rule of prior stowage for operation lines with small container amount.

Principle 3: concurrent operation of multiple container handling points in the yard

Each operation line generally consists of one quay crane, 1.5–2 yard cranes, and 5–7 trailers. The uniqueness of quay crane determines its important role in each operation line all the way. Any pause of quay crane will cause intermission of the whole operation line, hence, to ensure the continuity and efficiency of the quay crane operation is the key to ensure the vessel operation efficiency.

In order to ensure the continuity and efficiency of the quay crane operation, the stowage planner will try to ensure concurrent operation of multiple yard cranes, that is, to ensure that one bay in the vessel corresponds to multiple container handling points (usually 2–3 blocks) in the yard. Too many corresponding handling points will lead to frequent shifting of yard facilities, resulting in equipment pressure, operating efficiency reduction, and operating cost increase, while too few will reduce operation flexibility. Terminal operation condition is complex, and unexpected situations such as gantry crane failure, ETG rack failure, power transmission failure, and trailer failure occur from time to time. If a bay corresponds to only one container handling point, any emergency will inevitably affect the quay crane operations, which affects the operation line progress, and multiple points can greatly improve the stability of container delivery.

Principle 4: container handling priority for key blocks

The ideal mode of container collection in the yard is to evenly distribute the export containers of the vessel to several blocks; however, due to various reasons, it is difficult to realize in the actual operations, and uneven storage of containers is inevitable, resulting in concentrated stacking of containers in one or several key blocks, taking a high proportion of the export containers of the vessel. If special attention is not paid to this in actual stowage, operating speed will be gradually reduced, finally causing congestion in the blocks and equipment conflict. There are two ways to deal with such situation. One is to shift containers in the yard in advance, appropriately reducing the amount of containers stacking in the key blocks, which will cause unnecessary cost of container shifting and is not adopted in general. The other is to give container handling priority to the key blocks under the same conditions in actual stowage, and the proportion of export containers of the vessel in the key blocks shall be gradually reduced consciously.

Principle 5: container handling priority for conflicting bays

Unbalanced container collection in the blocks also exists. If several continuous bays are stacked with one type of containers, the conflicting bays will be generated due to the limitation of the yard crane operation techniques. Generally, the operation spacing of gantry crane is 8 small containers and 18 small containers for stacker. Bays within the operation spacing cannot be operated simultaneously, which will reduce loading flexibility. Its harmfulness to the operations will mitigate as the remaining number decreases in the actual stowage. Figure 6.13 shows a block with a gantry crane in the yard, and bays 10, 15, 18, 21, and 26 in the figure are considered to be conflicting. The priority of containers at these bays is much higher than those at the bays 44 and 56 in the figure in actual stowage. When selecting containers in the conflicting bays, the principle is to first select the containers in the middle and then on the sides. As

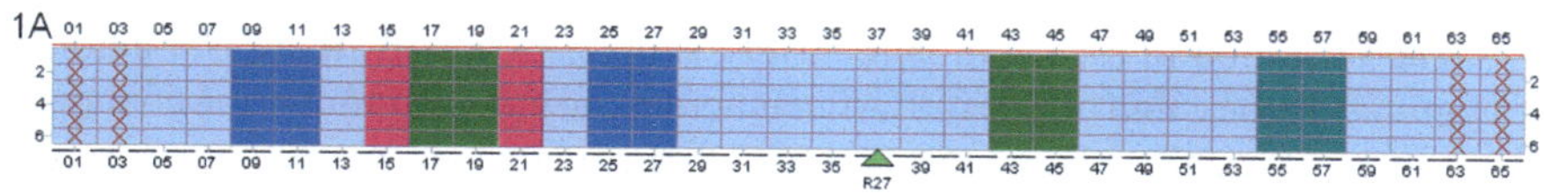

Fig. 6.13 Conflicting bays in the yard

shown in the figure, after the containers at bays 15, 18, and 21 are removed, the two bays on both sides will be divided into two independent container handling points, and the conflicting bays will not exist. The earlier the conflicting bays are broken up, the greater the freedom of actual stowage will be.

Principle 6: controlling the yard crane shifting

In stowage, measures should be considered to reduce the reshuffle operation of yard facility and minimize the shifting distance of the gantry and the trolley. So as to improve the efficiency of yards and vessel loading operations, the most important is not to make the gantry shift back and forth.

If the yard crane often shifts between multiple bays in the process of container handling, the operating efficiency will be greatly reduced and energy resources will be wasted. Figure 6.14 shows two different ways of stowage at a certain bay in the vessel. The stowed containers are stacked at bays M and N in the yard, respectively.

When handling the containers during vessel loading, the delivering process of stowage I will make the yard crane shift cross over two bays, the process is N → M → N → M → N; while the delivering process of stowage II is just N → M, that is, when container handling at bay N is finished, the yard crane moves to bay M to continue container handling.

Principle 7: loading sequence at bays in the vessel

The reasonable loading sequence in the vessel can reduce the operation difficulty of quay crane drivers, improve the operation smoothness, and eliminate the hidden threat of accidents. Therefore, it is also very important for stowage planner to optimize the loading sequence during and after the actual stowage and establish a reasonable loading sequence.

(1) Loading sequence in the cabin

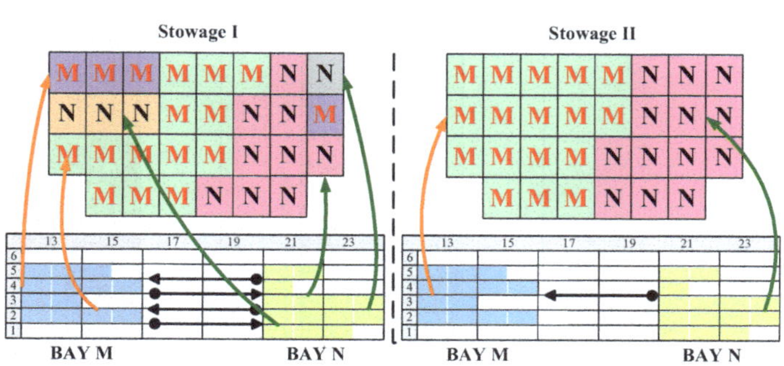

Fig. 6.14 Impact of stowage on the operation of yard facility

The operation in the cabin is relatively simple. Generally, there is a guiding groove in the cabin of a vessel. After entering the guiding groove, the container can be vertically placed into the target position along the guiding groove, which is easy for the quay crane drivers to operate. During the actual stowage in the cabin, the stowage planner only needs to meet the requirements of multiple container handling, delivering containers by grooves instead of by tiers. As shown in Fig. 6.15 (different colors indicating containers from different handling points), it is necessary to note that each handling point occupies independent grooves, so that even if there is an abnormal situation at one handling point, the containers from the other handling point can be loaded smoothly, so as to keep the quay crane operation continuous.

(2) Loading sequence on deck

During deck stowage, attention should also be paid to each handling point occupying independent groove just as in the cabin to meet the requirements of multiple container handling. The following points should also be noted.

(a) Loading direction

The deck guardrail of a vessel is generally 1-to-2 storey high. In order to ensure that the drivers of quay cranes will not collide with the guardrail of the vessel in the operation, the regulations of terminal operation will require the drivers to lift each container to the safety height in the deck operation. The actual stowage sequence shown in Fig. 6.16 is not appropriate. As the containers with sequence no. 7 and 8 are loaded before the container with sequence no. 9, the quay crane driver needs to raise the container of no. 9 above three tiers to ensure the safety of operation.

(b) Level difference

If the difference between the deck and the adjacent row level is too large, the possibility of containers tipping over, rubbing, and falling will be greatly increased in the operations. As shown in Fig. 6.17a, the container with sequence no. 3 is too high in a single row, leading to the increased

Fig. 6.15 Independent container delivery in the cabin

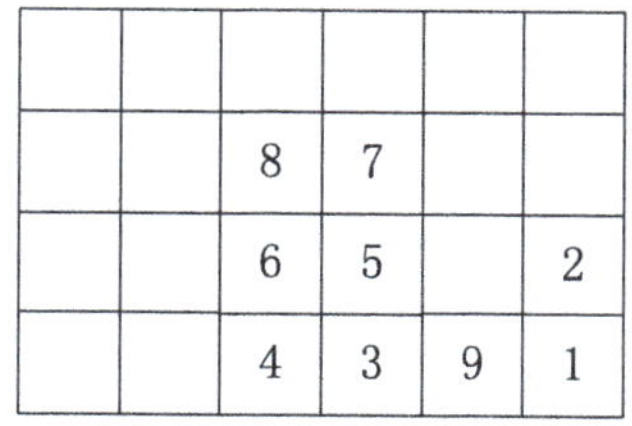

Fig. 6.16 Operation over certain height

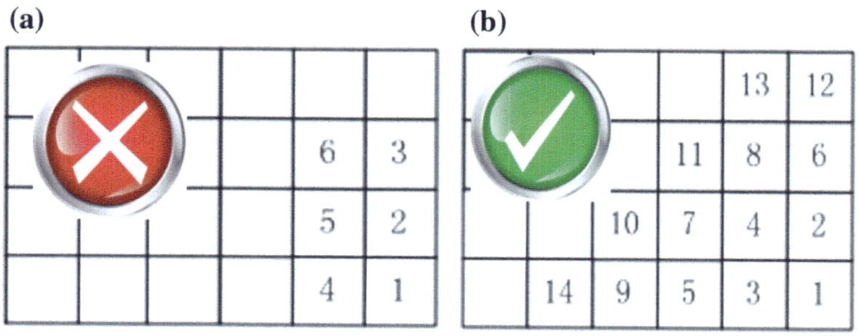

Fig. 6.17 Loading sequence in a right trapezoid style on deck

risk of collision and falling. The correct actual stowage sequence is shown in Fig. 6.17b, trying to ensure that the stacking is in such a right trapezoid.

(c) Avoiding container insertion

Container insertion will also increase the risk of collision in the operations, which needs to be avoided.

(3) Loading sequence of vessel with full manual lock structure

For the vessel with full manual lock structure, the top locks of the containers need to be installed manually. In order to avoid workers constantly climbing the containers, stowage of vessels with such structure needs to be made by tiers, as shown in Fig. 6.18.

There are certain conflicts between the vessel's principles and the terminal's principles, and an excellent stowage planner needs to weigh the advantages and

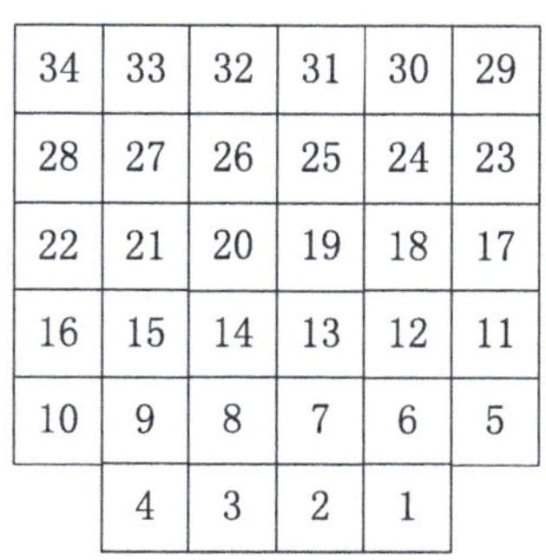

Fig. 6.18 Suggested loading sequence for hand-dropped lock structure

disadvantages of the two. The vessel's principles are the basis, and only when the seaworthiness conditions are satisfied, the subsequent operations of the terminal can be implemented. Therefore, how to meet the vessel's principles under the condition of guaranteeing the maximum benefit of the terminal is the main problem to be solved by the stowage planner of the terminal.

6.3.4 *Process of Stowage Operation*

1. Collecting and verifying stowage documents

As mentioned above, the stowage documents, including the vessel data and container data, are the original reference of stowage operation, which must be collected as complete as possible. Before the start of stowage operation, such data should be verified to avoid improper stowage or error caused by the inconsistency of the documents. The documents to be verified mainly include:

(1) The consistency of packing list and container load plan
Only containers with the container no. listed in the packing list with the "customs release seal" can be stowed and loaded for export; otherwise, it will constitute smuggling violation. Therefore, it is necessary for terminal stowage planner to check whether the packing list is consistent with the container load plan. At present, such verification is done by the shipping agencies in most container terminals in China, but it is also very important for stowage planner to verify it again to avoid mistakes.

(2) The collected data of export containers
The container terminal usually stops the container entry in 24 h before the vessel leaves. The stowage planner has to summarize and verify data of containers that have been collected in the terminal and released by the customs, including the amount of total export containers, the amount of 20 ft containers, the amount of 40 ft containers, the amount of special containers and their specific types, the amount of dangerous goods containers, ports of call, and the containers for discharging at each port. Usually, the container data can be summarized and verified in the order of discharging ports, respectively, including the amount of containers, size, container type, special containers, and dangerous goods containers for each port.

(3) The locations of export containers in the yard
The operation requirements in the terminal should be taken into account in stowage under the premise of meeting the shipping requirements. In order to meet the requirements of the shipment and loading sequence, the export containers are usually stacked according to four main principles after their entry, namely separation of containers with different discharging ports, separation of containers with different sizes, separation of containers with different tonnage levels, and separation of general containers and special containers. However, due to the limitation of yard space and unpredictable entry order of

the containers, these four main principles cannot be followed strictly, which requires the stowage planner to be aware of the actual stacking locations of the export containers, to avoid frequent container reshuffle in the yard caused by improper stowage.

2. Making the stowage BAPLIE

Making stowage BAPLIE is the important work of stowage planner, which is the reference for the vessel commander to instruct the loading operation, and also for the controlling office to coordinate the loading and unloading operations. According to the basic principles of stowage, the stowage BAPLIE can be established on the basis of the complete collected data. In fact, the same stowage data can lead to different stowage BAPLIEs by different stowage planners, which reflects the basic skills and professional ability of different stowage planners. The stowage BAPLIE is divided into two parts, namely the general overview of the stowage plan and the bay view of the stowage plan.

The stowage of export containers needs to give consideration of both the vessel and the terminal on the premise of meeting the shipping requirements and achieve the best balance. In addition, due to the special configuration requirements of the special containers which are not as interchangeable as the general containers in the vessel, the stowage planner should stow the special containers first, then the general containers, and the loading positions of the special containers should not be changed once their stowage plan is completed. Generally, only the code of discharging port and special container mark are indicated in the container positions at the general overview of the stowage BAPLIE. In addition, upon completion of the stowage, a statistical table should be made for the containers in the order of discharging ports to check whether the stowage is correct.

The bay view of the stowage plan is actually the enlarged image of the general overview of the stowage plan for each bay. The number of bay view stowage plan corresponds with the number of bays. There is more information indicated in the container position at the bay view of the stowage plan, including the discharging port and discharging port code, container no. and holder, container weight and mark of special container, etc. In some container terminals, the container location in the yard is also indicated for the convenience of container delivery in the yard.

3. Reviewing the stowage plan

After the stowage of export containers is completed, the stowage planner shall conduct a careful review to check whether the stowage plan meets the basic requirements of stowage and whether there is any improper or error in stowage, so as to correct in advance and avoid passive and undue losses in loading and unloading operations. The main contents of the review are:

(1) Whether the amount, size, and type of containers of the whole vessel are consistent with the data of export containers; whether the amount, size, and type of containers at each discharging port are consistent with the export data; and whether there are short-loading, over-loading, and misloading.

(2) Whether the stacking weight of containers at each bay exceeds the allowable specification of the vessel.
(3) Whether the configuration of dangerous goods containers, reefer containers, and other special containers is appropriate; whether the configuration of over-height and over-width special containers is appropriate; whether it complies with the stowage requirements of special containers and ship specifications; and whether the permit for dangerous goods is complete.
(4) Whether the arrangement of container positions for each discharging port is reasonable; whether the containers of previous discharging port are stacked under the containers of the latter ports; and whether screwed stowage of containers and excessively concentrated stowage of containers with the same discharging port are in existence.

4. Signing and issuing the stowage plan

After review, the stowage planner should sign on the cover of BAPLIE and record the date of stowage, so as to make clear responsibility, and send the stowage plan to the mate of the vessel for review. Once the mate confirms and signs on the stowage plan, the stowage plan will be sent to the vessel handler, the yard, the central controlling office, tally and other departments, as the reference of loading operations and legal documents to distinguish responsibility.

5. Stowage of the released shut-out containers

Due to the reasons of the vessel or the goods, there will be a number of shut-out containers for each voyage. The stowage planner should timely summarize and verify the amount of shut-out containers, container no., container type, bill of lading no., seal no. and other data, as well as the reasons for shut-out, and submit them to the relevant departments. For the containers that have been released again, the stowage planner should pay more attention to avoid short-stowage.

6.3.5 Data Required for Stowage

In order to make the stowage plan scientifically and rationally, it is necessary to obtain as many detailed stowage data as possible, which mainly include:

1. **Particularity of the vessel**

(1) Container loading capacity
The container loading capacity of a vessel refers to the maximum amount of containers loaded in the vessel, which is usually expressed by TEU. The container loading capacity can represent the size of a container vessel, and it is also the limit data that must be understood in stowage. Besides the container loading capacity, data related to it must also be mastered.

① Maximum loading capacity of 20 ft containers
This refers to the maximum amount of 20 ft containers that a container vessel can carry. In general, the maximum loading capacity of 20 ft containers is the same as the container loading capacity of a container vessel. But some container positions in certain container vessels are specially designed for 40 ft containers and the stowage of 20 ft containers is normally not allowed there. If the amount of 20 ft containers planned for this voyage exceeds the maximum loading capacity of 20 ft containers, the shipping agency must be contacted to return the extra 20 ft containers and replace them with 40 ft containers.
② Maximum loading capacity of 40 ft containers
This is the maximum amount of 40 ft containers that a container vessel can carry. This amount is not equal to half of the container loading capacity of the vessel. Due to the structure of the vessel, some container positions (such as in the cabin) can only hold 20 ft containers. In addition, some container vessels have fixed 20 ft container grids in the cabin, or leave lashing aisles on deck, so such positions can only hold 20 ft containers instead of 40 ft containers. Also, stowage planner must be aware that the amount of 40 ft containers in the stowage plan must not exceed the maximum loading capacity of 40 ft containers of the vessel.
③ Distribution of container positions
The stowage planner should also be aware of the maximum loading capacity in TEU on deck as well as that in the cabin (including the maximum amount of 20 and 40 ft containers that can be loaded), in order to meet the requirements on the container position distribution of the vessel, and improve the space utilization as well as stowage rationality.
④ Compatibility of 20 ft containers and 40 ft containers
Usually, a 40 ft container position can be loaded with two 20 ft containers; however, due to vessel structure, the two are not compatible in some container vessels, that is, a 40 ft container position cannot be loaded with two 20 ft containers, such as lashing aisles on deck, container positions in the cabin without reinforced base for 20 ft container, etc. The stowage planner should pay particular attention to these data, and avoid stowing two 20 ft containers in such 40 ft container positions; otherwise, the vessel loading is unable to run smoothly.
⑤ Distribution of reefer container areas
There are usually some reefer container areas equipped with power sockets configured in the vessel. The stowage planner should know the distribution of reefer container areas in the vessel, so as to ensure the normal power supply to the reefer containers after loading. The reefer container areas are usually configured on deck. In addition to the deck, some container vessels also have reefer container areas configured inside the cabin.
⑥ Stowage restrictions for dangerous goods containers
In the stowage of dangerous goods containers, in addition to complying with the international dangerous goods isolation requirements, the stowage

planner should also know the container positions in the vessel which are not allowed to load dangerous goods containers in the consideration of structure safety, so as to avoid reshuffling when violating the restrictions of the vessel.

(2) Stacking weight
Stacking weight refers to the maximum weight of containers allowed to be stowed on the container bases of the cabin frame, deck and hatchway cover of a container vessel, including stacking weight for 20 ft containers and stacking weight for 40 ft containers. During stowage, it must be noted that the total weight of each row of containers, no matter inside the cabin or on deck, shall not exceed the stacking weight of the vessel, especially when there are many full containers or overweighted containers, in order to avoid being refused by the vessel due to damage to the vessel structure.

(3) Length, width, and draft of the vessel
The vessel length in stowage data refers to the length between perpendiculars (LBP), that is, the horizontal distance between the leading edge of the stem and the rear edge of the stern, which is the necessary data to calculate the draft difference of the vessel. The vessel width usually refers to the maximum horizontal distance between the two sides of a vessel, which is the necessary data to calculate the rolling period and determine the height range of initial stability. Draft refers to the fully loaded draft, that is, the average draft of the vessel under the rated load, which is the limit draft that must be mastered in stowage. In stowage operation of the container terminal, it is necessary to master the length between perpendiculars (LBP).

(4) Hatchway covers and vessel cranes
The type and quantity of hatchway covers have a great influence on stowage and loading/unloading operations. Especially for foldable hatchway covers and translational hatchway covers, the stowage planner should have a detailed understanding of them to avoid the loss caused by port congestion and reshuffling of the stowed containers. Translational hatchway cover is to use jacking oil cylinder to lift the hatchway cover plate to the horizontal guiding track and then use gear rack or sprocket chain to translate the hatchway cover to the side of the hatch, to achieve the purpose of opening a cabin. Foldable hatchway cover is to use hydraulic cylinder to lift two foldable hatchway cover plates, standing on one side of the hatch, thus to achieve the purpose of opening the cabin. Some small and medium-sized container vessels are equipped with vessel cranes, and the stowage planner should also know their locations to avoid or reduce their impact on the loading operations.

(5) Weight of the empty vessel and vessel constant
The weight of the empty vessel refers to the weight of the vessel when delivered or after dock repair, excluding the weight of any oil, water, cargo, crew, etc. The constant refers to the weight of the oil scale in oil tank, the water scale in water tank and container lashing tools, etc. Since this part of weight is neither included in the loading weight, nor included in the weight of the empty vessel, and it is relatively constant within a certain period of time, so it is called the

vessel constant. These two data are necessary to calculate stability and draft difference.

(6) Stability calculation and draft difference calculation
After the construction and delivery of a container vessel, its size, shape, and structure have been determined. Relevant data can be calculated in advance according to the data of the vessel under different draft conditions and tabulated to facilitate and correctly calculate the stability and draft difference after stowage.

2. **Container data**

(1) Container load plan
The container load plan is a document that records the detailed information of the goods in the container, including the name, weight, package, number of pieces, etc. Meanwhile, it also provides necessary information for the stowage planner, such as container no., seal no, bill of lading no., container size, container type, total weight, name of vessel, voyage, discharging port, and port of destination. One container load plan corresponds to one container.

(2) Packing list
The packing list is the fifth of the ten copies of "dock receipt," which is mainly used for export customs declaration. The stowage planner must verify the packing list. Only the container with the container no. listed in the packing list with "customs release stamp" can be loaded for export. One packing list corresponds to one B/L.

(3) Pre-stowage list
The pre-stowage list is a list of containers consigned for this voyage made by the shipping agency according to the confirmation of booking information, including bill of lading no., goods name, number of pieces, weight, container size, container type, discharging port, etc., which can be used as stowage reference for stowage planner.

(4) Special container list
Special containers, such as reefer containers, open top containers, flat rack containers, and tank containers, have special requirements for stowage. Stowage planner can know about these special containers in advance through the list of special containers, so as to make full consideration and preparation for stowage.

(5) Dangerous goods container list and dangerous goods permit
The list of dangerous goods containers provides the stowage planner with the total amount of dangerous goods containers of the voyage, their type and size, as well as the name, weight, IMDG grade, and other information of the dangerous goods inside the containers. Accordingly, the stowage planner can know the stowage requirements and direct loading requirements of these dangerous goods containers.

A dangerous goods permit is the *certificate of dangerous goods shipped in containers* issued by the port supervision. The permit must be obtained before stowing dangerous goods in the container terminal.

(6) Data of export containers in the yard
After the entry of the export containers, the container terminal usually makes a list of export containers or displays the data of export containers directly through the computer screen, such as the total amount of containers, the amount of general containers, the amount of special containers, container weight, container type, size, discharging port, and the location of each container in the yard. For the stowage planner, he can take into account the loading requirements of the vessel and the operation requirements of the terminal by mastering various information and materials of the containers in the yard, so as to make scientific and reasonable stowage plan.

6.3.6 General Procedures of Stowage

6.3.6.1 Difference and Relationship of Pre-stowage and Stowage

Although "pre-stowage" and "stowage" are similar in the literal sense, they are actually two greatly different stowage plans."Pre-stowage" is not from the terminal, but issued by the shipping company to the terminal, which is the most important stowage reference for the stowage planner. The "stowage" from the terminal is "actual stowage," which is a real stowage. However, the stowage needs to satisfy the container release constraints defined in "pre-stowage." In simple words, pre-stowage defines the specific positions for a certain group of containers. The stowage defines the specific position in the vessel for a single container. Figure 6.19 shows the pre-stowage data at bay 13 of vessel GUOTA for a certain voyage.

As shown in the figure above, in the module of pre-stowage BAPLIE, the stacking area in the vessel for a group of containers with a certain discharging port can be defined. On this basis, the special properties of required containers can also be defined, such as container type, over-limited container or not, dangerous goods or not, and 45 ft container allowed or not.

There are paper version and EDI transmission version of pre-stowage BAPLIE or BAPLIE made in the stowage system. The pre-stowage BAPLIE should be properly handled according to different situations of the shipping company or shipping agency.

1. If pre-stowage BAPLIE is provided by the shipping agency

Professional shipping agency with high work quality will provide a pre-stowage BAPLIE before the terminal stowage, for the purpose of requiring the terminal to take into account the loading and unloading requirements of the whole route. For the voyage with tight operation time, many ports of call and large container amount, if the pre-stowage BAPLIE has been provided by the shipping agency, the stowage planner can skip the pre-stowage operation in the computer and directly make the stowage plan according to the pre-stowage principles from the shipping agency. In addition, since the pre-stowage BAPLIE is made by the shipping agency according to the vessel booking information, the actual containers collected in the terminal may be different

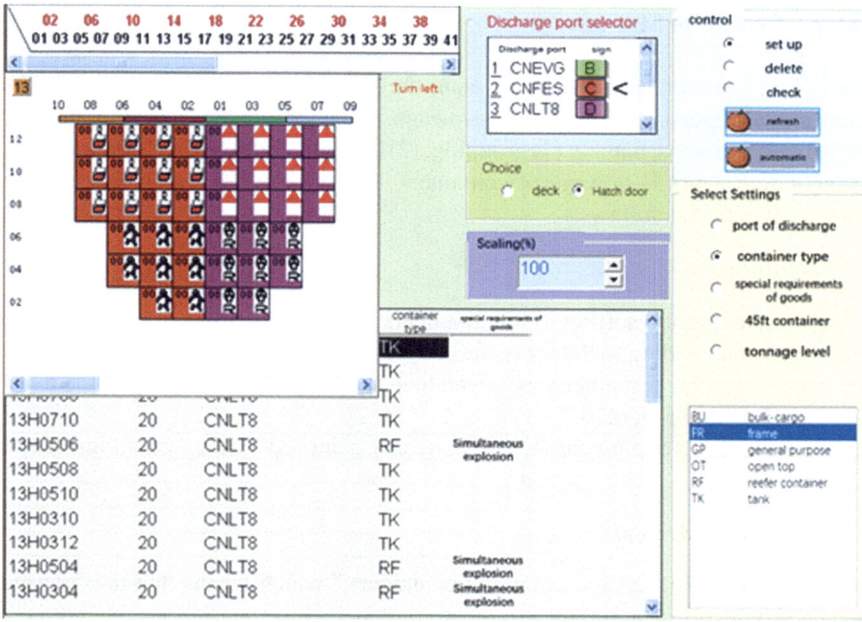

Fig. 6.19 Pre-stowage plan at bay view

from those in the BAPLIE, and sometimes, the difference could be extremely huge. Therefore, when making the stowage plan according to the pre-stowage BAPLIE, the stowage planner does not have to blindly follow the pre-stowage BAPLIE; instead, he should give comprehensive consideration to the actual containers collected in the terminal on the basis of the principles of pre-stowage BAPLIE and negotiate with stowage planner of the shipping agency; otherwise, troubles during loading in the terminal and delay in shipment may be caused.

2. If pre-stowage BAPLIE is not provided by the shipping agency

The purpose of the shipping agency providing the pre-stowage BAPLIE is to balance the loading and unloading operations of the whole route. For the voyage with short course and less ports of call, and short-range ocean vessels and feeder vessels with a small container volume, the shipping agency usually does not provide the pre-stowage BAPLIE, especially for the one-port route. In this case, the stowage operation is relatively simple, and pre-stowage BAPLIE is not necessary to be made in computer system. But for the voyage with long route, many ports of call, and a large container volume, if the shipping company fails to timely send the pre-stowage BAPLIE, it is necessary to make pre-stowage plan by the terminal according to the route and container volume, roughly determining the stowage principles and formulating a scientific and reasonable stowage BAPLIE.

6.3.6.2 Process of Stowage

After the completion of pre-stowage plan, the stowage planner shall carry out the actual stowage according to the pre-stowage. Stowage process can be generally divided into five steps, namely (1) grouping, (2) container retrieval, (3) bay selection, (4) sectioning, and (5) stowage of containers. The meanings and contents of each step are described as follows.

1. **Grouping**

Grouping refers to the sorting of all containers of the voyage meeting the stowage requirements, according to the properties of the containers such as discharging port, size, empty weight, container type, container height, and special requirements of goods, as shown in Fig. 6.20.

Of course, the grouping can be automatically achieved by the system instead of manual operation.

2. **Container retrieval**

Container retrieval refers to "searching containers," which means that the stowage planner should select a certain group of containers and then retrieve the storage locations of such containers in the yard (see Fig. 6.21), and check and analyze the yard status information such as the number of blocks where the containers are distributed and the concentration degree in each block, so as to assist the stowage planner in formulating the stowage plan.

Figure 6.21 has shown the retrieval results for containers of discharging port CNSHA and size 20 ft.

3. **Bay selection**

Bay selection refers to the selection of the bay in the vessel to load the retrieved group of containers. Before selecting the bay, the pre-stowage BAPLIE should be reviewed to determine all the bays that can be loaded for such group of containers and then

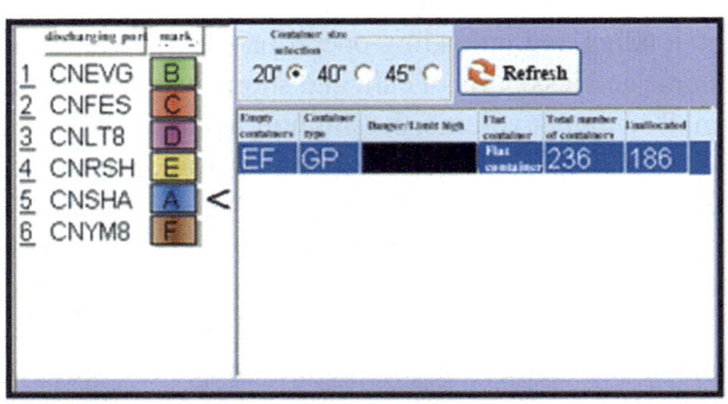

Fig. 6.20 Sorting of containers for stowage

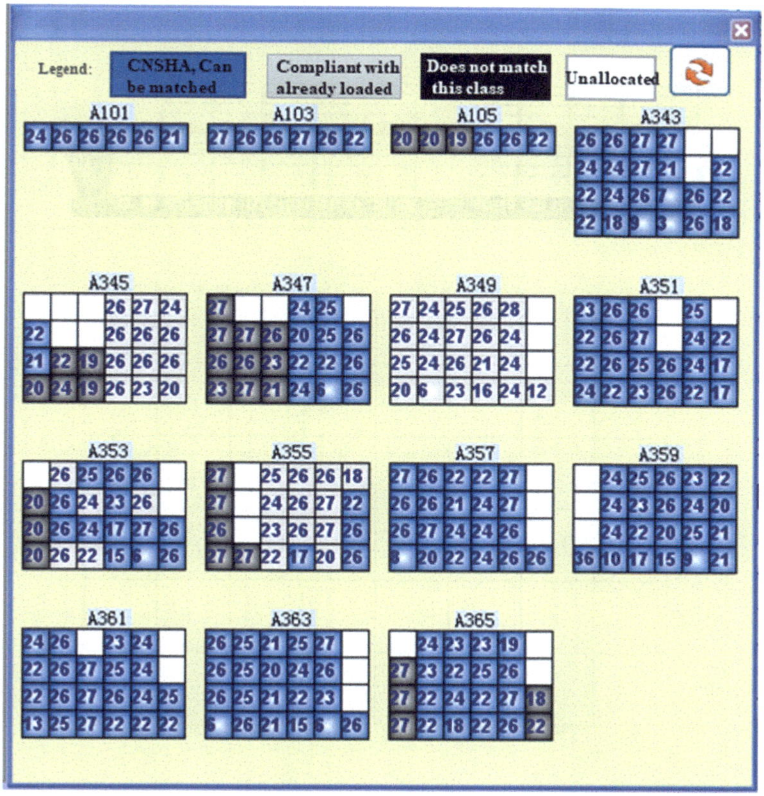

Fig. 6.21 Distribution of containers for stowage

check the loading capacity of each bay. Then, according to the storage situation of containers in the yard, one bay should be selected as the stowage target.

As shown in Fig. 6.22, bay 23 in the cabin is selected for stowage.

4. **Sectioning**

Sectioning refers to selecting a target stowage area within the selected bay, as shown in Fig. 6.23. After selecting the bay, it is not necessary to carry out uniform stowage within the whole bay. It is likely to need to divide the whole bay into two pieces or more, which is mainly related to the hatchway cover plate. If there are two pieces of translational hatchway covers at bay 23 pressed on both the left and right sides, respectively, in the operation process, the right hatchway cover has to be slide to the left to operate in rows 02, 03, 05, 07, and then, the two pieces of hatchway covers are slide to the right to operation in the left cabin. Therefore, bay 23 in operation process is divided into two sections, and then the stowage should also be formulated in two sections; otherwise, it will lead to container reshuffle in the yard.

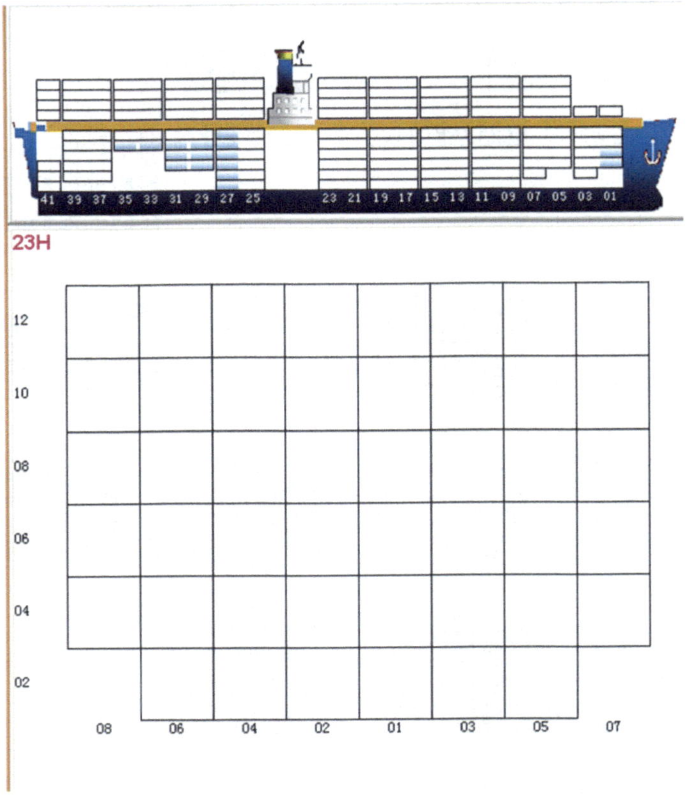

Fig. 6.22 Bay selection for stowage

5. **Stowage of containers**

The stowage of containers means that certain containers at each bay in the yard should be loaded in a certain sequence (stowage sequence) to the container positions in the vessel designated by the "sectioning" operation. The stowage sequence will have a definite influence on the operation in the yard in the future vessel loading, and it is also the main factor that dominates the vessel loading sequence. Therefore, a slight carelessness may lead to the reshuffle operation in the container delivery during vessel loading.

As the stowage results shown in Fig. 6.24, when the containers in the yard displayed on the right are stowed to the cabin displayed on the left, the system will automatically calculate the stowage sequence and positions of containers. Generally speaking, the containers are stowed from the sea side to the land side, from the bottom to the top, which is also related to the berthing direction of the vessel.

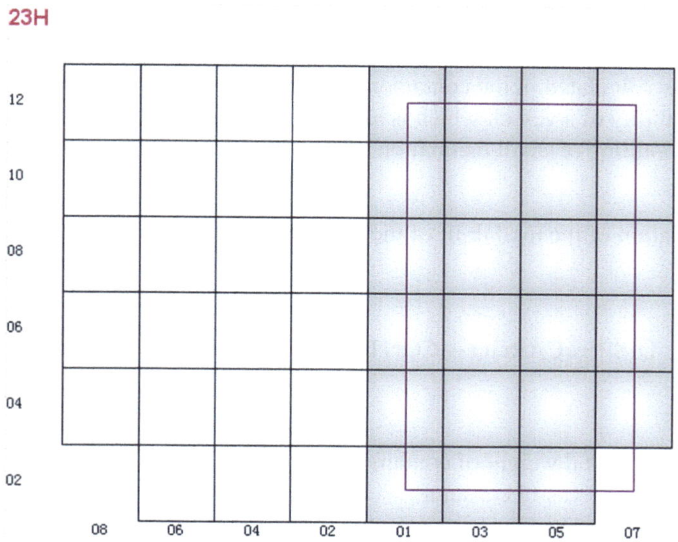

Fig. 6.23 Sectioning of the selected bay for stowage

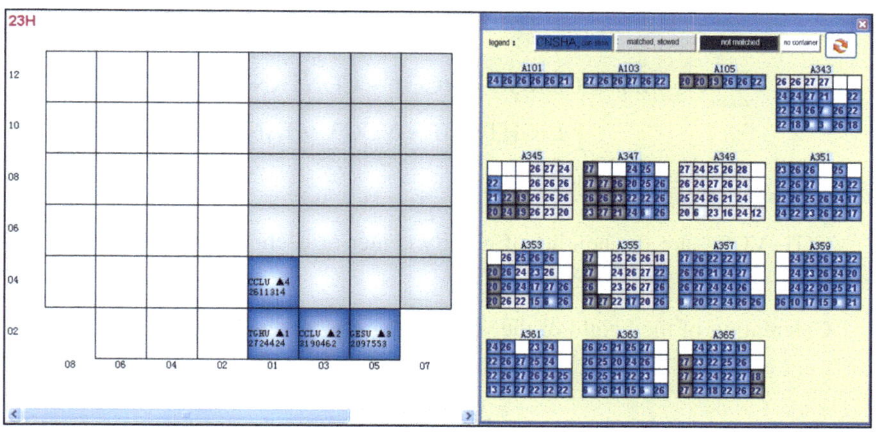

Fig. 6.24 Stowage of containers

6.3.7 Calculation of Stability and Draft Difference

At present, the stability and draft difference are not calculated after the completion of stowage in most terminals in China. Whether the vessel can meet the requirements of stability and draft difference is estimated according to experience, which may lead to passivity at work. In fact, the calculation of stability and draft difference is not only necessary but also easy for stowage planner, especially with the aid of computer stowage system.

6.3.7.1 Calculation of Stability

The stability calculated in the container terminal refers to the GM value, the height of initial stability of a vessel, which is an index to measure the resilience of a vessel under the condition of small inclination angle. The calculation steps and formula are as follows:

1. Calculation of the total draft of the vessel (DISP)

$$\text{DISP} = \text{CARGO} + \text{CONTANT} + \text{LIGHTSHIP} + \text{TANK}$$

where

CARGO: the total weight of container cargos;
CONTANT: the vessel constant;
LIGHTSHIP: the weight of the empty vessel;
TANK: the total weight of liquid in the tank.

2. Calculation of the sum of the vertical torques of each weight $\left(\sum V - \text{MT}\right)$

$$\sum V - \text{MT} = \text{CARGO} * \text{VCG1} + \text{CONTANT} * \text{VCG2} + \text{LIGHTSHIP} * \text{VCG3} + \text{TANK} * \text{VCG4}$$

where

VCG1–VCG4 are the heights of gravity center of each weight (vertical coordinate).

3. Calculation of the height of the vessel's gravity center (KG)

$$\text{KG} = \sum V - \text{MT}/\text{DISP}$$

Calculation of the height of the initial stability of the vessel (GM)

$$\text{GM} = \text{KM} - \text{KG}$$

where

KM is the metacentric height of the vessel, which can be directly obtained by DISP in the stability and draft difference calculation.

4. Calculation of the initial stability height after free liquid trimming (G_0M)

$$G_0M = GMGG_0$$

where,

$GG_0 = FSM/DISP$, where FSM is the sum of moments of free liquid level in the liquid tank of the vessel, which reduces GM value and is provided by the vessel.

6.3.7.2 Calculation of Draft Difference

The draft difference is mainly calculated by three indexes.

1. Calculation of the draft difference (TRIM)

$$\mathrm{TRIM} = \mathrm{DISP} * (\mathrm{LCG} - \mathrm{LCB})/\mathrm{MTC}$$

where

① LCB is the longitudinal coordinate of the vessel's buoyancy center, and MTC is the longitudinal moment per meter draft. If it is the longitudinal moment per centimeter draft, it will be converted into the longitudinal moment per meter draft by multiplied by 100. Both of these data can be directly inquired by DISP in the stability and draft difference calculation.
② LCG is the longitudinal coordinate of the vessel's gravity center, which needs to be calculated.

$\mathrm{LCG} = \sum L - \mathrm{MT/DISP}$, where, $\sum L - \mathrm{MT}$ is the sum of the vertical torques of each weight.

2. Calculation of the trim by head (F.DRAFT)

$$\mathrm{F.DRAFT} = \mathrm{DRAFT} + (\mathrm{LBP} - \mathrm{LCB})/\mathrm{LBP} * \mathrm{TRIM}$$

where,

① DRAFT is the average draft of the vessel, LCB is the longitudinal coordinate of the vessel's buoyancy center, which can be directly obtained from the stability and draft difference calculation by DISP.
② LBP is the length between perpendiculars of the vessel, provided in the vessel's data.

3. Calculation of the trim by stern (A.DRAFT)

$$A.DRAFT = DRAFT - LCB/LBP * TRIM$$

6.4 Facility Scheduling in Vessel Loading Operations

Similar to the vessel unloading operations, the scheduling of various loading and unloading facilities as well as transportation facilities before and during the vessel loading process is crucial. Since the vessel unloading is the first step in the import operations, while vessel loading is the last step of the export operations, and unloading process is from concentration to dispersion, while the process of loading is from dispersion to concentration, which makes the operating efficiency of vessel loading lower than that of vessel unloading at each port, and in facility scheduling during the process of vessel loading, more factors need to be considered compared with vessel unloading operations.

6.4.1 Quay Crane Scheduling During Vessel Loading Operations

Quay crane scheduling in vessel loading operations refers to the arrangement of designated quay cranes for the operation of the stowed export containers at certain bays. Together with the quay crane arrangement of unloading operations, it constitutes the operation line scheduling. See Sect. 3.3.2 for main contents, significance, and main principles of operation line scheduling.

The following aspects should be paid attention to in the scheduling process of the quay cranes.

1. Review the container amount distribution map and estimate the loading and unloading operation load for each bay.

As shown in Fig. 6.25, in order to ensure that all quay cranes can basically complete the loading and unloading tasks at the same time, the distribution of containers is crucial for the balance of the operation lines. The scheduler will generally multiply the amount of containers to be loaded by a coefficient, such as 1.3, when counting the operation load at each bay. The reason for this is that the operating efficiency of vessel unloading is often higher than that of vessel loading, so 1:1.3 is the speed ratio, which is an empirical value.

2. Determine which bay to load first

For each bay of the vessel, containers on deck should be unloaded first, then containers in the cabin, then containers in the cabin be loaded, and then containers on deck. So

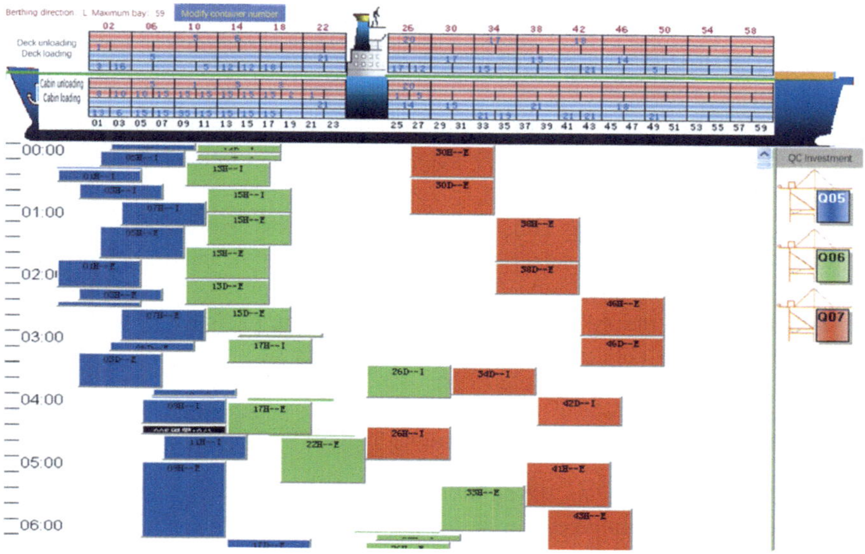

Fig. 6.25 Container amount distribution map

before beginning the vessel loading operations at a certain bay, whether containers at this bay or the corresponding large bay have been unloaded completely should be checked first. As shown in Fig. 6.25, taking the 39 containers to be loaded at bay 27 as an example, before beginning the loading task, 6 40 ft containers on deck at bay 26 should be unloaded first, as well as 20 ft containers at bay 25 and bay 27 on deck and 30 40 ft containers at bay 26 in the cabin.

3. Check whether the containers can be delivered normally

If the process of loading and stowage is reasonable, generally speaking, for loading operations at each bay, the containers can be delivered normally regardless the container sequence. However, the vessel control scheduler also needs to check whether the containers to be loaded at certain bay are stacked under others; that is, the containers stowed at this bay are stacked under other containers in the yard and the containers above are stowed to other bays. This situation should be avoided as much as possible by the stowage planner, but it also happens sometimes due to negligence or the non-ideal storage state in container collection. In this case, vessel control scheduler should try to load the containers at bays without container reshuffle.

6.4.2 Container Truck Scheduling in Vessel Loading Operations

All the containers that need to be loaded must be transported from the yard to the quay by container trucks before they can be loaded. As shown in Fig. 6.26, six container trucks are designated (bound) to Q05.

Container truck scheduling in the vessel loading process also includes the configuration of operation line, the selection of container truck pool, and the determination of the number of container collection trucks and the number of loop trucks. For the definition of operation line and container truck pool, please refer to 3.3.3. The number of container collection trucks and the number of loop trucks are different from the unloading process.

1. **Number of container collection trucks**

For vessel loading operations, the number of container collection trucks refers to the minimum number of container trucks waiting in the yard for container handling of a certain operation line. When the number of trucks is less than the given number of container collection trucks, the system will automatically search for appropriate idle trucks and assign them to collect containers in the yard.

2. **Number of loop trucks**

The number of loop trucks refers to the maximum number of container trucks for the whole operation line, including the number of trucks waiting for container handling in the yard and the number of trucks that have been transporting the containers to the quay. When the number of trucks for the operation line exceeds the given number

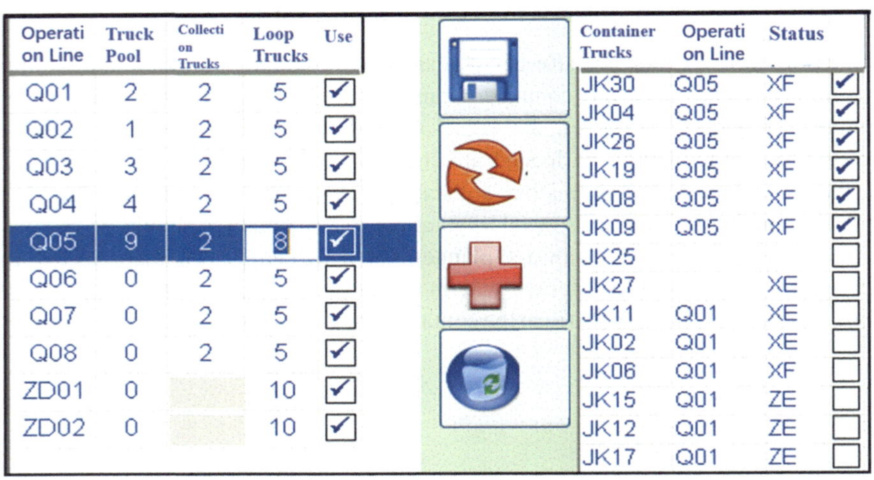

Fig. 6.26 Container truck scheduling

of loop trucks, the system will not arrange other trucks to deliver containers for the operation line.

Note: see Sect. 3.3.3.2 for common modes of container truck scheduling.

6.4.3 Yard Crane Scheduling in Vessel Loading Operations

In the vessel loading process, the yard cranes are responsible for the handling of containers in the yard. The yard crane scheduling here refers to designating yard cranes to certain blocks or areas for container handling of different bays of different vessels, so as to ensure that the yard cranes can deliver containers to the container trucks when they enter such blocks.

The yard crane scheduling of vessel loading is often more difficult than the yard crane scheduling in the process of vessel unloading. The main reason is that the containers to be loaded should be stacked according to the discharging port, with more groups and relatively dispersive storage. However, there are less groups of containers to be unloaded, and there is no need to consider the discharging port. Moreover, there is no one-to-one correspondence between the container position in the vessel and the storage location in the yard, so the utilization of yard cranes can be relatively concentrated.

For vessel loading operations, yard control scheduler generally needs to carry out the following operations to ensure reasonable scheduling of yard facilities.

1. Select the operating vessel and check the current loading bay of the corresponding quay crane, as shown in Fig. 6.27.
2. Further check the BAPLIE at bay view of vessel loading operations and determine the distribution status of containers in the yard according to the stowage results, mainly the distribution of blocks;
3. Arrange the yard cranes to ensure that each block is designated with a yard crane for container handling.

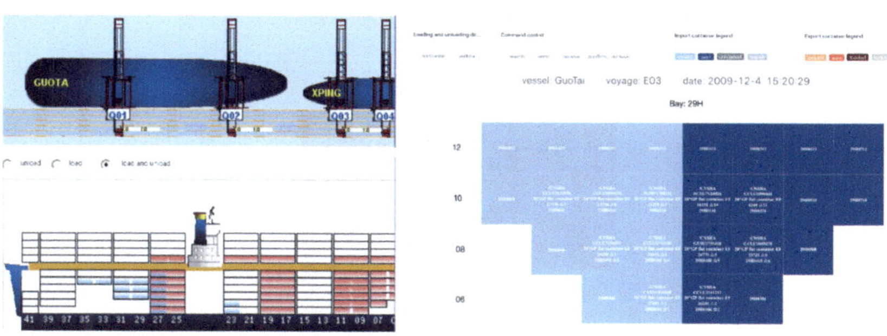

Fig. 6.27 Export container information of currently operating bay

6.5 Instruction Sending of Export Container Delivery in the Yard

1. Meaning of instruction sending of container delivery

Container delivery instructions are a set of serial numbers compiled by the vessel control scheduler, which can control the delivery sequence of export containers of a certain bay from the yard to the quay, as shown in Fig. 6.28.

Shown in Fig. 6.28 are the containers to be loaded at bay 09 in the cabin of vessel GUOTA, and the vessel control scheduler sends instructions via the bay view of BAPLIE shown in Fig. 6.28. The containers that have been given instructions are green, and a star "★" and a serial no. will be generated in the lower right corner of the container, which represents the sequence of each container at the operating bay. After the instruction of the container is sent, if there is an idle container truck in the operating line and the number of container collection trucks and the number of loop trucks do not exceed the upper limit, the yard crane driver in the corresponding block can see the container delivery task and the truck no. assigned by the system for the container, and the TPS terminal of the truck driver can also receive a message to collect the container in the block.

2. The difference and relationship between stowage sequence and container delivery sequence

Fig. 6.28 Instruction sending of container delivery

The stowage sequence is the serial no. generated by the stowage planner in the stowage process, which represents the idea of the stowage planner in container delivery. The final delivery sequence needs to be determined by the vessel control scheduler according to the stowage situation, the yard situation, and the facility situation. Therefore, stowage sequence does not represent the final loading sequence, but it is the most important reference for the vessel control scheduler to determine the container delivery sequence.

6.6 Actual Operations of Vessel Loading

6.6.1 Operation Process at Vessel Loading Scene

Operations at vessel loading scene refer to various operations involving port machinery, personnel, and other positions when loading export containers. Compared with loading system operations, there are more steps and more specific contents. In order to ensure the smooth running of the whole operation process, each position must ensure the quality of operation, cooperate, and coordinate with each other. The operation process at vessel loading scene is shown in Fig. 6.29.

As shown in Fig. 6.29, loading instructions are first sent by vessel control scheduler when loading export containers. Then, the container truck collects the container in the yard according to the instructions received and gears to the quay according to the TPS instructions. After the container truck is loaded, the container on the truck shall be provided with the twist lock by the mechanics after confirmation by the terminal. According to the container position provided by the quayside scheduler, the commander directs the quay crane driver to manipulate the crane to load the container into the vessel. Finally, the quayside scheduler completes the loading confirmation through the wireless terminal.

6.6.2 Generation of Container Delivery Task

Container delivery task refers to the operation task generated by the server according to the current loading situation, including a certain container truck no. and a container location in the yard. In the process of loading operations, the server monitors the loading operations at the current bay in real time, finds the corresponding container location in the yard according to the retrieved container position to be loaded, searches the available idle container trucks at the same time, and selects a container truck to assign the delivery task of the found container location.

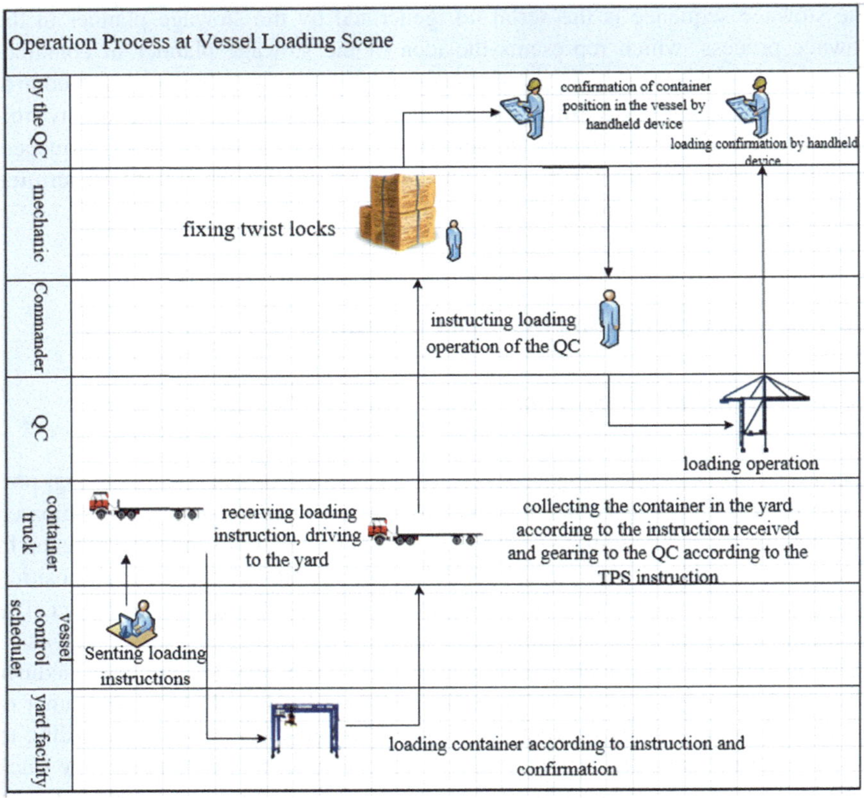

Fig. 6.29 Operation process at vessel loading scene

6.6.3 Container Delivery Confirmation by the Yard Crane

The container truck receives the task of container delivery generated by the server and goes to the storage yard to wait the yard crane driver for container delivery.

1. **Contents of container delivery confirmation**

The container delivery confirmation by the yard crane includes two parts, namely the truck no. and container location of the container for loading. In the container terminal, each container in the yard corresponds to a unique container location. Since the container no. cannot be seen when the container is in the yard, the container location is often used to search and confirm the container in the actual operations in the yard.

When the yard crane driver delivers the container, he should first confirm that the current container truck is the corresponding one in the container delivery task, as shown in Fig. 6.30. Then, he will lock the actual location in the yard and manipulate the yard crane to load the container onto the truck according to the corresponding

Fig. 6.30 Confirmation of container truck information

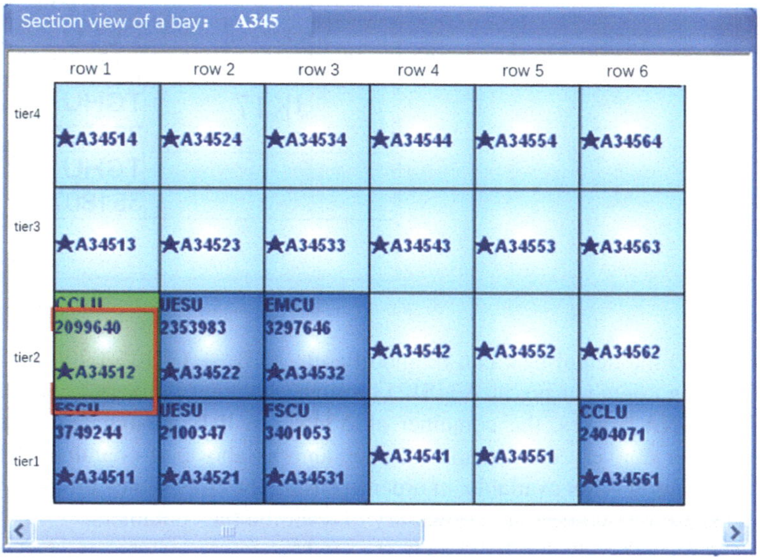

Fig. 6.31 Confirmation of container location

container location in the container delivery task. Container location confirmation is shown in Fig. 6.31.

2. **Significance of container delivery confirmation**

Container delivery confirmation of export containers plays a very important role in the effective and smooth loading operations of export containers, which can be summarized as follows.

(1) It can exchange the tasks of container delivery, avoiding container reshuffle

Due to factors such as road congestion or mechanical failure, the actual arrival sequence of container trucks may be inconsistent with the planned sequence during loading. Assuming that JKA and JKB correspond to the container locations A34512 and A34511 in the yard, respectively, in Fig. 6.32, namely JKA → A34512 and JKB → A34511. When actually entering the yard, JKB arrives before JKA. At this moment, the container at A34512 can be assigned to JKB

Fig. 6.32 Task list of vessel loading confirmation

Task	license plate no.	container no.	size
vessel loading	JK06	CCLU 2172783	20
		CCLU 2311404	20
	JK15	TTNU 3218951	20
		CCLU 2430949	20
	JK17	TGHU 3239721	20
		TGHU 3518074	20

first, and the actual task can be modified to JKB → CCLU2099640 (corresponding container no. at A34512) through container delivery confirmation. When JKA arrives, the container at A34511 will be designated to it, and container delivery confirmation is the same as above. If container delivery confirmation is not available, in order to ensure the consistency of information and smooth operation, it is necessary to reshuffle the containers.

(2) It can release container location in the yard

It is known that there is a one-to-one correspondence between the container location and the container. When a certain container is delivered and confirmed, the current storage location in the yard is released at the same time, and the current storage status in the yard is timely updated.

(3) It can the generate vessel loading task
The vessel loading task refers to a record containing the container truck no. and the corresponding container no. When container delivery is confirmed, the system will "bind" the container truck with the container no. of the container it carries, that is, the vessel loading task. Then, the tally clerk at the quayside will prepare for loading operation according to loading instructions received.

6.6.4 Vessel Loading Confirmation by the Quay Crane

1. **Contents of vessel loading confirmation**

The vessel loading confirmation by the quay crane refers to the process in which the quay crane driver manipulates the quay crane to transfer the container for loading from the container truck to the specific container position in the vessel, which mainly includes the confirmation of container no., container position in the vessel, and container truck no.

2. **Steps in vessel loading confirmation**

(1) Selecting the task
After the container truck arrives at the quay, the vessel loading task is selected according to the container truck no. (i.e., the container no.). The vessel loading task list is shown in Fig. 6.32.
(2) Checking and confirming container position in the vessel
According to the container position reported by the tally clerk at the quayside, the commander directs the quay crane to execute the loading operation. Generally, the containers can always be smoothly loaded, but sometimes due to the failure of the container truck, the sequence of the truck arrival at the yard is changed, which causes the situation of "suspended container" similar to that in the yard. Meanwhile, the terminal operators should contact the mate in time. If the properties of the containers above and below are not extremely different, the loading positions of the containers can be exchanged after the mate's approval. If the positions cannot be exchanged due to the extreme difference in the properties of the containers above and below, the loading of the current container can only be carried out after the loading of the container below.

6.7 Comprehensive Case of Vessel Loading Operations

The vessel loading operation process in the container terminal, from information collection to vessel loading of all the containers and the print of related statistical reports, is mainly introduced in this chapter. In this section, the vessel GUOTA, E03 is taken as an example, to illustrate the whole process of vessel loading operations, helping readers better understand and master the vessel loading operations of export containers in the container terminal operation system.

1. **Review of export container information**

The review of export containers is one of the most important steps in vessel loading operations of export containers. Before completing the collection operations of export containers, the information of the containers must be reviewed before the actual vessel loading. To review the export containers of the vessel GUOTA, E03, it is necessary to log in the module of query and review of export container load plan, and select "review" to display the information of the containers of this voyage that fail to pass the review. As shown in Fig. 6.33, there is inconsistency in the container with the no. of "BHCU3039777."

At this moment, the export container information of this voyage needs to be verified via the "automatic verification," as shown in Fig. 6.34. The discharging port of the container in the main file is not consistent with that in the export manifest, so the review is not passed. After verification, it is found that the main file input was wrong, and the discharging port of the container is indeed "CNYM8," so the discharging port in the main file will be corrected and saved. Meanwhile, there are still 11 over-loaded containers in the export manifest of this voyage, which may be caused by the fact that there are still export containers of this voyage not arrived and this will not affect the review. After modifying the error information and saving it,

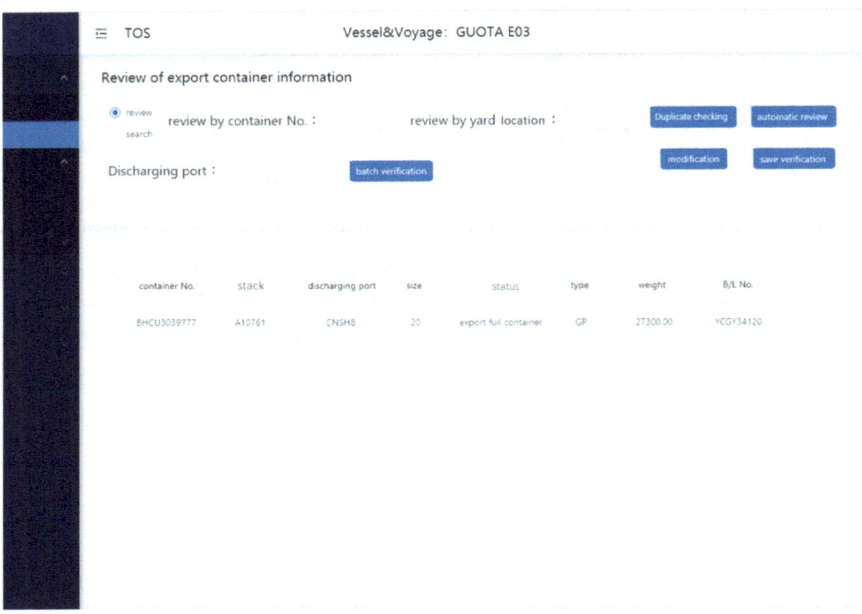

Fig. 6.33 Review of export container information

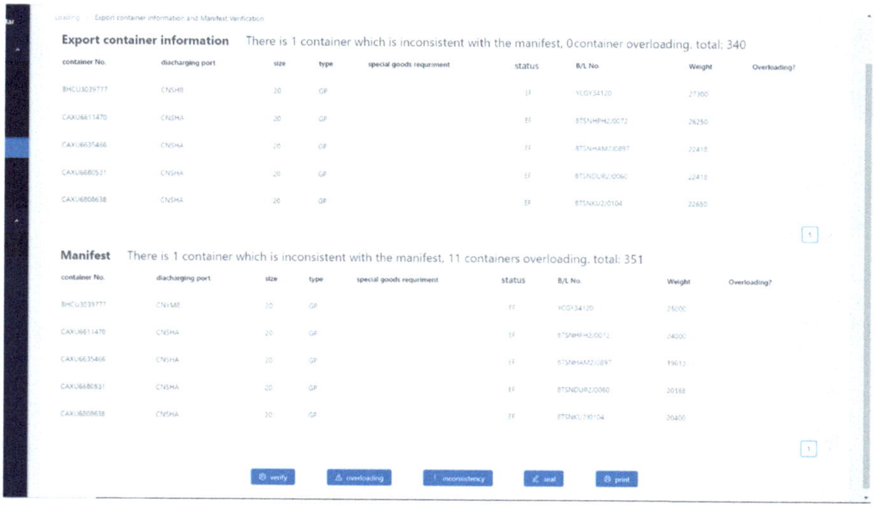

Fig. 6.34 Verification of export container information

the information will be rechecked. The containers that failed in the last review have passed, which means the completion of the review of export container information.

2. **Dock receipt and customs release**

After the review of export container information, the containers can only be stowed when they are released by the customs, and the dock receipt is the reference of customs release. According to paper document, the dock receipt of the vessel GUOTA, E03 is input, as shown in Fig. 6.35.

Click the button of "add the whole shipment," in the edit box of "bill of lading no." enter the bill of lading no. "141588023501," then click the button of "add the whole container" and in the edit box of "container no.," enter the container no. corresponding to the bill of lading. Since there are several containers in the bill of lading, the operation of "add the whole container" should be repeated. Input the four container no. in turn as shown in Fig. 6.35 and click the save button to save the records.

As shown in Fig. 6.36, click and select a container in the window of "containers of a B/L with complete dock receipts," such as "CCLU3030037," then click the button of "release," and the container will be released by the customs and displayed in the sub-window of "the released." In order to facilitate the operation, multiple containers can be released at one time. Select multiple containers continuously with "shift" and left mouse button, or select multiple containers at intervals with "ctrl" and left mouse button, and perform release operation.

3. **Stowage of export containers**

With the data preparation for the stowage, such as information review and customs release, the stowage of this export voyage can be executed as shown in Fig. 6.37.

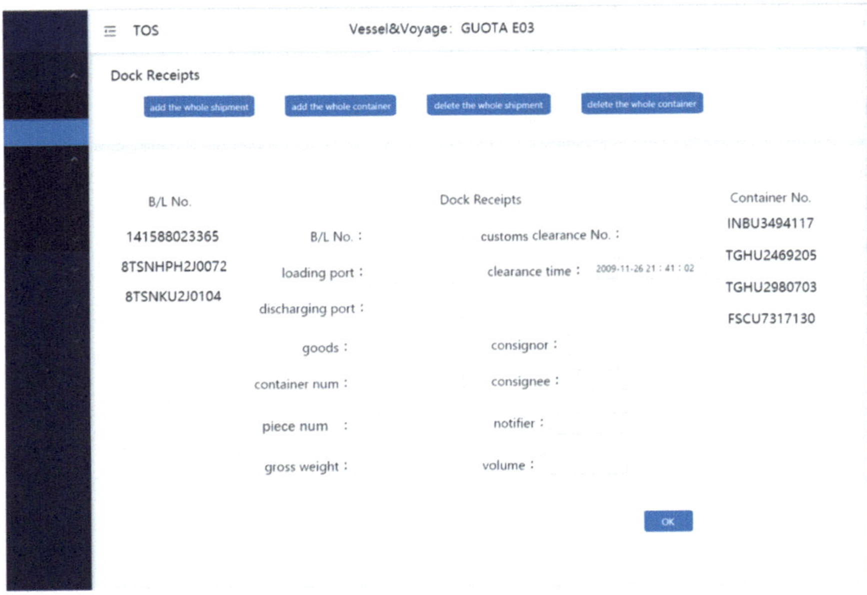

Fig. 6.35 Data input of dock receipt

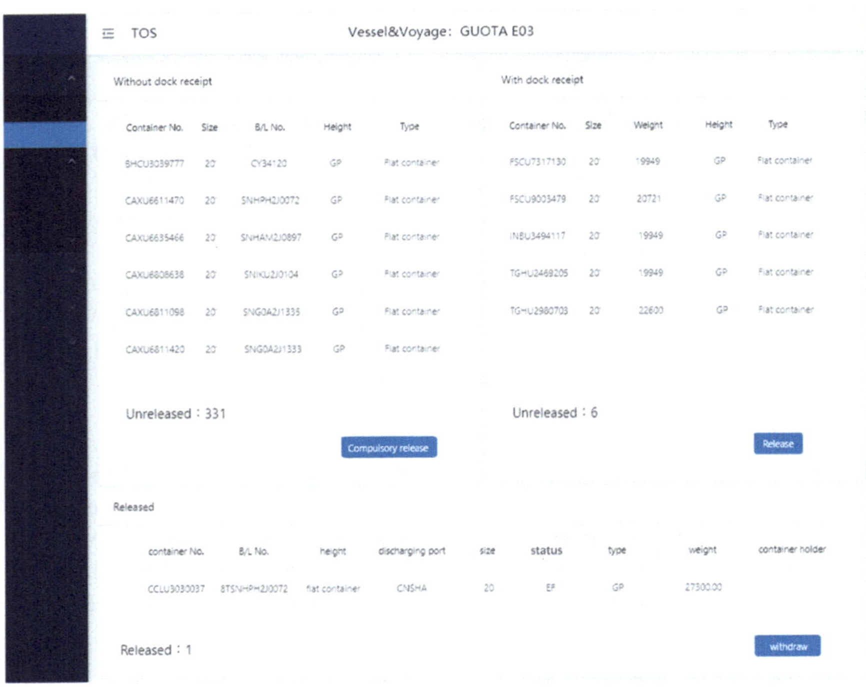

Fig. 6.36 Customs release

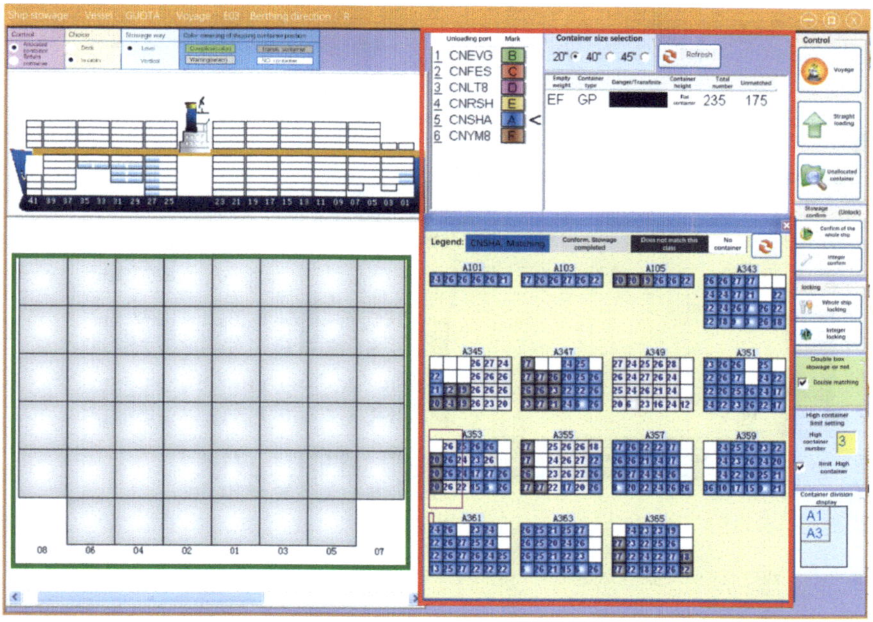

Fig. 6.37 Stowage of export containers (before stowage)

The 20 ft containers with the discharging port "CNSHA" are taken as an example, such container group are distributed in blocks A1 and A3 in the yard, and most of them in block A3, which will be the first target in the stowage.

The specific steps and precautions for stowage planner to make stowage by referring to the pre-stowage BAPLIE from the shipping agency are as follows.

(1) Display the bay view. In this example, bay 22H is selected for stowage. Click the position of bay 22H in the vessel's profile map to display the bay view.
(2) Select the container positions for stowage. Select the check box of "stowage" to place the bay in stowage state. Since there are a large number of containers distributed in block A3, drag all the container positions at the current bay at bay view with the mouse and set their state as "to be stowed," as shown in Fig. 6.37.
(3) Stow the containers, namely to select containers for stowage in the corresponding locations in the yard. When selecting containers, the principle of no heavy containers stacking above the light ones as well as the avoidance of frequent shifting of the yard crane between different bays should be taken into account. According to the distribution characteristics of the export containers in block A3 shown in Fig. 6.37, the weight distribution of the export containers in block A3 is relatively uniform. Therefore, the containers at bay 43 in the block, namely from A343 to A365, can be stowed in order. According to the weight and stacking characteristics of containers at bay 43, the containers weighing above 20 ton will be stowed first. The containers are selected from top

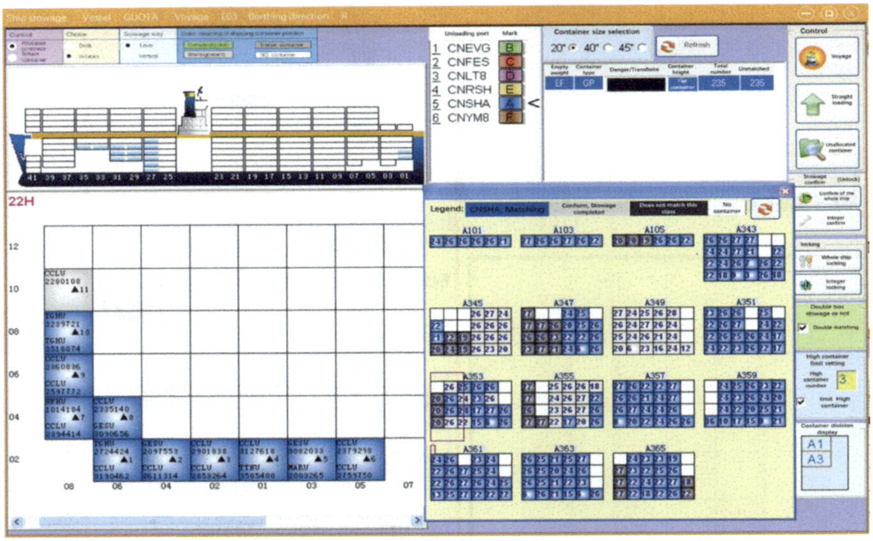

Fig. 6.38 Stowage of export containers (during stowage)

to bottom in the order of decreasing weight. The system automatically stows the containers from left to right and bottom to top, as shown in the container position—in Fig. 6.38. Considering the weight of the other containers, stack the four containers weighing 18, 9, 7, and 3 ton in the left row from bottom to top. During stowage, when there are only a few containers with small weight among the export containers, a special row or two are often configured to stack these containers. Select the containers for stowage that have not been stowed yet, and cancel their waiting state. Then select the three container positions in the leftmost row, and select the remaining three containers in A343 in turn according to the principle of decreasing weight. The stowage results are shown in Fig. 6.38.

The stowage method of other bays in the yard is the same. When the weight difference of all containers at a certain bay is not significant, which does not affect the rules of stacking light containers above the heavy ones during loading, all containers at this bay can be directly selected instead of choosing one by one, so as to improve stowage efficiency. Bay view of bay 22H after stowage is shown in Fig. 6.39.

4. **Facility scheduling in vessel loading operations**

The facility scheduling of export vessel loading is the same as that of import vessel unloading, including three parts: quay crane scheduling, container truck scheduling, and yard crane scheduling, and the scheduling method is the same. As shown in Figs. 6.40 and 6.41, the operating quay crane is Q01, and the corresponding container

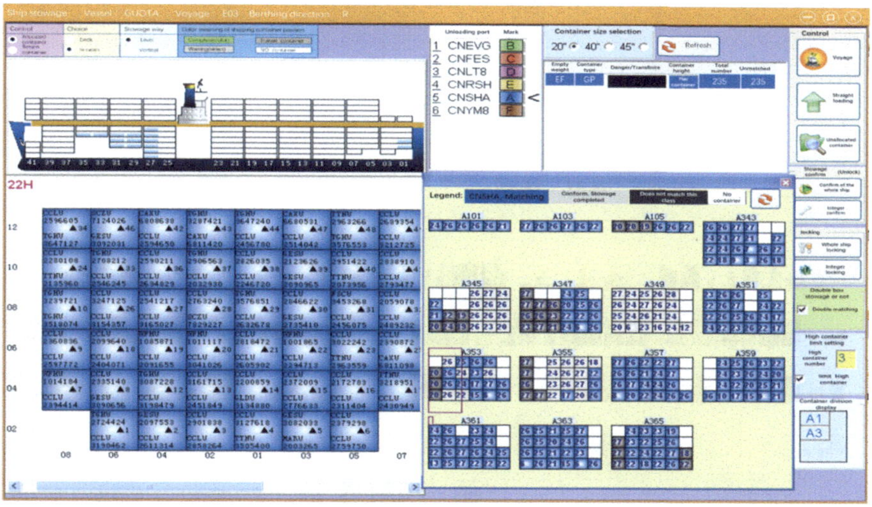

Fig. 6.39 Stowage results

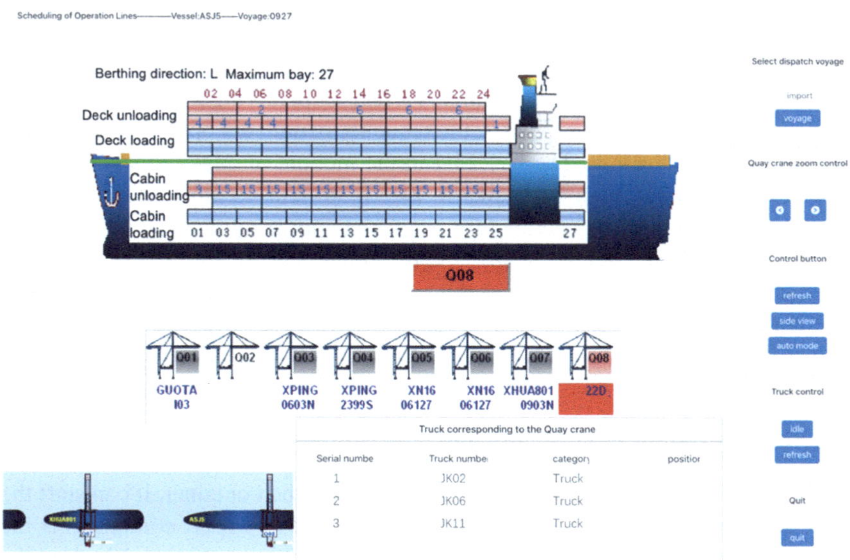

Fig. 6.40 Quay crane scheduling in vessel loading

trucks are JK02, JK06, JK11, JK12, JK15, and JK17. The operating facility in the yard is L04.

5. **Delivery instruction of export containers**

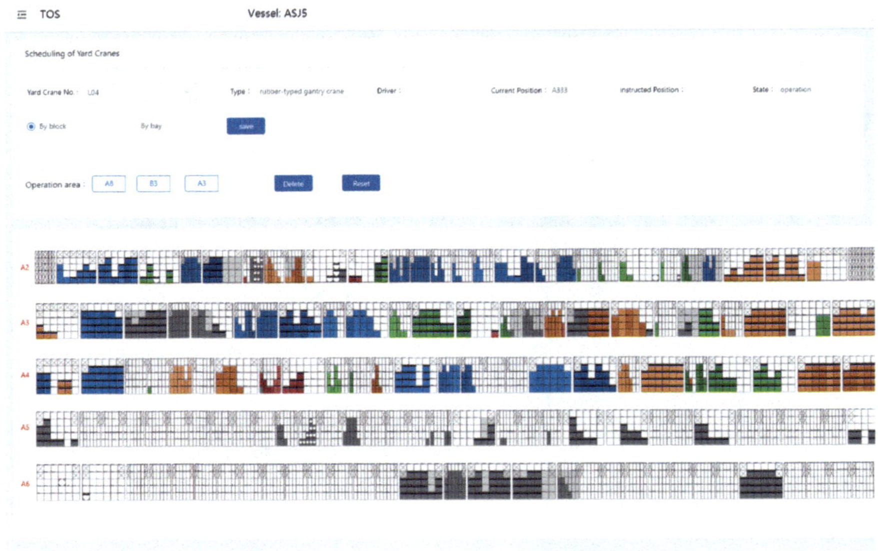

Fig. 6.41 Yard crane scheduling in vessel loading. *Note* During actually loading, all export containers should be stowed before carrying out facility scheduling

The delivery instruction of export containers refers to specifying a delivery sequence for export containers in the yard, that is, the sequence of delivery operation of the yard crane during vessel loading. The bay 23H in the module of quay crane scheduling should be right-clicked first, as shown in Fig. 6.38. During stowage, a sequence no. has been automatically generated for each container position at such bay. Here, the control instruction is set as "send," and the containers are selected successively according to the sequence no., such as the serial number of the first container is ★1, and the container position in the vessel is displayed in green, where the corresponding container delivery instruction has been completed, as shown in container positions 1–9 in Fig. 6.42.

6. **Actual operations of vessel loading**

After the yard crane receives the instruction to deliver the container, it can start the actual operation. The module of yard crane's operation of export containers is the same as the import operations. In this example, by clicking the "current" button, the task list of L02 can be checked including three containers from the gatehouse for collection and four containers to deliver for vessel loading, as shown in Fig. 6.43. The operation tasks can be screened out by selecting the loading and delivery tasks from the drop-down list of task categories. Then click the first task, namely the container "TGHU2724424," and click the "OK" button. The system will automatically complete the delivery task and jump to the next task, as shown in Fig. 6.44. When the container truck is fully loaded, the next task will be selected. The system will

Fig. 6.42 Sending instructions of export container

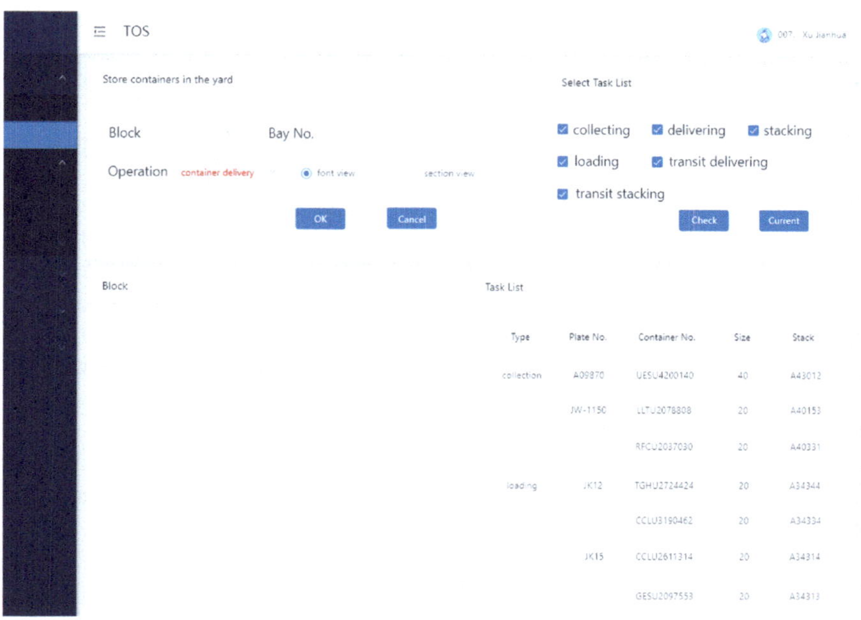

Fig. 6.43 Container delivery for vessel loading

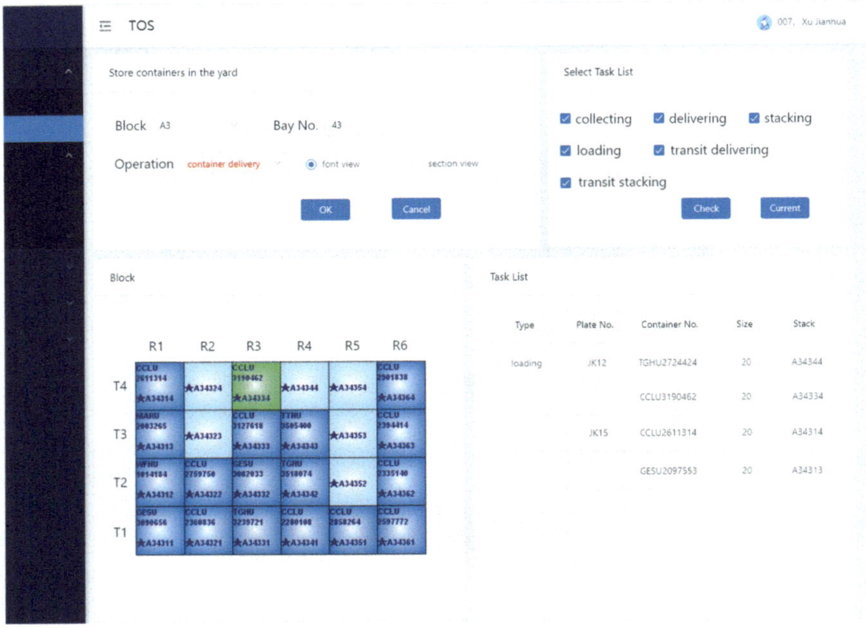

Fig. 6.44 Process of container delivery for vessel loading

automatically send the container truck to the quay for vessel loading, and meanwhile, the handling operation of the next container truck will begin.

When the container truck is fully loaded, it leaves the yard and goes to the apron for the loading and stacking operation by the quay crane. When it arrives at the apron, the container truck will be automatically displayed in the list of loading tasks of the quay crane. Different from the import operations, by clicking one of the tasks in the loading task list, such as "CCLU3190463" and confirming, the quay crane will complete the loading and stacking of the container, and the loading task will automatically jump to the other container on the container truck. After confirming the stacking position, the empty container truck will return to the yard for a new round of container delivery. The quay crane can continue the vessel loading operation by clicking the next task, as shown in Fig. 6.45.

The vessel loading operations of the containers with the instructions of delivery can be completed through the cyclic operations of yard crane, quay crane, and container trucks.

7. **Operation of directly loaded container**

The operation of directly loaded container does not need to go through the container collection, and its data are directly input in the module of data input of directly loaded container. In this voyage, there is a directly loaded container with the no. of "CCLU6637836." A record can be added in the interface shown in Fig. 6.46, by inputting relevant information according to written data and saving.

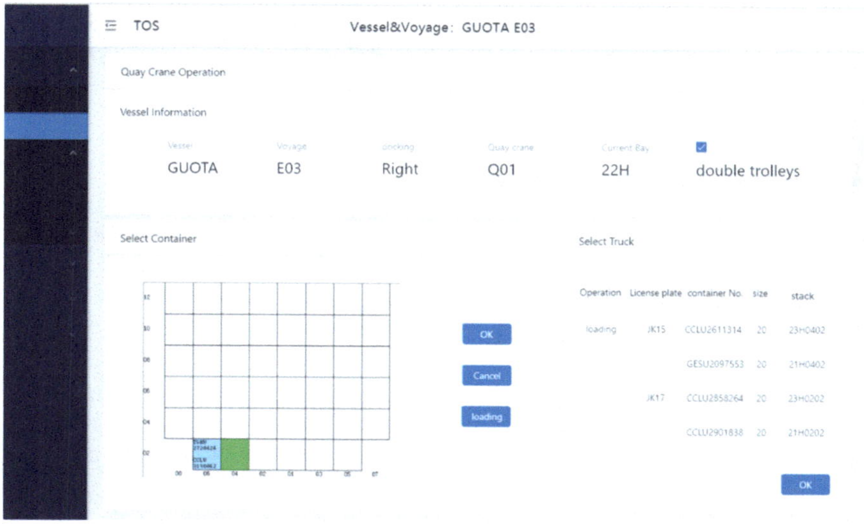

Fig. 6.45 Vessel loading by the quay crane

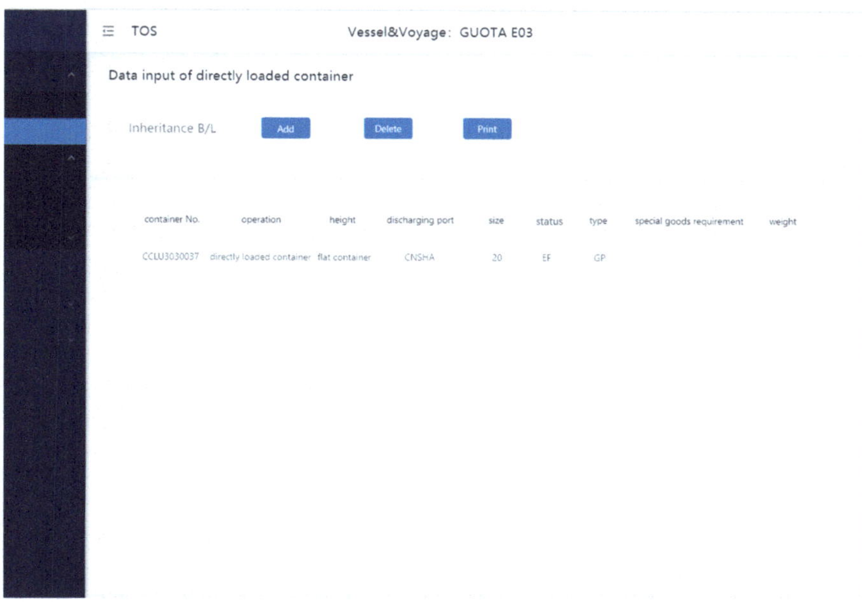

Fig. 6.46 Data input of directly loaded container

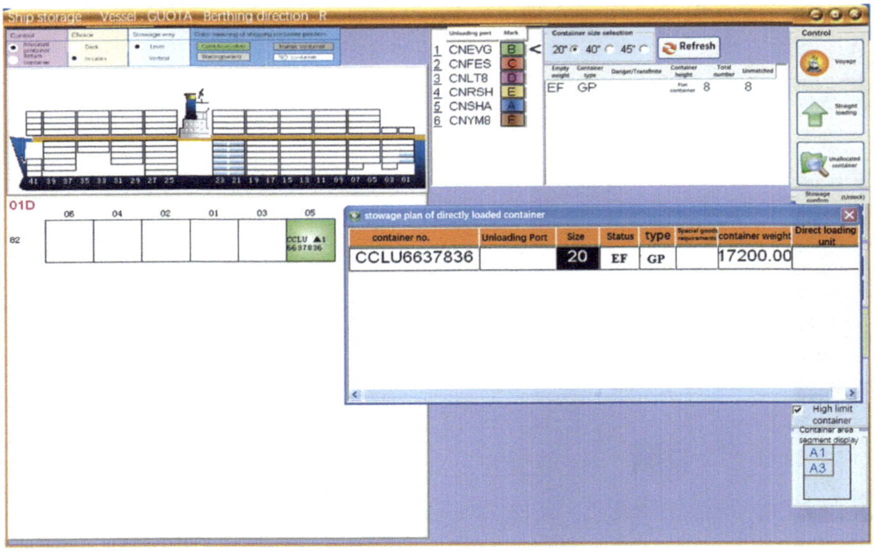

Fig. 6.47 Stowage of directly loaded container

The same as non-directly loaded containers, stowage cannot begin until the container is released by the customs according to dock receipt after data input of directly loaded container. The method of dock receipt input and customs release is the same as non-directly loaded containers. Select "directly loaded container" in the vessel stowage module, and stow the container to slot 01D05 as shown in Fig. 6.47.

The directly loaded container will be directly transported to the quay for loading after entering the terminal without operation in the yard, so it is not necessary to arrange the yard crane and container truck for it. The scheduling of quay crane of directly loaded container is the same as that of non-directly loaded, and in this example, the loading task of the directly loaded container will be assigned to Q01.

After the completion of the operation line scheduling, the directly loaded container will enter the terminal. The entry operation of the directly loaded container will be carried out in the module of entry through gatehouse, which mainly includes three steps.

(1) Truck selection. The container truck for entry can be selected through the "next vehicle" drop-down list, and the truck "JB-34429" is selected here;
(2) Container selection. The container no. "CCLU6637836" can be input in the edit box of "container no."
(3) Entry confirmation. The entry of the directly loaded container can be confirmed by clicking the "entry data of export containers" and selecting the record of such container in the sub-window of directly loaded container data, shown in Fig. 6.48.

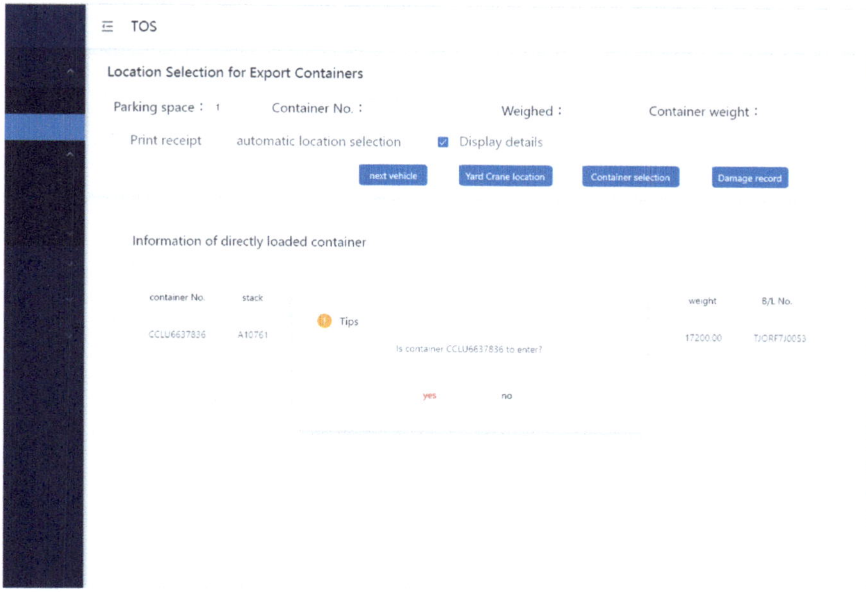

Fig. 6.48 Entry of directly loaded container

After entry, the directly loaded container is transported to the quay for loading. The vessel loading operation process of the quay crane is the same as that of non-directly loaded containers, by selecting the vehicle and confirming in the module of quay crane operation.

8. **Statistical reports of vessel loading**

Similar to the unloading of import containers, there are also statistical reports of loading of export containers, such as the bay view of BAPLIE, the general overview of export BAPLIE, vessel loading summary, and the summary of the single vessel for export. The operator can check and print these data at any time. The vessel loading summary is taken as an example shown in Fig. 6.49.

In this example, 22 20 ft containers have been loaded on board for this voyage, so the actual number of loaded containers in the vessel loading summary is 22 (20 ft full containers), and there are another 257 20 ft containers and 62 40 ft containers waiting for loading, as shown in Fig. 6.49.

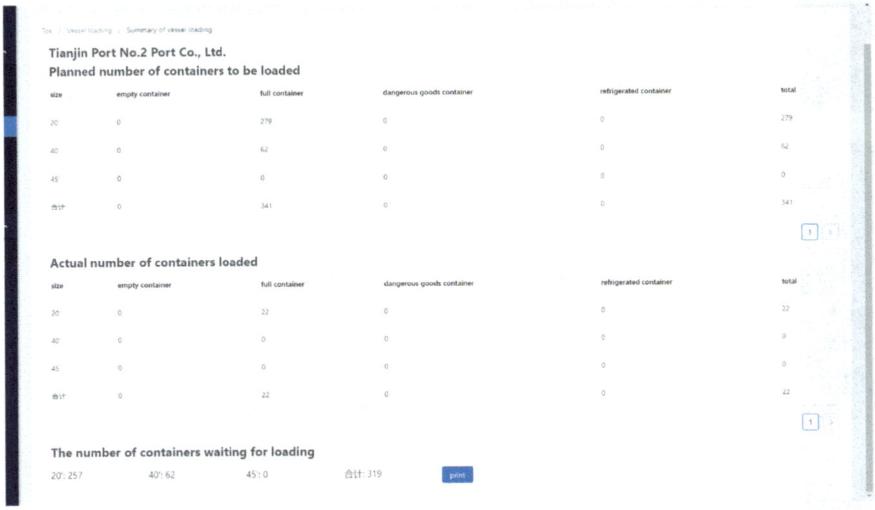

Fig. 6.49 Vessel loading summary

9. **Closure of export voyage**

When all the vessel loading operations of this voyage are completed, the terminal should close the voyage in time to prohibit the modification of the operational data of the vessel, timely release the occupied resources and archive the operational data.

6.8 Intelligent Stowage System

6.8.1 Definition

Intelligent stowage refers to the decision-making processes of automatically stowing the scheduled export containers to the target vessel according to the requirements of pre-stowage BAPLIE from the vessel, seaworthiness and terminal operation, via simulating the thought and method of stowage planner of the container terminal by the artificial intelligence algorithm, which comprehensively considers factors such as facilities, operation distribution, and stacking state.

6.8.2 Significance

(1) Greatly improving stowage efficiency
The efficiency of automatic stowage is about 8–10 times that of manual stowage. Taking the stowage of 2000 unit as an example, the time consumption of automatic stowage is 15 min, while manual stowage is about 2–3 h.
(2) Reducing work intensity
For ultra-large vessels, it can greatly reduce the work intensity of employees and gradually transform the role of stowage planner from operator of repeated work to rule maker.
(3) Solidifying stowage planner experience
The retirement or resignation of experienced stowage planner is a loss to the terminal, and the stowage quality can be steadily and effectively improved by continuously absorbing and solidifying the stowage operation experience of the stowage planner through the computer automatic stowage system.
(4) Improving night stowage quality
Another characteristic of stowage system is the stable stowage quality. The super computing ability of computer can effectively avoid the unfavorable conditions such as the decrease of stowage quality caused by manual fatigue at night.
(5) Reducing container reshuffle and improving stowage coverage
The container reshuffle rate is closely related to the customs release condition of containers during stowage. As customs are not closed at some large seaports, the later the stowage time is, the higher the customs release rate of export containers will be, which can not only improve stowage coverage, but also effectively control the effect of customs release on container reshuffle rate.
(6) Demonstrating the process of container loading with animation, effectively connecting the thoughts of stowage planner and vessel control scheduler
The system supports the animation demonstration of container loading process of the stowage and can step through to view the process of container loading at each loading point and the loading process inside the bay on board, so as to effectively connect the thoughts of the stowage planner and the vessel control scheduler.

6.8.3 Developing History and Current Situation

In recent years, the construction of informatization and intelligence has been promoted in port enterprises in China, introducing Internet thinking, Internet of Things technology, big data technology, cloud computing technology, GIS, operation management system, and other advanced technical means to promote the independent loading and unloading capacity and "self-thinking" capacity of the terminals. Up to now, certain achievement has also been obtained correspondingly. Under the background of "smart port," port facilities are becoming increasingly large-scale, digitalized and integrated, and the level of informatization, automation, and intelligence of container terminals has been continuously improved. Therefore, the operation and management mode of container terminal is required to transform to the intelligent, predictive, and controllable lean management.

The operation and management of container terminal mainly include yard management, facility and equipment management, vessel loading and unloading operation management, berth date management, and so on. Among them, the management of vessel loading and unloading operation directly affects the operating efficiency of the terminal, and the stowage operation of container vessel is one of the important steps of vessel loading and unloading, which is to stow export containers to the appropriate positions in the vessel in a reasonable order. Therefore, the lean management of container vessel stowage has become an important way to reduce costs, energy consumption and improve efficiency of the terminal, and the effective technical means of lean management of container vessel stowage is to realize intelligent stowage of container vessels by exploiting the knowledge base in the management information system.

The intelligent container vessel stowage is not only an important part of lean management of container terminal, but also an inevitable requirement of the development of container terminal industry.

Throughout the history of container vessel stowage, its development can be divided into three stages: manual stowage, computer-aided stowage, and intelligent stowage. Before 2000, the stowage mode in container terminals was mainly manual stowage with paperboards. After 2000, computer-aided stowage system appeared, which is gradually developing toward intelligent stowage in recent years.

(1) Manual stowage
At the early stage, most of the transportation vessels in the international shipping industry were small vessels, while the container transportation industry in China was in its initial stage with few terminals and small scale. The operation mode of container terminals was mainly recording the container storage plan and vessel stowage plan manually. Among them, "T card management" was a typical manual operation mode. In this mode, plane record board was usually hung on the wall to represent different locations in the yard, and each "T card," recording important parameters such as container size, container type, and weight, was used to represent a container.

(2) Computer-aided stowage
With the development of international trade, the proportion of container transportation has increased rapidly, the number of containers carried by a single vessel in container terminals is constantly increasing, and the original manual mode cannot meet the demand of operation. The management information system has been developed successively in each terminal to assist the operation and improve the efficiency and flexibility of terminal plans and gradually developed into the current terminal operation system (TOS).

(3) Intelligent stowage
In recent years, the construction of automated container terminals in the world has resulted in large vessels with complicated structures, demanding for refined and intelligent operations in the terminal. In addition, in order to improve competitiveness, the conventional terminals even propose "non-closing" to improve service quality and attract customers. The increase in the number of medium-sized or above vessels in the terminal and the increase in the total container volume of import and export require the terminal to improve operating efficiency and lower labor cost, which stimulates the demand for intelligent stowage.

At present, most of the vessel stowage plans at ports in China are mainly completed by manual operation with computer assistance. With the proposal of the concept of "smart port" and port enterprises' requirements for themselves, the management mode of manual stowage based on the personal experience of stowage planners has become a major bottleneck in the current container vessel operations. Although a lot of relevant experience has been mastered through continuous practice in the actual process, manual stowage is easily affected by personal experience. Therefore, how to combine the stowage experience with the computer and how to apply the computer to make a reasonable stowage plan quickly and effectively have important research significance.

Container vessel stowage is an important step in vessel loading operations in the terminal. It connects the operation process of the quay crane at apron and the yard crane in the backup yard, which directly affects the operating efficiency of the terminal. Therefore, the realization of reasonable and efficient stowage by computer is an essential and important way under the trend of large-scale vessels and has practical significance.

6.8.4 Common Techniques of Intelligent Vessel Stowage in Container Terminals

At present, common stowage techniques in container terminals can be divided into the following four types.

(1) Intelligent stowage technique based on manual rules
This technique mainly refines the basic stowage rules into the rule base and recalls the manually refined rule base during the stowage process. Its advantage is that the stowage conditions within the range of rule base are better and the stowage can be solved quickly. However, with the continuous evolution of large-scale vessels in recent years, the increase of the stowage rules is not in a geometric growth, and more and more stowage skills are difficult to be

described with simple manual rules. As a result, there is little room for further improvement.

(2) Intelligent stowage technique based on mathematical models
This technique mainly abstracts the vessel stowage problem into a mathematical model and applies some intelligent algorithms to solve it. Compared with manual rule-based technique, this technique can solve more complicated stowage problems. But at the same time, the complexity of the model also seriously affects the solution efficiency of stowage. It is difficult to get a satisfactory solution in a reasonable time for large and ultra-large vessels.

(3) Combined intelligent stowage technique
The combined stowage technique is a new one which combines the advantages of rule-based technique and model-based technique. This technique usually applies rule-based method for stowage under simple operation conditions, and intelligent algorithm to solve the problem under complex conditions, so as to improve the solution efficiency of stowage to a certain extent. However, when the container amount for stowage reaches a certain number, its solution efficiency is also unsatisfactory.

(4) Intelligent stowage technique based on learning
The technique is like the combined stowage technique, but the core is completely different. From the aspects of rules extraction, the neural network deep learning algorithm is adopted, and thus, the normal human rules have been raised to the level of stowage features, to overcome the problem that most abstract stowage strategy cannot be described with the structured artificial rules. Meanwhile, the intelligent algorithm is also applied in the process of stowage solution, but above the algorithm, a layer of workflow engine is constructed for fast recalling stowage feature database, so as to boost the stowage solution speed. Finally, this technique adopts the way of preview to realize the stowage evaluation system, which is not available in other techniques.

6.8.5 Examples of Technical Architecture for Intelligent Stowage

Technical architecture: deep leaning, workflow engine, preview evaluation
With the introduction of the concept of smart port, the manual stowage mode based on the personal experience of stowage planner has become a major bottleneck of the current stowage operations in the container terminal. The problem of intelligent stowage is mainly solved by applying intelligent optimization method, including optimization theory, system analysis and decision science, artificial intelligence, and other research directions. The following is an example of classic architecture for a learning-oriented intelligent stowage system. This architecture decomposes the stowage decision-making process into three stages: learning, decision-making, and evaluation. In each stage, deep learning, workflow engine, and operation preview techniques are adopted to solve the three key problems of difficult knowledge storage,

low solving speed, and poor evaluation effect in the stowage decision-making process, so as to ultimately realize the intelligent stowage (Fig. 6.50).

The technical architecture has the following features.

Feature 1: deep leaning network

This architecture realizes two deep learning networks of stowage strategy and stowage condition, breaks through the technical bottleneck that the traditional learning mode is difficult to automatically extract data features, and thus solves the problem of accurate extraction and storage of stowage knowledge and complex operation conditions.

The operation conditions faced in the process of stowage decision-making in the container terminal are ever-changing. Experienced stowage planner can accurately determine the current stowage conditions according to the pre-stowage BAPLIE and container distribution in the yard, but it is difficult to describe these operation conditions accurately in words. Similarly, the stowage strategies adopted by the stowage planner under different operation conditions cannot be described with simple rules. Stowage condition and stowage strategy are the most important basis of the whole stowage decision-making process. Therefore, in order to solve the problem of refining, expression, and storage of these two bases, two deep neural networks are trained, respectively. Among them, the stowage condition network transforms the historical stowage data of each bay of different vessels into a sparse matrix with a constant size and extracts more than 100 feature mapping matrices through the convolutional neural network. Finally, through a series of processing such as pooling, a fully deep network is connected to construct the stowage condition network. The stowage condition network is shown in Fig. 6.51.

The stowage strategy network is designed to take the basic stowage principles summarized as the basic input and abstract them to form more complex stowage strategies. Therefore, based on the historical stowage steps and stowage reference as

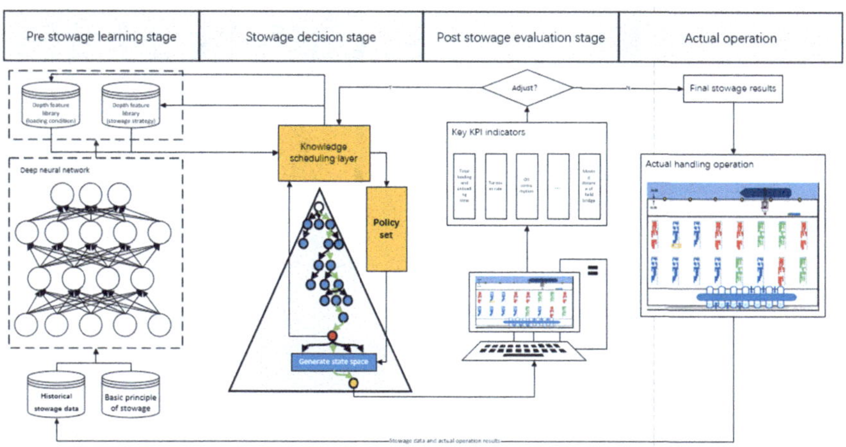

Fig. 6.50 Architecture of the intelligent stowage system based on leaning orientation

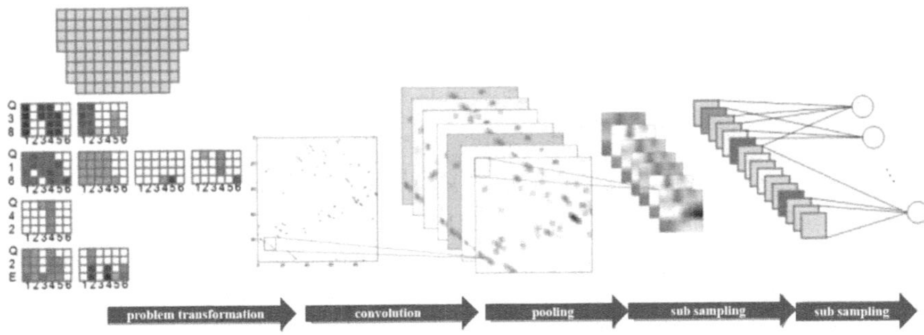

Fig. 6.51 Stowage condition network

input, the stowage strategy network is constructed by establishing a fully connected limited Boltzmann machine in the form of cascade self-coding machine. The stowage strategy network is shown in Fig. 6.52.

The results of training by these two networks provide the core decision reference for realizing intelligent stowage. Meanwhile, when the existing knowledge is unable to make effective decisions on the current vessel, by manually adjusting the stowage results and using the adjusted results to enhance the training of the network, the two major stowage networks can be further improved, so as to realize the promotion of intelligent stowage techniques.

Feature 2: Rule scheduling based on workflow engine

The architecture provides a reasonable knowledge network scheduling plan for complex conditions of intelligent stowage decision-making by applying a method of

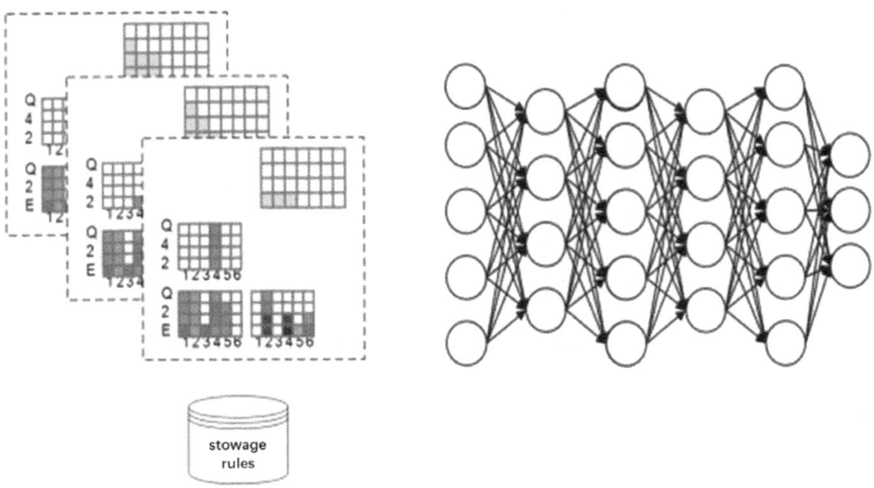

Fig. 6.52 Stowage strategy network

rule scheduling based on workflow engine, which has overcome the technical difficulty of impossible implementation of efficient decision in the process of decision-making of large scale due to the exponential growth of solution space required, and solved the key problems of pruning strategy arrangement of different conditions in the process of intelligent stowage.

The engine prunes the current decision space by pattern matching for the current decision condition, identifying and scheduling different rules. For some non-simple conditions, the engine could select and reconstruct the knowledge stack suitable for the current condition from the knowledge network, optimize and solve the state nodes to be searched, and greatly reduce the decision space.

The traditional methods for solving stowage problems are mainly intelligent algorithms (such as genetic algorithms, tabu search, and hybrid algorithms) and heuristic algorithms. Since the search space of stowage problem is very large, the constraints are very complex and usually different under different operation conditions. Therefore, the traditional solving methods have the following limitations.

(1) The constraint rules are complex, and different algorithms or heuristic methods have to be designed for different operation conditions, so the reusability is low.
(2) The solution space is large, the convergence efficiency is low, the solving time is long, and it is difficult to get satisfactory solution.

Therefore, an efficient method is needed to organize decision rules, reduce search space by pruning, and improve solution efficiency and effect.

The rule scheduling method is designed based on the workflow engine and can effectively deal with the correlation between operation flow and solution flow. In the solving process, according to the features of the current solution node, through the analysis of historical data and process of the workflow engine, the corresponding operation conditions are matched to obtain the constraint set of the current node, generate the current sub-search node, and identify the required solution strategy. If there is no condition matched in the historical condition database, it indicates that the current node is a new condition. At this point, the workflow engine analyzes the features of the new condition and the operation process to obtain the corresponding solution strategy. After the strategy extraction is completed, the knowledge stack is reconstructed according to the strategy priority to form a strategy set for the current node, and each sub-node is searched. After the solving is completed, the strategy and the effect of pruning are analyzed in reverse according to the optimal path finally selected for the solution tree constructed. Through the difference between the actual estimation and the actual value of the sub-node in each step of the solving process, residual learning is carried out to correct the corresponding parameters of the strategy network, and the modified knowledge is interpreted through workflow analysis. Then, the operation features and strategy features of the newly added operation condition in the solving process are updated, in order that once the same operation condition is encountered again, the features can be directly matched for solving.

Feature 3: operation preview

The architecture adopts the loading and unloading operation preview system of the container terminal to dynamically simulate the loading and unloading process of the whole vessel after the intelligent stowage decision and estimates the operating efficiency and other key performance indicators of the terminal under the target decision-making plan, so as to break through the technical bottleneck that the traditional stowage plan cannot be reasonably evaluated before the actual loading and unloading operations.

The stowage plan can only determine the location of each container in the vessel and the yard, as well as the operation sequence of the quay crane and the yard crane, so the actual efficiency and cost of subsequent loading and unloading operations cannot be known. Meanwhile, due to the complexity of the evaluation factors of stowage plan, it is difficult to describe the actual total evaluation index with static functions. Therefore, it is necessary to develop dynamic preview and evaluation method to comprehensively analyze the actual effect of stowage. Through simulating the actual loading and unloading operation process of the container terminal, the actual operating efficiency and cost of the stowage plan can be precisely evaluated.

The preview and evaluation system takes the stowage plan as input, including the berthing position of the vessel, the number of quay cranes in operation, the operating order of each quay crane, and the operating position of the quay crane and the yard crane for each operation task. The loading tasks may appear in the form of task group. The loading tasks within the same group centralize at the adjacent bays of the same block and are performed successively by the same quay crane and yard crane. The unloading tasks may also form an unloading task group, whose definition is similar to that of the loading task group.

For a given stowage plan, the preview module can produce two types of output. The first type of output is used to support decisions in stowage planning, including vessel loading and unloading time, quay crane operating efficiency, and total loading and unloading cost. The second type of output is used to propose the direction of improvement of the stowage plan, including the waiting time of each quay crane and yard crane, which is sorted and made statistics according to different task groups. The first type of output is the main index to evaluate the stowage plan. It is generally believed that the stowage plan with short loading and unloading time, balanced operating efficiency of quay cranes, and low total loading and unloading cost is always better. The second type of output is used to indicate unreasonable parts of the stowage plan. If the quay crane waiting time of a task group is longer than zero, it means that the quay crane operating efficiency of the task group still has room for improvement. If the yard crane has a longer waiting period for the container trucks, appropriate adjustments can be made to the corresponding part of the stowage plan related to the task group.

6.8.6 *Application of Intelligent Stowage Techniques in China*

6.8.6.1 CMICT, Ningbo

CMICT is the first container terminal in China to apply intelligent stowage techniques. Up to now (December 2018), the number of intelligently stowed vessels (with loading capacity greater than 300 UNIT) has been over 1000, and the application rate in the large and ultra-large vessels is about 90%. The average efficiency of single vessel using intelligent stowage techniques has been significantly improved, and the average number of operation lines has been reduced compared with the same period of previous years, which is expected to reduce cost of more than 10 million yuan for each year. Intelligent stowage techniques have greatly improved the efficiency of the stowage plan. Stowage for 1000 UNIT can be finished in 10 min. When the company's throughput reaches 300 WTEU, the staffing of planning posts can remain unchanged. Especially for large vessels with short closing period, this technique can advance the start time of loading operations by 3–4 h on average and significantly reduce the operation costs of terminals while saving energy and reducing emissions.

6.8.6.2 Shanghai International Port Group

Shanghai International Port Group implements the "non-closing" service, and the closing time is greatly postponed. Due to the long decision-making time of previous manual stowage, it is necessary to make the first stowage plan according to the entry of export containers in a few hours before vessel docking, and the second and third subsequent stowage plans according to the entry of export containers just before vessel docking. After the application of intelligent stowage techniques, the time needed for stowage decision-making has been significantly shortened; therefore, the first stowage decision can be made according to the entry and customs release of the export containers in advance in several hours before vessel docking, and the second decision can be made just before vessel docking for the remaining export containers. The time point for making the first decision has been greatly postponed, which has reduced the number of export containers released by the customs after the first decision and improved the decision-making efficiency and level.

6.9 Intelligent Vessel Control System

6.9.1 *Definition*

The intelligent control system of vessel loading instructions refers to the computer-aided decision system which stimulates the control principle and method of the vessel control scheduler with computer programs and comprehensively considers the key operation lines, facility conditions, storage conditions, and other factors to send vessel loading instructions intelligently instead of manually. The intelligent system can dynamically calculate the container loading point and loading order and automatically activate the containers waiting for loading in the yard according to the real-time loading order, so as to realize the instructed and intelligently controlled vessel loading operations.

6.9.2 Developing History and Current Situation

In recent years, in the context of intelligent manufacturing 2025, there has been a trend of construction of automated terminals in China. Meanwhile, the intelligent transformation of the conventional artificial terminals also becoming increasingly fierce, which has put forward new and higher requirements for daily operation planning, operation scheduling, and other aspects of container terminals.

The loading operation is the last stage of the export operations of the container terminal, for which the previous container collection and vessel stowage are prepared. The sending of loading operation instructions is the first procedure in the loading stage. The quality of instruction control directly affects the overall operating efficiency of the terminal. How to reasonably control loading operation instructions is the core issue to improve the operating efficiency of the terminal. Therefore, the realization of intelligent control of instructions of export container loading in the container terminal is the most critical technical step in the process of intelligent transformation of the conventional artificial terminal.

At present, the intelligent loading instructions control in the container terminal is still at the initial stage. Most terminals still use manual methods to control loading instructions. Currently, there are mainly the following problems.

(1) The central control schedulers make the instruction control depending too much on manual experience, and schedulers of different levels have different understanding degrees of the instruction rationality.
(2) Different central control schedulers make decisions on the container loading orders of different vessels independently. The lack of communication between them may lead to unreasonable facility deployment and reduce the efficiency of loading operations.
(3) The lack of a mechanism to balance the efficiency of the whole operation lines at the quay makes it liable to cause conflicts among operation instructions.

In conclusion, the current instruction control mode relies too much on manual experience, the schedulers lack of effective means of communication, and have not enough time to reasonably consider and deploy related resources, so that the overall operating efficiency of the terminal cannot always maintain a high level. The intelligent instruction control system can balance the operating efficiency of each operation line in real time according to the current operation situation from a global perspective, so as to improve the overall loading and unloading operating efficiency of the terminal.

6.9.3 Significance

The next generation of intelligent control system of vessel loading instructions in the container terminal should be mainly aimed at guiding operation and improving the operating efficiency of the terminal. The significance of realizing the intelligent control system of vessel loading instructions for export containers is as follows.

(1) The computer is capable of coordinating and optimizing the container loading process among multiple operation lines. A big advantage of the next generation of intelligent control system of vessel loading instructions in the container terminal is to take over the decision of the container loading instruction control of all the vessels currently under operation along the terminal coastline and compute the container loading point in the yard for each operation line from a more global perspective, thus to ensure the overall efficiency of the vessel loading operation.

(2) The loading technique modes are refined.
The computer can make specific instruction control rules according to different voyage and different loading technique modes. In recent years, the continuous emergence of ultra-large vessels has brought more challenges to terminal operations. In order to better serve the ultra-large vessels, operators in the terminal have also put forward many new handling techniques, such as loading and unloading at the same time. And these new handling techniques all have raised new requirements for the instruction control of loading operations. By refining the operation modes, the configurable instruction control rules are implemented to ensure the efficiency of loading operations in the container terminal under different operation modes.

(3) The human experience is solidified and labor intensity reduced
Another major advantage of intelligent instruction control is that it eliminates many complicated operations that are limited by human experience and working state, and the computer automatically takes over the relevant work by solidifying the human experience through intelligent calculation. The terminal operators can be liberated from trivial and complicated work and have more time to find problems, summarize rules, and further put forward ideas to optimize the system, so as to help optimize the decision-making results of the intelligent system.

It can be seen that the next generation of intelligent control system of vessel loading instructions in the container terminal is one of the key technologies to realize the intelligent transformation of the conventional container terminal, which is of great significance to the improvement of operating efficiency and intelligent level of the terminal.

6.9.4 *Examples of Technical Architecture for Intelligent Instruction Control Systems*

There are many methods to realize the intelligent instruction control of vessel loading in the container terminal. The basic structure and operation principle of this intelligent decision-making system are illustrated by taking the technical architecture as an example. The basic architecture of the technical plan is shown in Fig. 6.53.

(1) Modules of the system

The system is composed of operation database, information extraction module, information processing module, dynamic modeling module, model solving module, and plan feedback module, among which:

The operation database is used to store the data needed to run the operation system.

The information extraction module, which is a data extractor, obtains the original operation data needed to build and solve the container loading sequence model of the relevant bay in vessel loading from the operation database, including: stowage information, container information of the bay, container storage information in the yard, and container position information within the bay. In order to obtain complete and accurate information, view tools can be applied to extract specific data when necessary.

Since the extracted data from the operation database are two-dimensional table data, the information processing module is applied to matrix the extracted data. For each parameter matrix, the required data are mapped according to the dimensions defined in the model by merging and combining, so as to get the corresponding parameter matrix required by the model.

The dynamic modeling module uses the parameter matrix formed by information processing module to construct the corresponding constraint conditions and objective function according to the actual operation requirements, so as to build the corresponding model of container loading sequence for vessel loading, which can be used for the decision of container loading. In order to generate correct and efficient loading sequence, the dynamic modeling module also includes several sub-modules for constructing constraint conditions and objective function, mainly including the sub-module for preventing the suspension of loading, the sub-module for controlling reshuffle in the yard, the sub-module for controlling yard crane shifting, the sub-module for unique constraint, and the sub-module for model induction.

The model solving module is realized by calling CPLEX solver to solve the model.

The plan feedback module, which also needs to interact with the operation database, applies the SQL statements to write the solution of vessel loading

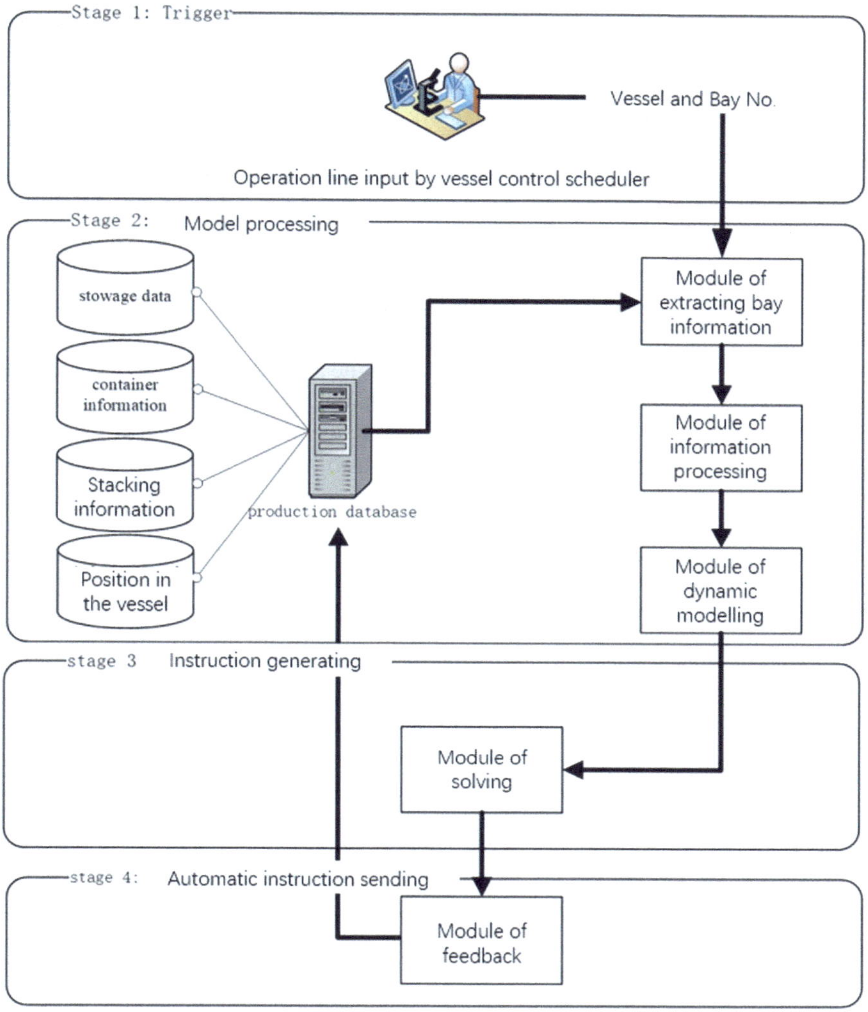

Fig. 6.53 Architecture of the intelligent control system of vessel loading instructions

instructions into the operation database, so that the operation system can automatically load containers according to the loading serial no. of containers at the corresponding bay.

(2) Operation principles

According to the decision-making system formed by the above technical plan, its operation process is shown in Fig. 6.53. The five functional modules of the system coordinate with each other, and each operation needs to go through four stages: event triggering, model processing, instruction formation, and automatic instruction sending.

First, the vessel control scheduler puts the quay crane into the designated operation bay and arranges the mechanical facilities such as container trucks, yard crane, and so on in the operation line, and then triggers the module of automatic generation of vessel loading instruction. After receiving the vessel call sign and bay no., the information extraction module extracts the corresponding information, and then the information processing module performs matrix processing, and then dynamic modeling. After the model and known data are formed, the model solving module will automatically arrange the sequence of vessel loading instructions. After the arrangement is formed, it will be sent to the feedback module for writing into the operation database, which means the completion of the automatic sending of vessel loading sequence.

The decision-making system formed by the above technical plan is capable of performing decision analysis of the container loading sequence for vessel loading according to the information of containers for loading, container locations in the yard, container positions within the bay, etc., to minimize the number of container reshuffle in vessel loading and reduce shifting frequency of the yard crane, so as to optimize the final decision and maximize the vessel loading efficiency of the container terminal.

Correction to: Digital Management of Container Terminal Operations

Correction to:
N. Zhao et al., *Digital Management of Container Terminal Operations*, https://doi.org/10.1007/978-981-15-2937-5

The original version of the book was inadvertently published with incorrect figures in Chapters 1, 3, and 6, which have now been corrected as follows:

The correction chapter and the book have been updated with the changes.

The updated version of these chapters can be found at
https://doi.org/10.1007/978-981-15-2937-5_1
https://doi.org/10.1007/978-981-15-2937-5_3
https://doi.org/10.1007/978-981-15-2937-5_6

N. Zhao et al., *Digital Management of Container Terminal Operations*,
https://doi.org/10.1007/978-981-15-2937-5_7

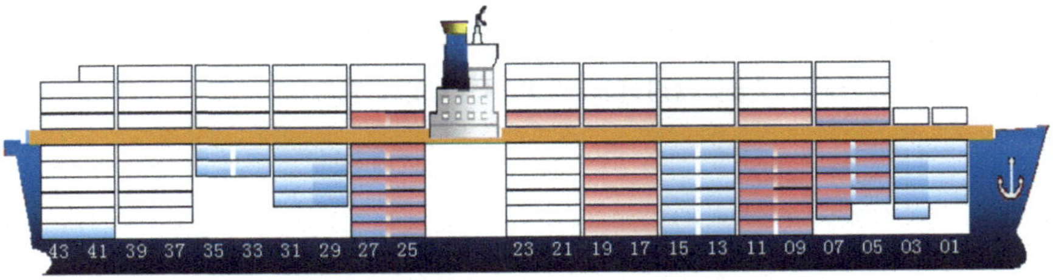

Fig. 1.18 Bays distribution on the vessel

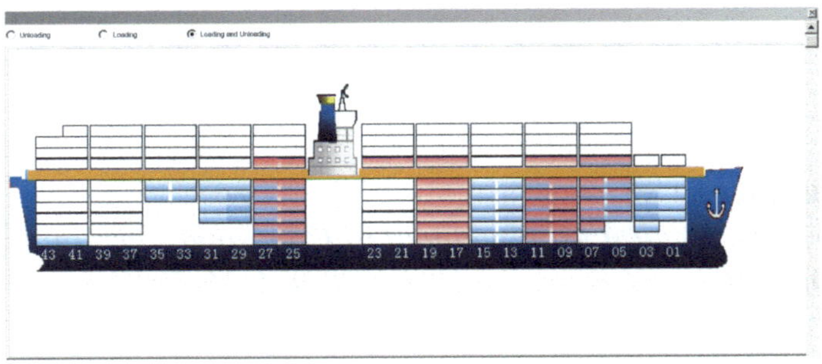

Fig. 3.9 Side profile of the vessel

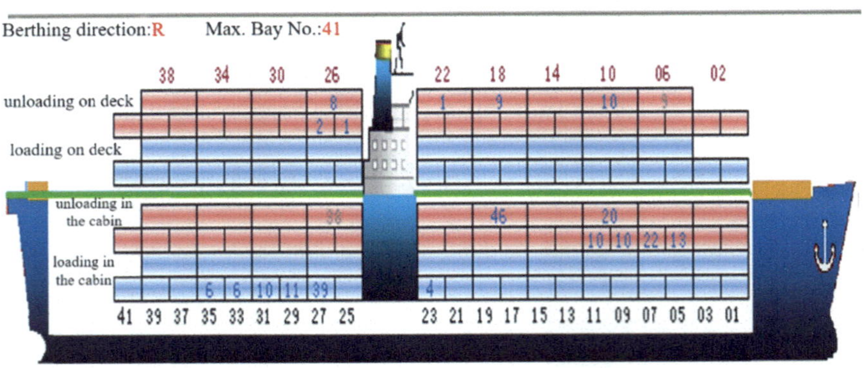

Fig. 3.10 Container volume distribution map

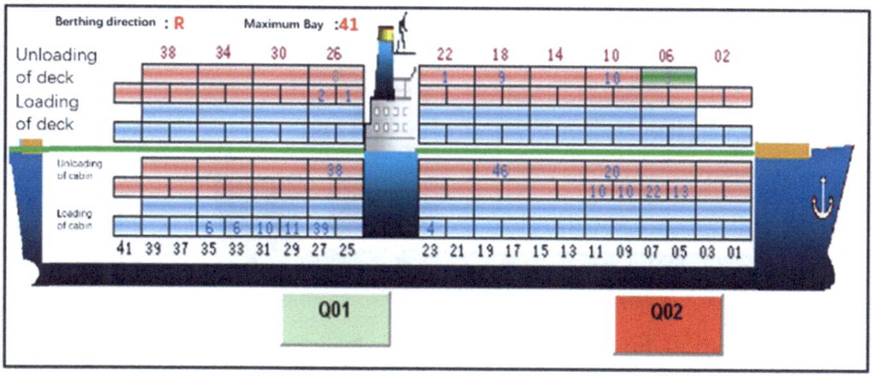

Fig. 3.21 Containers distribution in loading and unloading operations

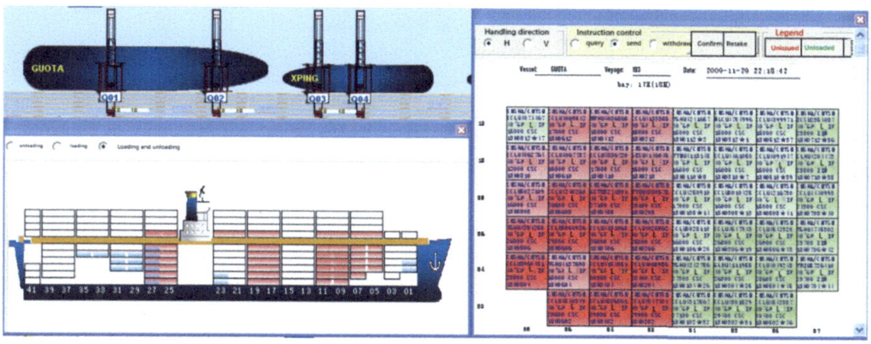

Fig. 3.25 Current information of containers for unloading checked by the yard control scheduler

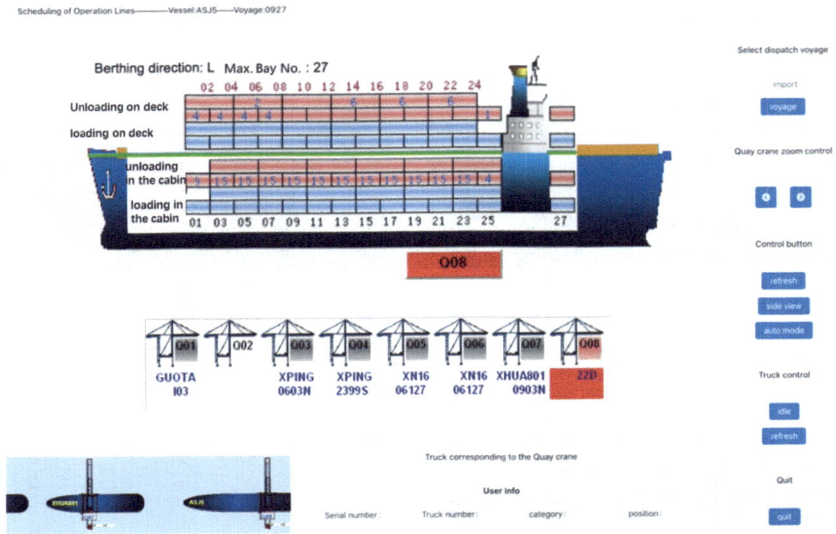

Fig. 3.50 QC scheduling of vessel unloading

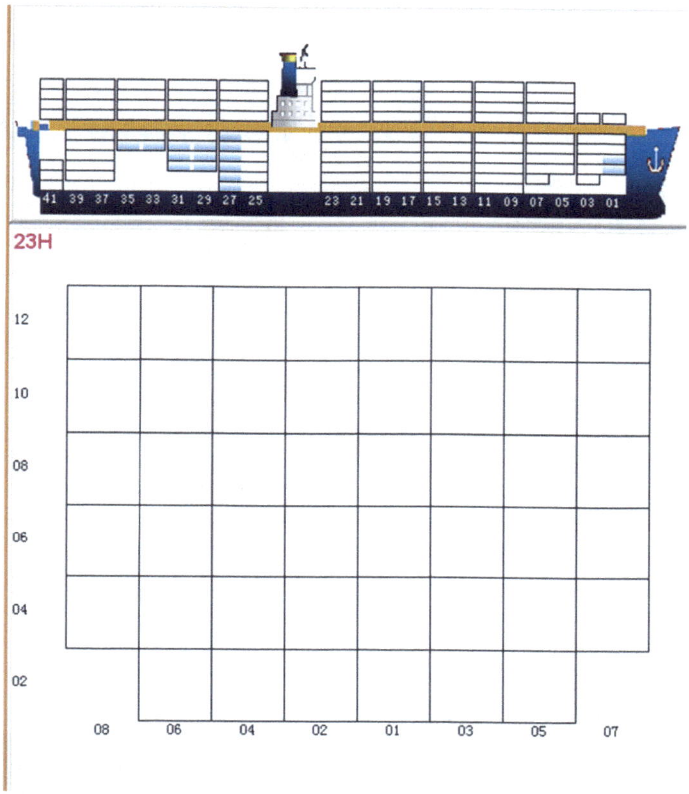

Fig. 6.22 Bay selection for stowage

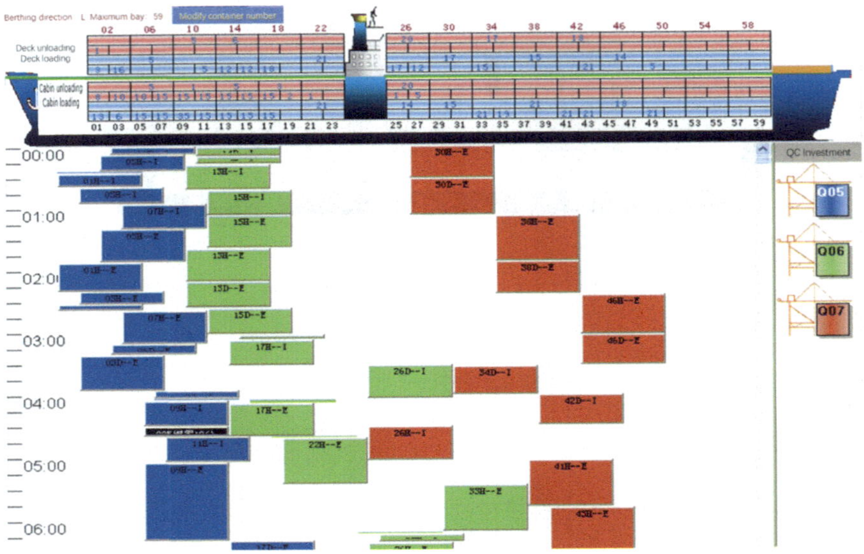

Fig. 6.25 Container amount distribution map

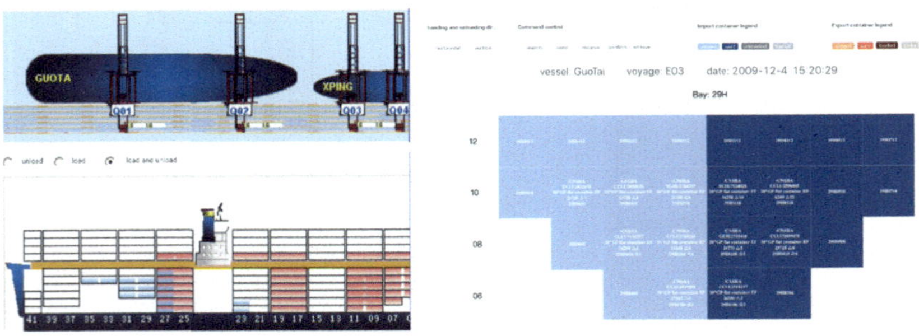

Fig. 6.27 Export container information of currently operating bay

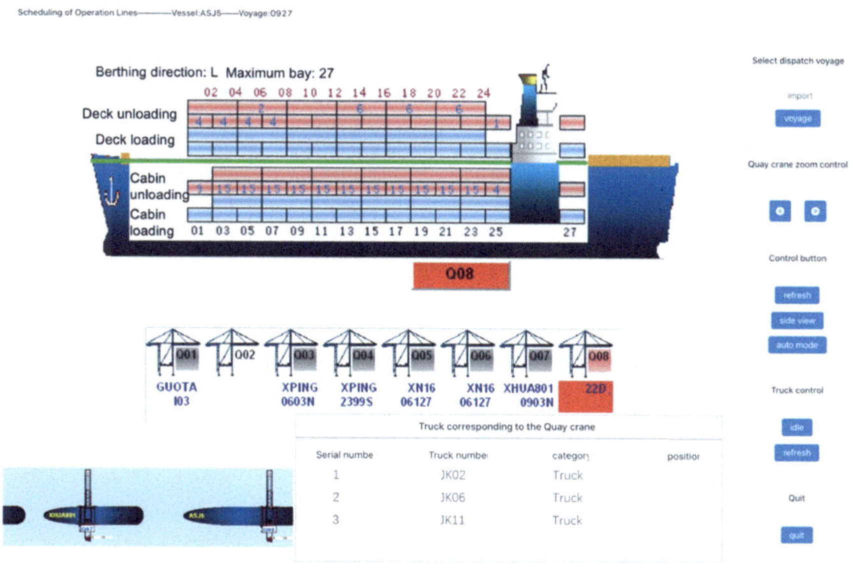

Fig. 6.40 Quay crane scheduling in vessel loading